MODERN CONTROL SYSTEMS:
A Manual of Design Methods

MODERN CONTROL SYSTEMS:
A Manual of Design Methods

JOHN A. BORRIE

Prentice/Hall International

Englewood Cliffs, N.J. London Mexico New Delhi
Rio de Janeiro Singapore Sydney Tokyo Toronto Wellington

Library of Congress Cataloging-in-Publication Data

Borrie, John A., 1937–
Modern control systems.

Bibliography: p.
Includes index.
1. Automatic control. 2. Control theory. I. Title.
TJ213.B6 1986 629.8 85–19270

ISBN 0-13-590290-8

British Library Cataloguing in Publication Data

Borrie, John A.
Modern control system: a manual of design methods.
1. Control theory
I. Title
629.8′312 QA402.3

ISBN 0-13-590290-8
ISBN 0-13-590282-7 Pbk

Prentice-Hall Inc., *Englewood Cliffs, New Jersey*
Prentice-Hall International (UK) Ltd, *London*
Prentice-Hall of Australia Pty Ltd, *Sydney*
Prentice-Hall Canada Inc., *Toronto*
Prentice-Hall Hispanomericana S.A., *Mexico*
Prentice-Hall of India Private Ltd, *New Delhi*
Prentice-Hall of Japan Inc., *Tokyo*
Prentice-Hall of Southeast Asia Pte Ltd, *Singapore*
Editora Prentice-Hall do Brasil Ltda, *Rio de Janeiro*
Whitehall Books Ltd, Wellington, *New Zealand*

Printed and bound in Great Britain for
Prentice-Hall International (UK) Ltd,
66 Wood Lane End, Hemel Hempstead, Hertfordshire, HP2 4RG
by A. Wheaton and Company Ltd, Exeter

2 3 4 5 90 89 88 87

ISBN 0-13-590290-8
ISBN 0-13-590282-7 PBK

CONTENTS

Preface xi

1 Basic Definitions and Mathematical Techniques 1

1.1 Introduction 1
1.2 Ordinary Differential Equations 1
    CAD Facility 3
1.3 The Laplace Transform 7
  (i) The Laplace Transform 7
    (a) Linearity 8
    (b) Delayed Function 8
    (c) Initial Value Theorem 8
    (d) Final Value Theorem 8
    (e) Differentiation with respect to t 8
    (f) Integration with respect to t 9
  (ii) The Inverse Laplace Transform 9
1.4 The Solution of Differential Equations using Laplace Transforms 10
    CAD Facility 11
1.5 System Transfer Functions and Dynamic Behavior (Continuous Time) 14
1.6 Transfer Functions of Some Common Circuits 17
1.7 Discrete Time Functions 17
1.8 The z Transform 22
  (i) The z Transform 22
    (a) Linearity 24
    (b) Initial Value Theorem 24
    (c) Final Value Theorem 24
    (d) Shift in Time 24
    (e) Multiplication by e^{an} 24
  (ii) The Inverse z Transform 24
    (a) The General Formula 24

(b) Table Look-up 28
(c) Power Series 29
CAD Facility 30
1.9 The Solution of Difference Equations Using z Transforms 30
1.10 System Transfer Functions and Dynamic Behavior (Discrete Time) 31
1.11 Sampled Data Systems 35
1.12 Relationship Between s and z Plane Poles 38
1.13 Block Diagrams and Composite Systems 40
1.14 Modified z Transforms 40
1.15 The Fourier Transform (Continuous Functions of Time) 42
(a) Linearity 43
(b) Shifting in Time 43
(c) Shifting in Frequency 43
(d) Differentiation of $f(t)$ with respect to t 43
(e) Differentiation of $F(j\omega)$ with respect to ω 43
(f) Real and Imaginary Parts of $F(j\omega)$ 43
(g) Amplitude and Phase Characteristics 44
(h) Common Fourier Transforms 45
1.16 System Frequency Behavior 46
1.17 The Fourier Transform (Discrete Functions of Time) 48
(a) Linearity 49
(b) Shifting in Time 49
(c) Shifting in Frequency 49
(d) Amplitude and Phase Characteristics 49
(e) Common Fourier Transforms 49
1.18 Shannon's Sampling Theorem 50
1.19 Discrete Time System Frequency Behavior Analysis 51
1.20 The Discrete Fourier Transform and its Inverse 52
(i) Calculation of the Discrete Fourier Transform 53
CAD Facility 55
(ii) Calculation of the Inverse Discrete Fourier Transform 57
1.21 Some Properties of Continuous Time Stochastic Processes 58
1.22 Some Properties of Discrete Time Stochastic Processes 65
References 70

2 Classical Techniques for Continuous and Discrete Time Systems 71
2.1 Introduction 71
2.2 Bode Plots (Continuous Time) 71
2.3 Root Locus Diagrams (Continuous Time) 77
2.4 Nyquist Diagrams (Continuous Time) 83
2.5 Nichols Charts (Continuous Time) 86
2.6 Inverse Nyquist Diagrams (Continuous Time) 89
2.7 Servomechanism Design Using Bode Plots, Root Locus Diagrams, Nyquist Diagrams, and Nichols Charts 91

(i) Bode Plot Design (Continuous Time) 92
(ii) Root Locus Diagram Design (Continuous Time) 94
(iii) Nyquist Diagram Design (Continuous Time) 96
(iv) Nichols Chart Design (Continuous Time) 98
(v) Inverse Nyquist Diagram Design (Continuous Time) 98
2.8 CAD Facility 102
Comment 109
2.9 Bode Plots (Discrete Systems) 109
2.10 Root Locus Diagrams (Discrete Systems) 115
2.11 Nyquist Diagrams (Discrete Systems) 119
2.12 Nichols Charts (Discrete Systems) 122
2.13 Digital Servomechanism Design Using Bode Plots, Root Locus Diagrams, Nyquist Diagrams, and Nichols Charts 123
(i) Design for a Short Sampling Period 124
(a) Backward Difference Approximation 124
(b) Approximation by Bilinear Transformation 125
(ii) Bode Plot Design (Discrete Systems) 126
(iii) Root Locus Design (Discrete Systems) 130
(iv) Nyquist Diagram Design (Discrete Systems) 131
(v) Nichols Chart Design (Discrete Systems) 132
2.14 CAD Facility 134
2.15 Proportional Integral Differential Controllers 134
(i) The Reaction Curve Method 136
(ii) The Continuous Cycling Method 137
Comment 141
References 141

3 Continuous Time State Space Design Techniques 142
3.1 Introduction 142
3.2 State Equations 142
3.3 Properties of Linear System State Equations 146
(i) Solution of the State Equations 146
CAD Facility 147
(ii) Similar Systems 149
(iii) Solution of the State Equations by Laplace Transforms 152
CAD Facility 153
3.4 Properties of LTI Systems 155
(i) Controllability 155
(ii) Observability 157
(iii) Stability of LTI Systems 158
3.5 Realization of State Equations 160
(i) Single-input, Single-output Systems 160
(ii) Multivariable Systems 163
(a) A General Formula 163

CAD Facility 164
(b) A Non-general Formula 167
CAD Facility 168
3.6 Pole Shifting by State Feedback 170
(i) Single-input Systems 172
CAD Facility 173
Comment 176
(ii) Multi-input Systems – Dyadic Feedback 176
3.7 Optimal Control Strategies 180
(i) Linear Quadratic Optimal Control 181
CAD Facility 182
Comment 188
(ii) Switching Curve Strategy 189
Comment 193
3.8 State Estimators for LTI Systems 193
(i) The Asymptotic State Estimator 194
CAD Facility 195
(ii) The Reduced-order Estimator (Luenberger Observer) 195
CAD Facility 196
Comment 201
3.9 Modeling and Behavior of Stochastic Systems 201
(i) Systems Affected by White Noise 201
(ii) Systems Affected by 'Colored' Noise 202
(iii) The Mean and Covariance of the State and Output 203
3.10 Multivariable System Design Methods 204
CAD Facility 207
Comment 214
References 214

4 Discrete Time State Space System Models 215
4.1 Introduction 215
4.2 State Equations 215
4.3 Properties of LTI System State Equations 216
(i) Solution of State Equations 216
CAD Facility 217
(ii) Similar Systems 220
(iii) Solution of State Equations by z Transforms 221
CAD Facility 222
4.4 Properties of LTI Systems 222
(i) Reachability (or Controllability) 222
(ii) Observability 222
(iii) Stability of LTI Systems 222
4.5 Realization of State Equations 223
(i) Single-input, Single-output Systems 223

(ii) Multivariable Systems 224
4.6 Sampled Data Systems 224
CAD Facility 226
4.7 Pole Shifting by State Feedback 228
4.8 Optimal Control by State Feedback 232
Comment 235
4.9 State Estimators for LTI Systems 235
(i) The Asymptotic State Estimator 235
(ii) The Reduced-order Estimator (Luenberger Observer) 236
4.10 Modelling and Behavior of Stochastic Systems 238
(i) Systems Affected by Colored Noise 238
(ii) The Mean and Covariance of the State and Output 239
(iii) Sampled Data Stochastic Systems 240
4.11 The Kalman Filter 241
(i) Basic Kalman Filter for Linear Time-variant Systems 242
CAD Facility 243
(a) A Suboptimal Kalman Filter 247
(b) An Optimal Filter 248
(ii) The 'Split' Kalman Filter 250
(iii) The Extended Kalman Filter 252
Comment 255
References 256

5 Useful Computer Techniques 257
5.1 Introduction 257
5.2 General Statement of the Optimization Problem 257
5.3 The Simplex Method 258
Comment 259
CAD Facility 259
5.4 Alternating Variable Methods 261
CAD Facility 264
5.5 Path of Steepest Ascent or Descent 266
Comment 270
CAD Facility 270
5.6 Some Useful Two-dimensional Object Functions 271
(i) Booth's Function 272
(ii) Zettl's Function 272
(iii) Rosenbrock's Function 273
5.7 System Identification by Computer Program 273
5.8 Least Squares Estimation of System Parameters 274
CAD Facility 278
Comment 279
5.9 Maximum Likelihood Estimation of System Parameters 281
CAD Facility 283
Comment 284

5.10 Other Methods of System Parameter Identification 285
References 285

Appendix **A Notes on Vectors, Matrices, and Determinants 286**
A.1 Vectors 286
A.2 Matrices – Basic Definitions 288
A.3 The Determinant of a Square Matrix 289
A.4 The Minors, Cofactors, Adjoints, and Trace of a Square Matrix 290
A.5 The Rank of a Matrix 291
CAD Facility 294
A.6 The Solution of Simultaneous Equations 294
CAD Facility 295
A.7 The Inverse of a Square Matrix 296
CAD Facility 298
A.8 Some Properties of Matrices and Determinants 298
A.9 The Eigenvalues and Eigenvectors of a Square Matrix; the Cayley–Hamilton Theorem 300
(i) Eigenvalues and Eigenvectors 300
(ii) The Cayley–Hamilton Theorem 301
CAD Facility 301
A.10 Similar Matrices 303
A.11 Definite and Semidefinite Quadratic Forms 304

Appendix **B Notes on Probability Functions 306**
B.1 Basic Definitions 306
B.2 Mean, Variance, and Standard Deviation 307
B.3 Joint and Conditional Probability Density Functions 309
B.4 The Central Limit Theorem 310
B.5 Pseudo-random Signal Generation 310
(i) Pseudo-random Binary Signals (PRBS) 311
(ii) Evenly Distributed Random Numbers 312
(iii) Random Numbers with Gaussian Distribution 313

Appendix **C Notes on Delta Functions 314**
C.1 The Dirac Delta Function 314
C.2 The Kronecker Delta Function 315
References 315

Index 317

PREFACE

This book contains an ordered presentation of practical modern control engineering techniques with explanations, formulas, and examples, but without mathematical proofs. Many detailed suggestions are made for the construction of computer aided design (CAD) algorithms suited to readily available microcomputers supporting BASIC. At the end of each chapter a limited but carefully selected bibliography helps the reader to explore further. Continuous and discrete time systems are given equal emphasis.

Chapter 1 sets the mathematical background with topics such as differential and difference equations, Laplace, z, Fourier transforms, and stochastic system definitions. It includes CAD algorithm designs which are useful in themselves and as subroutines for larger programs. Chapter 2 deals with 'classical' techniques for continuous and discrete time systems. Chapters 3 and 4 describe modern control ideas for linear systems including pole shifting, state estimation, stochastic systems, and Kalman filters. Chapter 5 is devoted to computing methods for function optimization and system identification since these are keys to the future development of this subject. The appendixes contain basic definitions, methods, and algorithms.

This material, which has been warmly welcomed by many short-course students, is suited to professional engineers, postgraduates, and advanced undergraduates. It is not intended as a first introduction to control engineering.

The author is grateful to the considerable number of Cranfield students who have investigated the methods and algorithms outlined in this book. He is also indebted to the staff members who have developed the laboratory experiments. These have provided valuable insight without being unduly complex, and can perhaps be copied by the interested reader.

J. A. B.

MODERN CONTROL SYSTEMS:
A Manual of Design Methods

1 BASIC DEFINITIONS AND MATHEMATICAL TECHNIQUES

1.1 INTRODUCTION

Some basic definitions and mathematical techniques are set out in this chapter in a compact form, together with suggestions for the design of CAD algorithms. Topics covered include linear differential and difference equations, Laplace, z, and Fourier transforms and basic stochastic system definitions. The reference books listed at the end provide thorough introductions to these topics from a fairly elementary level.

1.2 ORDINARY DIFFERENTIAL EQUATIONS

The behavior of many dynamical systems can be modeled by one or more differential equations of the form:

$$F(y, y^{(1)}, y^{(2)}, \ldots, y^{(m)}, t) = 0 \tag{1.1}$$

where t represents time, y represents some aspect of the system, typically its output, and

$$y^{(1)} = \frac{\mathrm{d}y}{\mathrm{d}t}; \quad y^{(2)} = \frac{\mathrm{d}^2y}{\mathrm{d}t^2}; \quad \ldots \quad ; \quad y^{(m)} = \frac{\mathrm{d}^m y}{\mathrm{d}t^m}.$$

Equation (1.1) is mth order (the highest order of derivative) and ordinary (only ordinary derivatives involved).

The general solution, i.e. the function $y(t)$ which satisfies Eq. (1.1), is of the form:

$$y(t) = y(t, c_0, c_1, \ldots, c_{m-1})$$

where $c_0, c_1, \ldots, c_{m-1}$ are constants which can usually be determined if values of $y, y^{(1)}, y^{(2)}, \ldots, y^{(m-1)}$ are known for some specific value of t, say t_0. Typically, 'initial' values $y(0), y^{(1)}(0), \ldots, y^{(m-1)}(0)$ are known.

EXAMPLE

Consider the first-order differential equation:

$$\frac{dy}{dt} + 2yt - e^{-t^2} = 0. \qquad (1.2)$$

This has the general solution:

$$y = e^{-t^2}(t + c_0)$$

where c_0 is a constant.

Given the 'initial' value $y(0) = 1$, it is easy to show by substitution that the solution is:

$$y = e^{-t^2}(t + 1)$$

While some useful types of differential equation do have analytic solutions (ref. 1), numerical solutions can very often be found for these and less tractable cases. This topic is of interest here.

If Eq. (1.1) can be rewritten with the highest-order derivatives on the left-hand side (LHS) – and it usually can – it can be recast as two equations, one first and one $(m-1)$th order. Repeating this process, Eq. (1.1) can finally be cast as a set of m first-order equations which in vector format (App. A.1) is:

$$\dot{\mathbf{x}} = \mathbf{f}(\mathbf{x}, t) \qquad (1.3)$$

with initial conditions:

$$\mathbf{x}(0) = \mathbf{x}_0$$

where

$$\mathbf{x} = (x_1 x_2 x_3 \dots x_m)^{\mathrm{T}}.$$

EXAMPLE

Consider the third-order differential equation:

$$\frac{d^3y}{dt^3} + 9\frac{d^2y}{dt^2} + 26\frac{dy}{dt} + 24y - 1 = 0 \qquad (1.4)$$

with initial conditions:

$$y(0) = 4, \quad y^{(1)}(0) = 3, \quad y^{(2)}(0) = 2.$$

By setting $x_1 = y$, rewriting Eq. (1.4) with d^3x_1/dt^3 on the LHS and substituting $x_2 = dx_1/dt$, two equations are generated, and by repeating the process with $x_3 = dx_2/dt$, Eq. (1.4) is recast:

$$\dot{\mathbf{x}} = \begin{bmatrix} \dot{x}_1 \\ \dot{x}_2 \\ \dot{x}_3 \end{bmatrix} = \begin{bmatrix} x_2 \\ x_3 \\ -9x_3 - 26x_2 - 24x_1 + 1 \end{bmatrix}$$

$$y = x_1$$

with initial conditions:

$$\mathbf{x}(0) = \begin{bmatrix} 4 \\ 3 \\ 2 \end{bmatrix}.$$

This is in the form of Eq. (1.3)

Once a differential equation, or set of equations, has been cast in this form, it can be fed to a fairly simple CAD algorithm to yield a numerical solution.

CAD Facility

Equations in the form of Eq. (1.3) with $\mathbf{x}(0) = \mathbf{x}_0$ can be solved numerically; i.e. the value $\mathbf{x}(nh)$, $n = 0, 1, 2, \ldots$, h a calculation interval, can be found by several well-known methods.

An interactive CAD algorithm is shown in the flow diagram of Fig. 1.1. The operation of the algorithm is as follows.

BLOCK 1

The differential equation is input in the form of Eq. (1.3) or in a manner allowing easy conversion to this format. It is displayed and corrected if necessary.

BLOCK 2

The calculation interval h, a printout ratio, R, and the total number of iterations required, N, are input.

BLOCK 3

$\mathbf{x}(nh)$, $n = 0, 1, 2, \ldots, N$, are calculated. Two types of mathematically stable method are outlined here.

(a) *Runge–Kutta Methods* (*refs.* 5, 7). These are based on the Taylor expansion:

$$\mathbf{x}((n+1)h) = \mathbf{x}(nh) + h\mathbf{x}^{(1)}(nh) + \frac{h^2}{2!}\mathbf{x}^{(2)}(nh) + \frac{h^3}{3!}\mathbf{x}^{(3)}(nh) + \cdots. \quad (1.5)$$

Commonly, four terms in this series are used to derive the fourth-order Runge–Kutta algorithm:

Calculate:

$$\mathbf{k}_1 = h\mathbf{f}(\mathbf{x}(nh), nh)$$

$$\mathbf{k}_2 = h\mathbf{f}(\mathbf{x}(nh) + \tfrac{1}{2}\mathbf{k}_1, nh + \tfrac{1}{2}h)$$

$$\mathbf{k}_3 = h\mathbf{f}(\mathbf{x}(nh) + \tfrac{1}{2}\mathbf{k}_2, nh + \tfrac{1}{2}h)$$

$$\mathbf{k}_4 = h\mathbf{f}(\mathbf{x}(nh) + \mathbf{k}_3, nh + h).$$

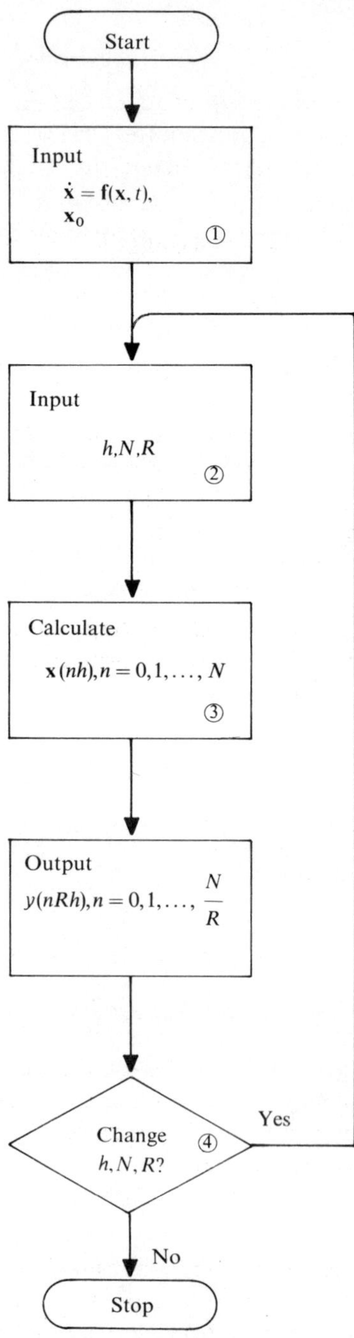

Fig. 1.1 CAD program for solving differential equations.

Then:

$$\mathbf{x}((n+1)h) = \mathbf{x}(nh) + \tfrac{1}{6}(\mathbf{k}_1 + 2\mathbf{k}_2 + 2\mathbf{k}_3 + \mathbf{k}_4).$$

Higher accuracy, at the cost of added complexity, can be achieved using a fifth-order Runge–Kutta algorithm (ref. 7). For simple cases, only two terms need be considered to yield the second-order algorithm:

Calculate:

$$\mathbf{k}_1 = h\mathbf{f}(\mathbf{x}(nh), nh)$$

$$\mathbf{k}_2 = h\mathbf{f}(\mathbf{x}(nh) + \tfrac{1}{2}\mathbf{k}_1, nh + \tfrac{1}{2}h).$$

Then:

$$\mathbf{x}((n+1)h) = \mathbf{x}(nh) + \mathbf{k}_2.$$

In each of these algorithms, since $\mathbf{x}((n+1)h)$ can be found from $\mathbf{x}(nh)$, the starting data $\mathbf{x}(0)$ are sufficient to allow the algorithm to proceed.

The Runge–Kutta method yields estimates of calculation errors only with some difficulty (ref. 7). Predictor–corrector methods are better in this respect, and one of these is outlined next.

(b) *Predictor–Corrector Methods.* Perhaps the simplest such method, due to Adams–Moulton (ref. 5), uses the information $\mathbf{x}(nh)$, $\mathbf{x}((n+1)h)$, to predict $\mathbf{x}((n+2)h)$ by linear extrapolation.

$$\bar{\mathbf{x}}(n+2)h = \mathbf{x}(nh) + \frac{h}{2}\{3\mathbf{f}[\mathbf{x}((n+1)h, (n+1)h] - \mathbf{f}(\mathbf{x}(nh), nh)\}.$$

This is then used in the correction formula:

$$\mathbf{x}((n+2)h) = \mathbf{x}((n+1)h) + \frac{h}{2}\{\mathbf{f}[\mathbf{x}((n+1)h), (n+1)h]$$

$$+ \mathbf{f}[\bar{\mathbf{x}}((n+2)h), (n+2)h)]\}.$$

The second equation can be used a number of times to improve $\mathbf{x}((n+2)h)$ until a stable value is obtained.

More sophisticated algorithms of this kind are based on nonlinear extrapolation formulas using three or more initial known values, typically $\mathbf{x}(0)$, $\mathbf{x}(h)$, $\mathbf{x}(2h)$. The initial information required by such algorithms is usually generated by Runge–Kutta methods.

BLOCK 4

The calculation interval may be shortened after a run, and the process repeated. If compatible results are obtained, the solution may be judged satisfactory and the longer interval selected for further runs. Otherwise the process is repeated.

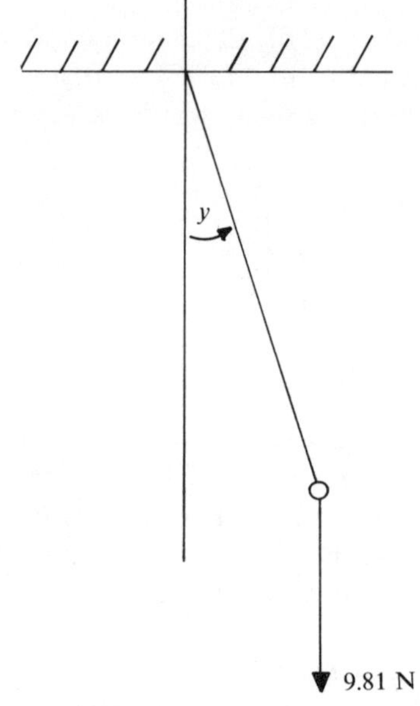

(a)

t	y	t	y	t	y
		3.6000	−0.39111	7.4000	−0.98938
0.00000	1.0000	3.8000	0.20125	7.6000	−0.93718
0.20000	0.83732	4.0000	0.71921	7.8000	−0.57453
0.40000	0.39007	4.2000	0.98663	8.0000	−0.63817E-02
0.60000	−0.20023	4.4000	0.93162	8.2000	0.56450
0.80000	−0.71649	4.6000	0.56747	8.4000	0.93354
1.0000	−0.98298	4.8000	0.96057E-04	8.6000	0.99306
1.2000	−0.92784	5.0000	−0.56767	8.8000	0.72872
1.4000	−0.56446	5.2000	−0.93251	9.0000	0.21087
1.6000	0.12048E-02	5.4000	−0.98804	9.2000	−0.38555
1.8000	0.56677	5.6000	−0.72082	0.4000	−0.84112
2.0000	0.92967	5.8000	−0.20239	9.6000	−1.0113
2.2000	0.98391	6.0000	0.39150	9.8000	−0.85346
2.4000	0.71631	6.2000	0.84242	10.000	−0.40682
2.6000	0.19882	6.4000	1.0076		
2.8000	−0.39284	6.6000	0.84547		
3.0000	−0.84104	6.8000	0.39667		
3.2000	−1.0038	7.0000	−0.19743		
3.4000	−0.84008	7.2000	−0.71855		

(b)

Fig. 1.2 (a) Pendulum. (b) Runge–Kutta model results.

EXAMPLE

The behavior of a simple undamped pendulum of mass 1 kg and length 1 m is modeled by the second-order differential equation:

$$\ddot{y} + 9.81 \sin y = 0$$

where y is the angular displacement of the pendulum from the vertical as shown in Fig. 1.2(a).

Following the rules outlined above, this is cast as two first-order differential equations:

$$\dot{x}_1 = x_2$$
$$\dot{x}_2 = -9.81 \sin x_1$$

and

$$y = x_1.$$

Given initial conditions $x_1(0) = y(0) = 1$ rad, $x_2(0) = y^{(1)}(0) = 0$ rad sec^{-1}, a calculation interval $h = 0.05$, a printout ratio $R = 4$, and $N = 200$ calculation intervals, and using a second-order Runge–Kutta algorithm:

$$\mathbf{k}_1 = 0.05\begin{bmatrix} 0 \\ -8.254 \end{bmatrix} = \begin{bmatrix} 0 \\ -0.413 \end{bmatrix}$$

$$\mathbf{k}_2 = 0.05\begin{bmatrix} -0.207 \\ -8.254 \end{bmatrix}$$

$$\mathbf{x}(h) = \begin{bmatrix} 1 \\ 0 \end{bmatrix} + \begin{bmatrix} -0.0104 \\ -0.4127 \end{bmatrix} = \begin{bmatrix} 0.989 \\ -0.413 \end{bmatrix}.$$

The algorithm proceeds, calculating $\mathbf{x}(nh)$, $n = 0, 1, \ldots, 200$, and printing $y(nRh)$, $R = 4$, as shown in Fig. 1.2(b).

1.3 THE LAPLACE TRANSFORM (REF. 9)

The Laplace transform is used to transform a function of the time variable, t, into another function of the 'complex frequency' variable, $s = \sigma + j\omega$.

(i) The Laplace Transform

The Laplace transform of $f(t)$, a function of t, is defined:

$$F(s) = \mathscr{L}\{f(t)\} = \oint_0^\infty e^{-st} f(t) \mathrm{d}t. \tag{1.6}$$

The conditions of convergence of this integral are of some theoretical interest (ref. 9), but are not explored here.

EXAMPLES

(a) Consider a step function:

$$f(t)=\begin{cases}0, & -\infty \leqslant t<0 \\ a, & 0 \leqslant t \leqslant +\infty .\end{cases}$$

Then

$$F(s)=\int_0^{\infty} e^{-st} a \, \mathrm{d}t=\frac{a}{s}.$$

(b) Consider an exponential 'decay' function:

$$f(t)\begin{cases}0 & -\infty \leqslant t<0 \\ e^{-at}, & 0 \leqslant t \leqslant +\infty .\end{cases}$$

Then

$$F(s)=\int_0^{\infty} e^{-st} e^{-at} \, \mathrm{d}t=\frac{1}{s+a}.$$

The Laplace transforms of some common functions of time are listed in Fig. 1.13.

Some important properties of Laplace transforms are:

(a) Linearity

$$\mathscr{L}\{a_1 f_1(t)+a_2 f_2(t)\}=a_1 F_1(s)+a_2 F_2(s) \tag{1.7}$$

where a_1, a_2 are constants, $\mathscr{L}\{f_1(t)\}=F_1(s)$, $\mathscr{L}\{f_2(t)\}=F_2(s)$.

(b) Delayed Function

$$\mathscr{L}\{f(t-a)\}=e^{-as} F(s). \tag{1.8}$$

(c) Initial Value Theorem

$$\lim_{t \to 0}\{f(t)\}=\lim_{s \to \infty}(sF(s)). \tag{1.9}$$

(d) Final Value Theorem

$$\lim_{t \to \infty} f(t)=\lim_{s \to 0}(sF(s)). \tag{1.10}$$

(e) Differentiation with respect to t

$$\mathscr{L}\left\{\frac{\mathrm{d}}{\mathrm{d}t}(f(t))\right\}=sF(s)-f(0). \tag{1.11}$$

$$\mathscr{L}\left\{\frac{\mathrm{d}^n}{\mathrm{d}t^n}(f(t))\right\} = s^n F(s) - s^{n-1} f(0) - s^{n-2} f^{(1)}(0) - \cdots - f^{(n-1)}(0). \quad (1.12)$$

(f) Integration with respect to t

$$\mathscr{L}\left\{\int_0^t f(\lambda)\mathrm{d}\lambda\right\} = \frac{F(s)}{s} - \frac{f^{(-1)}(0)}{s} \quad (1.13)$$

where λ is a dummy variable.

(ii) The Inverse Laplace Transform

The inverse Laplace transform of $F(s)$ is $f(t)$. A formula, the *Bromwich–Wagner integral*, is available:

$$\begin{aligned} f(t) &= \mathscr{L}^{-1}(F(s)) \\ &= \frac{1}{2\pi j}\int_{c-j\infty}^{c+j\infty} e^{st}F(s)\,\mathrm{d}s. \end{aligned} \quad (1.14)$$

Equation (1.14) includes a line integral in the s plane equal to that round a closed D-shaped contour enclosing all the 'poles' of $F(s)$ (s for which $F(s) = \infty$). $f(t)$ can be found by summing the residues of $e^{st}F(s)$ at these poles.

$$f(t) = \sum \mathscr{R}[e^{st}F(s)] \quad (1.15)$$

where the residue of $e^{st}F(s)$ at an rth-order pole $s = s_0$ is:

$$\mathscr{R}[e^{st}F(s)] = \lim_{s\to s_0}\left\{\frac{1}{(r-1)!}\frac{\mathrm{d}^{r-1}}{\mathrm{d}s^{r-1}}((s-s_0)^r e^{st}F(s))\right\}. \quad (1.16)$$

The sign $\sum$ represents summation over all the poles of $[e^{st}F(s)]$. In the usual case, $r = 1$, this becomes:

$$f(t) = \sum \mathscr{R}[e^{st}F(s)] = \sum \lim_{s\to s_0}[(s-s_0)e^{st}F(s)].$$

EXAMPLE

Consider:

$$F(s) = \frac{(s+1)}{(s+2)(s+3)}$$

$$f(t) = \sum \mathscr{R}\left[\frac{e^{st}(s+1)}{(s+2)(s+3)}\right].$$

In this case there are poles at $s = -2$, $s = -3$.

$$\begin{aligned} \mathscr{R}_1[e^{st}F(s)] &= \left.\frac{e^{st}(s+1)(s+2)}{(s+2)(s+3)}\right|_{s=-2} \\ &= (-1)e^{-2t} \end{aligned}$$

$$\mathscr{R}_2[e^{st}F(s)] = \left.\frac{e^{st}(s+1)(s+3)}{(s+2)(s+3)}\right|_{s=-3}$$

$$= (+2)e^{-3t}.$$

Hence $f(t) = 2e^{-3t} - e^{-2t}$.

$\mathscr{L}^{-1}(F(s))$ can often be found by manipulating $F(s)$ into the sum of expressions listed in the table of Fig. 1.13.

EXAMPLE

Consider:

$$F(s) = \frac{(s+1)}{s(s+2)(s+3)(s+4)}.$$

By partial fractions:

$$F(s) = \frac{1}{24s} + \frac{1}{4(s+2)} - \frac{2}{3(s+3)} + \frac{3}{8(s+4)}.$$

From Fig. 1.13,

$$f(t) = \frac{u^*(t)}{24} + \tfrac{1}{4}e^{-2t} - \tfrac{2}{3}e^{-3t} + \tfrac{3}{8}e^{-4t}.$$

where u^* is a unit step function, defined in the footnote to Fig. 1.13.

1.4 THE SOLUTION OF DIFFERENTIAL EQUATIONS USING LAPLACE TRANSFORMS

A useful class of differential equations (Eq. (1.1)), namely ordinary linear differential equations with constant coefficients, can be solved using Laplace transforms. These are of the form:

$$\alpha_m \frac{d^m y}{dt^m} + \alpha_{m-1}\frac{d^{m-1}y}{dt^{m-1}} + \cdots + \alpha_0 y = \beta_m \frac{d^m u}{dt^m} + \beta_{m-1}\frac{d^{m-1}u}{dt^{m-1}} + \cdots + \beta_0 u \quad (1.17)$$

where $u(t)$ is a known function of t.

A solution can be found if initial conditions (or conditions at some time t_0), $y(0)$, $y^{(1)}(0), \ldots, y^{(m-1)}(0)$, are specified.

By taking the Laplace transform of Eq. (1.17) and applying Eq. (1.12), an algebraic equation is formed which can be solved for $Y(s)$, the Laplace transform of $y(t)$. $\mathscr{L}^{-1}(Y(s))$ gives the solution $y(t)$.

EXAMPLE

Consider Eq. (1.4):

$$\frac{d^3y}{dt^3} + 9\frac{d^2y}{dt^2} + 26\frac{dy}{dt} + 24y = 1$$

with initial conditions (in this case): $y(0) = y^{(1)}(0) = y^{(2)}(0) = 0$. Taking Laplace transforms and applying Eq. (1.12):

$$s^3 Y(s) + 9s^2 Y(s) + 26s Y(s) + 24Y(s) = \frac{1}{s}.$$

where

$$Y(s) = \mathscr{L}\{y(t)\}$$

$$Y(s) = \frac{1}{s(s+2)(s+3)(s+4)}$$

$$= \frac{1}{24s} - \frac{1}{4(s+2)} + \frac{1}{3(s+3)} - \frac{1}{8(s+4)}.$$

From Fig. 1.13, the solution is:

$$y(t) = \mathscr{L}^{-1} Y(s) = \frac{u^*(t)}{24} - \tfrac{1}{4}e^{-2t} + \tfrac{1}{3}e^{-3t} - \tfrac{1}{8}e^{-4t}$$

where $u^*(t)$ is a unit step function at $t = 0$.

The solution of equations by this method often depends on an ability to factorize polynomials in s. In the above example:

$$(s^3 + 9s^2 + 26s + 24) = (s+2)(s+3)(s+4).$$

In general this is not an easy task, and a CAD facility is a considerable asset.

CAD Facility

The algorithm outlined here, based on the Bairstow–Hitchcock method (ref. 8), factorizes a polynomial:

$$G(s) = a_0 s^m + a_1 s^{m-1} + \cdots + a_m. \tag{1.18}$$

First, a quadratic factor is found:

$$F(s) = s^2 + ps + q.$$

Then:

$$G(s) = F(s)[b_0 s^{m-2} + b_1 s^{m-3} + \cdots + b_{m-2}].$$

The roots of $F(s)$ are found by the quadratic formula:

$$s = \tfrac{1}{2}[-p \pm \sqrt{p^2 - 4q}]. \tag{1.19}$$

The process is repeated until all the roots have been found.

An interactive algorithm, effective for most polynomials without repeated quadratic factors (ref. 8), is shown in Fig. 1.3, and operates as follows.

BLOCK 1

The parameters $a_0, a_1, \ldots, a_m$ are input, displayed, and corrected if necessary.

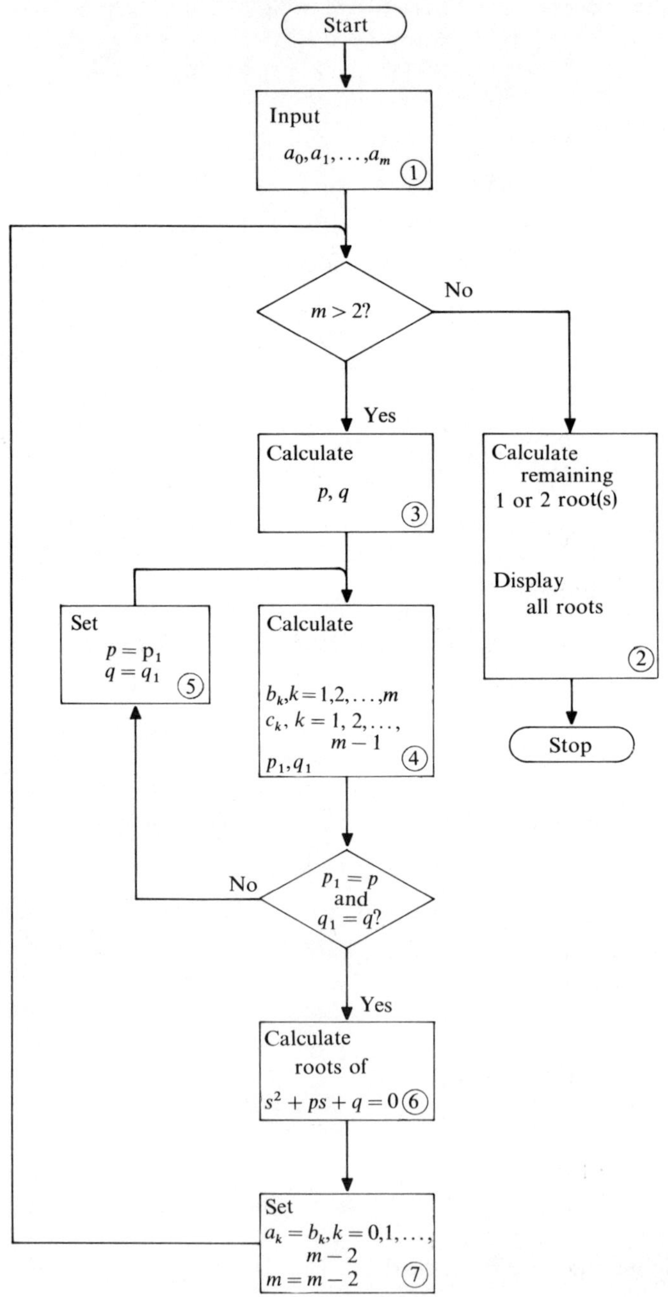

Fig. 1.3 The Bairstow–Hitchock method of finding polynomial roots.

BLOCK 2
If $m \leqslant 2$, the case is trivial and the solution is calculated and displayed. Any roots previously calculated are also displayed.

BLOCK 3
As a first approximation, set:

$$p = \frac{a_{m-1}}{a_{m-2}}$$

$$q = \frac{a_m}{a_{m-2}}.$$

BLOCK 4
Set $b_{-2} = 0$, $b_{-1} = 0$, and calculate:

$$b_k = a_k - pb_{k-1} - qb_{k-2}, \qquad k = 0, 1, 2, \ldots, m.$$

Set $c_{-2} = 0$, $c_{-1} = 0$, and calculate:

$$c_k = b_k - pc_{k-1} - qc_{k-2}, \qquad k = 0, 1, \ldots, m-1.$$

From these calculate:

$$\Delta p = \frac{b_{m-1}c_{m-2} - b_m c_{m-3}}{c_{m-2}^2 - (c_{m-1} - b_{m-1})c_{m-3}}$$

$$\Delta q = \frac{c_{m-2}b_m - (c_{m-1} - b_{m-1})b_{m-1}}{c_{m-2}^2 - (c_{m-1} - b_{m-1})c_{m-3}}.$$

Set

$$p_1 = p + \Delta p$$

$$q_1 = q + \Delta q.$$

BLOCK 5
The process is repeated until p and q are found as accurately as the computer resolution allows.

BLOCK 6
Find the roots of the quadratic:

$$s^2 + ps + q = 0$$

(Eq. (1.19)) and store them.

BLOCK 7
Find the new reduced polynomial:

$$G(s) = b_0 s^{m-2} + b_1 s^{m-3} + \cdots + b_{m-2}.$$

Update m to $(m-2)$ and repeat the process.

1.5 SYSTEM TRANSFER FUNCTIONS AND DYNAMIC BEHAVIOR (CONTINUOUS TIME)

The methods outlined in Secs. 1.3 and 1.4 can readily be applied to linear system dynamics and linear circuit analysis (refs. 3, 9).

A single-input, single-output system S is illustrated in Fig. 1.4.

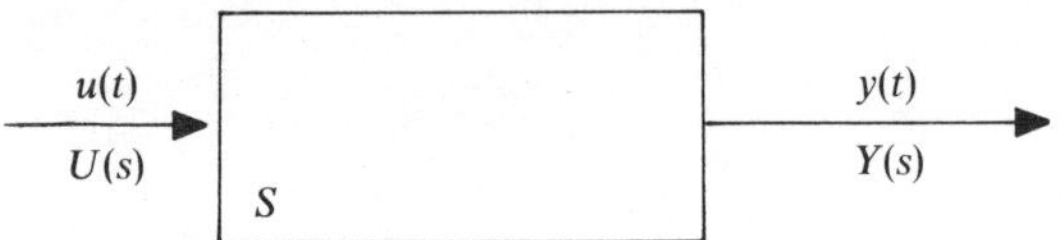

Fig. 1.4 Single-input, single-output system.

If the behavior of this system can be described by a differential equation of the type Eq.(1.17), the system is described as linear 'time invariant' – a term referring to the constant coefficients $\alpha_0, \ldots, \alpha_m, \beta_0, \ldots, \beta_m$.

$u(t)$ is the input to the system S, and $y(t)$ is the output.

Taking Laplace transforms of Eq. (1.17), and assuming zero initial conditions, it is easy to derive:

$$G(s) = \frac{Y(s)}{U(s)} = \frac{\beta_m s^m + \beta_{m-1} s^{m-1} + \cdots + \beta_0}{\alpha_m s^m + \alpha_{m-1} s^{m-1} + \cdots + \alpha_0}. \tag{1.20}$$

$G(s)$, the *transfer function* of the system S, is the ratio of the Laplace transform of the output to that of the input, assuming zero initial conditions.

$G(s)$, which is independent of the input, $U(s)$, and the output, $Y(s)$, describes the natural modes or responses of the system, which occur whenever an input is applied, in addition to the response particular to that input.

By factorizing the numerator and denominator of $G(s)$, the *zeros* and *poles* of S can be found, and it is instructive to consider cases where these occur in pairs:

$$G(s) = \frac{k(s + a_{n1} + jb_{n1})(s + a_{n1} - jb_{n1})(s + a_{n2} + \cdots)\ldots}{(s + a_{d1} + jb_{d1})(s + a_{d1} - jb_{d1})(s + a_{d2} + \cdots)\ldots}. \tag{1.21}$$

The zeros are the values of s for which $G(s) = 0$:

$$s = -a_{n1} \mp jb_{n1}, \quad -a_{n2} \mp jb_{n2}, \ldots.$$

The poles are the values of s for which $G(s) = \infty$:

$$s = -a_{d1} \mp jb_{d1}, \quad -a_{d2} \mp jb_{d2}, \ldots.$$

The poles are of particular interest, since the inverse transform of the output,

$$y(t) = \mathscr{L}^{-1}[Y(s)] = \mathscr{L}^{-1}[G(s)U(s)]$$

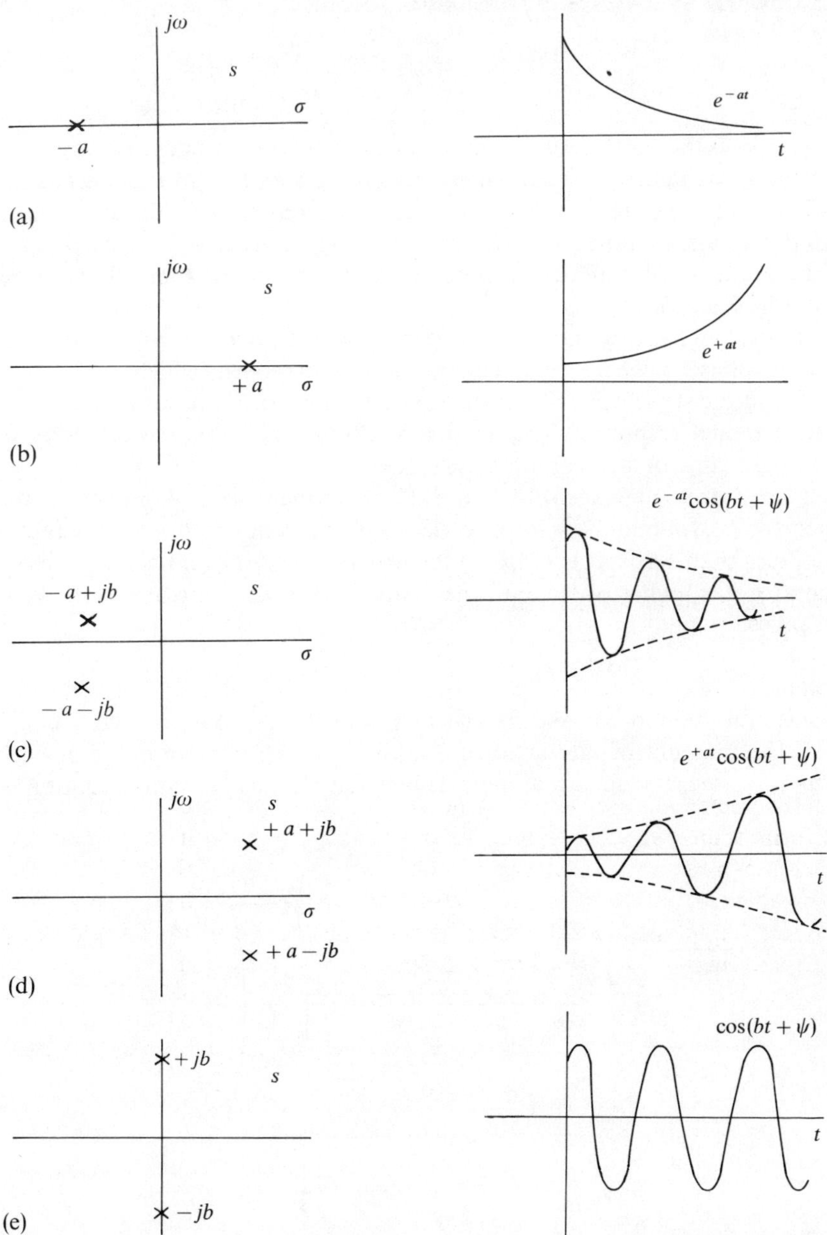

Fig. 1.5 Pole diagrams and corresponding time domain modes. (a) Single pole, real axis, left half plane. (b) Single pole, real axis, right half plane. (c) Complex conjugate pole pair, left half plane. (d) Complex conjugate pole pair, right half plane. (e) Complex conjugate pole pair, imaginary axis.

almost invariably involves expressions of the form:

$$e^{(-a_{d1}+jb_{d1})t}+e^{(-a_{d1}-jb_{d1})t}=2e^{-a_{d1}t}\cos(b_{d1}t),$$

$$e^{(-a_{d2}+jb_{d2})t}+e^{(-a_{d2}-jb_{d2})t}=2e^{-a_{d2}t}\cos(b_{d2}t).$$

The poles therefore describe the 'form' of the natural modes of the system, while the zeros indirectly describe the relative amplitudes of these modes.

The poles and zeros may conveniently be represented on an s plane Argand diagram, or 'pole–zero' diagram. The values of $a_{d1}, a_{d2}, \ldots, b_{d1}, b_{d2}, \ldots$ determine the natural modes as indicated above; some diagrams illustrating this are in Fig. 1.5.

If a system has any zeros in the open right half s plane, it is termed 'non-minimum phase.' Such a system has a negative gain mode(s) which responds to an input in the sense opposite to that expected (a positive input generating a negative model response). This feature naturally increases the difficulty of designing a control strategy for that system.

The system described by Eq. 1.20 is termed *strictly proper* if the numerator polynomial is of lower order than the denominator, i.e. there are more poles than zeros. It is *proper* if the orders are equal, and *improper* if the numerator has higher order than the denominator, i.e. there are more zeros than poles.

EXAMPLE

Consider the torsion bar and flywheel (with friction) shown in Fig. 1.6(a).

If the moment of inertia of the flywheel is 1 kg m^2, damping constant 8 N m rad^{-1} sec, and spring stiffness 25 N m rad^{-1}, the equation of motion is:

$$\frac{d^2y}{dt^2}+8\frac{dy}{dt}+25y=25u$$

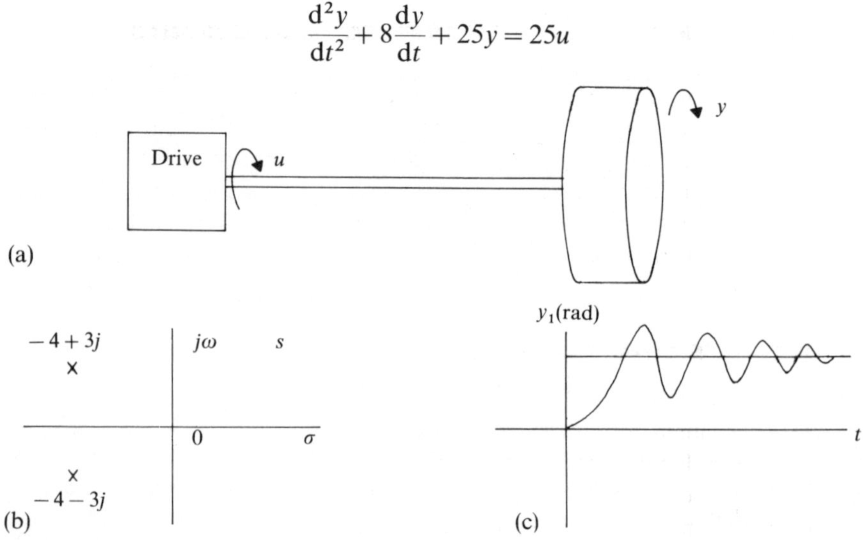

Fig. 1.6 Torsion bar and flywheel. (a) System. (b) Pole–zero diagram. (c) Step response.

u is the input position (rad) to the torsion bar, and y is the position of the flywheel (rad).

Assuming initial conditions to be zero, i.e. $y(0) = y^{(1)}(0) = 0$, and taking Laplace transforms:

$$G(s) = \frac{Y(s)}{U(s)} = \frac{25}{s^2 + 8s + 25}$$
$$= \frac{25}{(s + 4 + 3j)(s + 4 - 3j)}.$$

The pole–zero diagram and corresponding natural mode are in Fig. 1.6(b), and implicitly, 1.6(c).

Considering the particular case of a step function input of 1 rad applied to this system at $t = 0$, the output is readily found (using Fig. 1.13):

$$\begin{aligned} y(t) &= \mathscr{L}^{-1}(Y(s)) \\ &= \mathscr{L}^{-1}(G(s)U(s)) \\ &= \mathscr{L}^{-1}\left\{\frac{25}{s(s + 4 + 3j)(s + 4 - 3j)}\right\} \\ &= u^*(t) + 1.67e^{-4t} \sin(3t - 0.715). \end{aligned}$$

where u^* is a unit step function, defined in the footnote to Fig. 1.13.

This is illustrated in Fig. 1.6(c). The natural mode and the mode particular to the input are worth noting.

1.6 TRANSFER FUNCTIONS OF SOME COMMON CIRCUITS

The transfer functions of circuits are found by treating the impedances of resistance R, capacitance C, and inductance L, as $R, 1/sC$, and sLrespectively. Some common circuits and their transfer functions are in Fig. 1.7.

The transfer functions of composite systems are found by multiplying transfer functions of series-connected and adding those of parallel-connected systems, as illustrated in Fig. 1.8.

1.7 DISCRETE TIME FUNCTIONS

Discrete time functions are defined only at discrete, usually equally spaced, values of the independent variable time.

The usual notation for such a function, y, typically generated by sampling a continuous time function $y(t)$ (cf. Sec. 1.11), is:

$$y(n) = y(nT), \qquad n = 0, 1, 2, \ldots, \infty$$

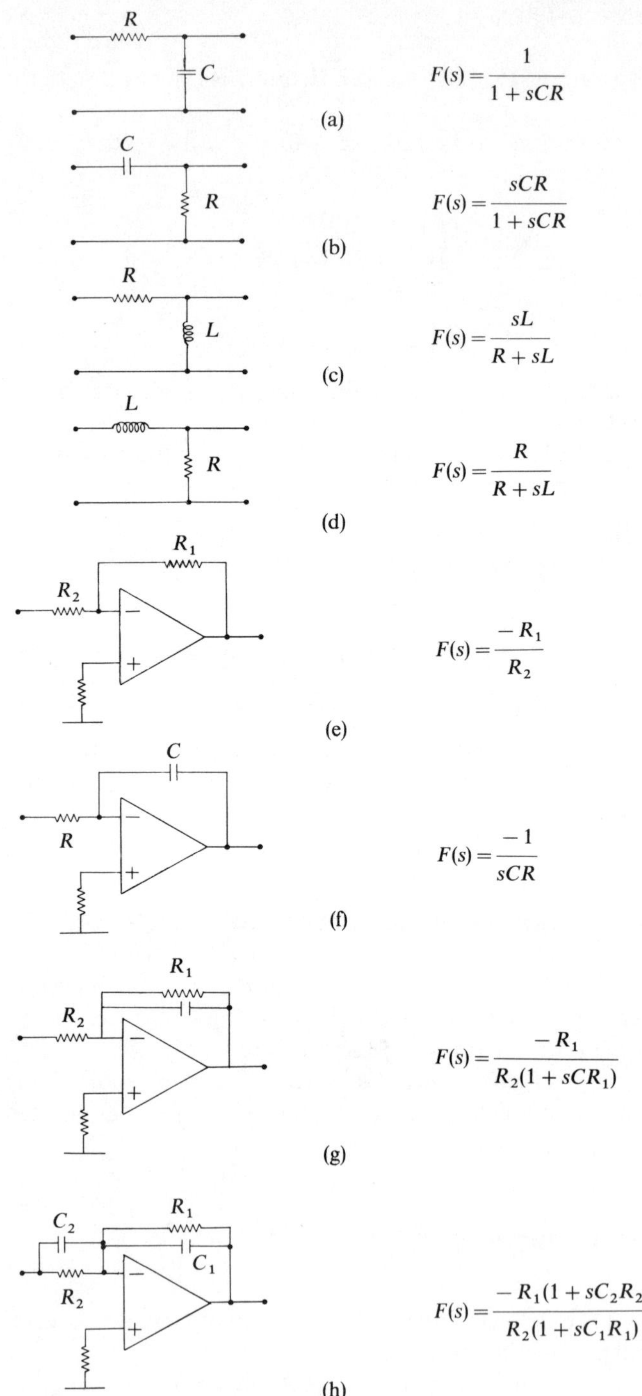

Fig. 1.7 Some common circuits and their transfer functions.

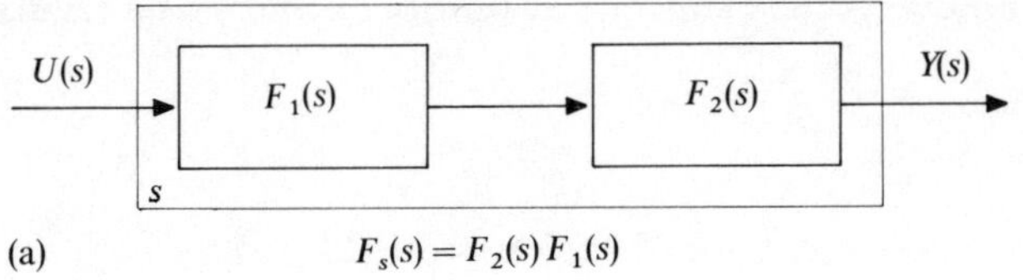

(a) $F_s(s) = F_2(s)\,F_1(s)$

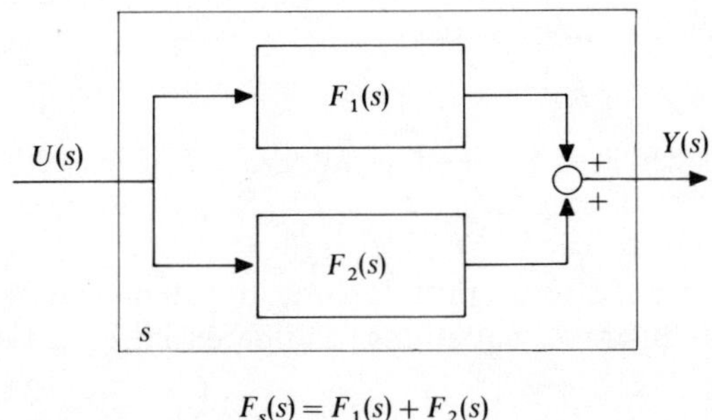

(b) $F_s(s) = F_1(s) + F_2(s)$

Fig 1.8 Composite system transfer functions. (a) Series connected systems. (b) Parallel connected systems.

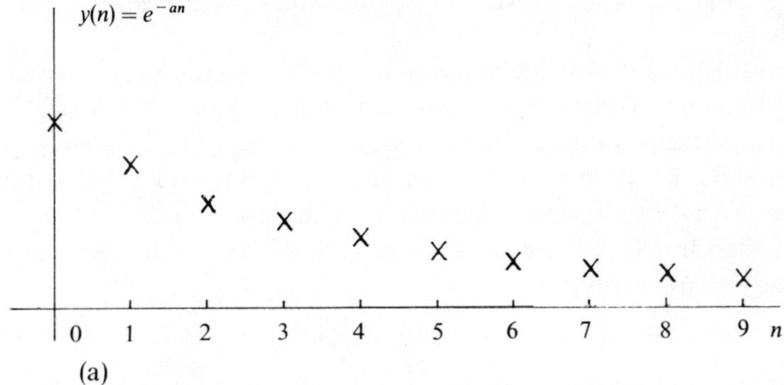

(a)

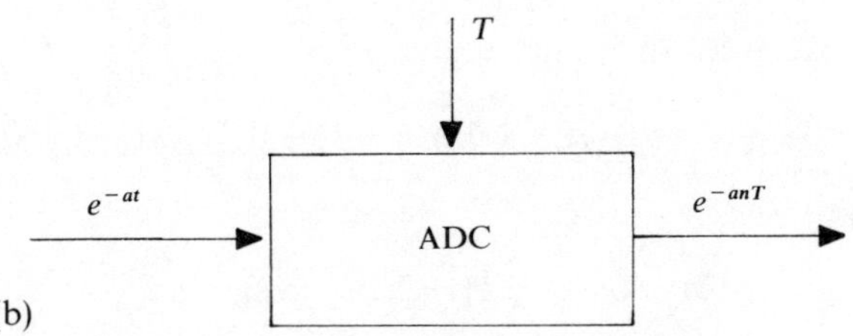

(b)

Fig. 1.9 Discrete time function (example). (a) Discrete signal. (b) Analog to digital converter.

where $y(n)$ represents the value of y at time (nT), and T is a regular sampling interval.

EXAMPLES

$$y(n) = e^{-an}, \qquad n = 0, 1, 2, \ldots, \infty.$$

This represents the sequence illustrated in Fig. 1.9(a).

The function shown could be generated by an analog to digital converter (ADC) to which the input signal

$$y = e^{-at}$$

is applied, and which is 'clocked' at intervals T (Fig. 1.9(b)),

$$y(n) = y(nT) = e^{-anT}.$$

The behavior of discrete time dynamical systems can often be represented by one or more difference equations (cf. differential equations, Sec. 1.2) of the general form:

$$F(y(n), y(n+1), \ldots, y(n+m), n) = 0 \tag{1.22}$$

where n represents discrete time $0, 1, \ldots, \infty$, and $y(n)$ represents some aspect of the system, typically the output.

Equation (1.22) is mth order since the largest discrete time difference involved is m.

Closed-form solutions can be found for certain classes of such equations, but an exploration of this subject is outside the scope of this book.

A direct solution is often easily found by computer. Setting $x_1 = y$, if Eq. (1.22) can be rewritten with the highest-order terms on the LHS, it can be recast as two equations, one first and one $(m-1)$th order. If this process can be repeated, it can finally be cast as a set of m first-order difference equations which, using vector format, is:

$$\mathbf{x}(n+1) = \mathbf{f}(\mathbf{x}(n), n). \tag{1.23}$$

Given initial conditions $\mathbf{x}(0)$, $\mathbf{x}(1)$ can be found directly from Eq. (1.23). From this $\mathbf{x}(2)$ can be found, thence $\mathbf{x}(3)$, and so on, to give the direct solution.

EXAMPLE

Consider the third-order difference equation:

$$y(n+3) - 1.8y(n+2) + 1.07y(n+1) - 0.21y(n) - 1 = 0, \tag{1.24}$$

with initial conditions:

$$y(0) = 1, \qquad y(1) = 2, \qquad y(2) = 3.$$

By setting $x_1(n) = y(n)$, rewriting Eq. (1.24) with $y(n+3)$ on the LHS, and substituting $x_2(n) = x_1(n+1)$, two equations are generated. By repeating this

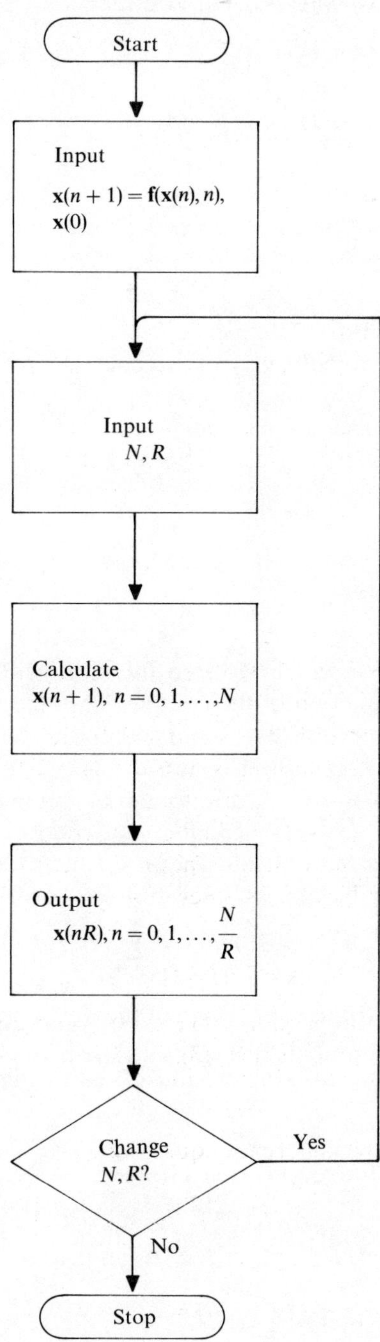

Fig. 1.10 CAD program for solving difference equations.

process with $x_3(n) = x_2(n+1)$, Eq. (1.24) is recast:

$$\mathbf{x}(n+1) = \begin{bmatrix} x_1(n+1) \\ x_2(n+1) \\ x_3(n+1) \end{bmatrix} = \begin{bmatrix} x_2(n) \\ x_3(n) \\ 1.8x_3(n) - 1.07x_2(n) + 0.21x_1(n) + 1 \end{bmatrix}$$

with initial conditions:

$$\mathbf{x}(0) = \begin{bmatrix} 1 \\ 2 \\ 3 \end{bmatrix}.$$

This is in the form of Eq. (1.23).

Clearly, starting with $\mathbf{x}(0)$, $\mathbf{x}(1)$ may be calculated. Hence $\mathbf{x}(2), \mathbf{x}(3), \ldots$ can be found.

It is easy to construct a CAD facility to yield such solutions and one is illustrated in Fig. 1.10 (cf. Fig. 1.1). The operation of the algorithm is self-explanatory.

1.8 THE z TRANSFORM

The z transform is used to transform a function of the discrete variable n (usually time) into a function of the complex variable $z = re^{j\beta}$, a measure of 'complex frequency'.

(i) The z Transform

The z transform of a function of n, $f(n)$, is:

$$\bar{F}(z) = Z\{f(n)\} = \sum_{n=0}^{\infty} f(n)z^{-n}. \tag{1.25}$$

The convergence conditions for $\bar{F}(z)$ are of theoretical interest (ref. 4) but this topic is outside the scope of this book.

Some simple functions, $f(n)$, are illustrated in Fig. 1.11.

EXAMPLES

(a) Consider the step function (Fig. 1.11(a)):

$$f(n) = \begin{cases} 0, & -\infty \leqslant n < 0 \\ 1, & 0 \leqslant n \leqslant +\infty \end{cases}$$

$$\bar{F}(z) = \sum_{n=0}^{\infty} 1z^{-n}$$

$$= z^0 + z^{-1} + z^{-2} + \cdots.$$

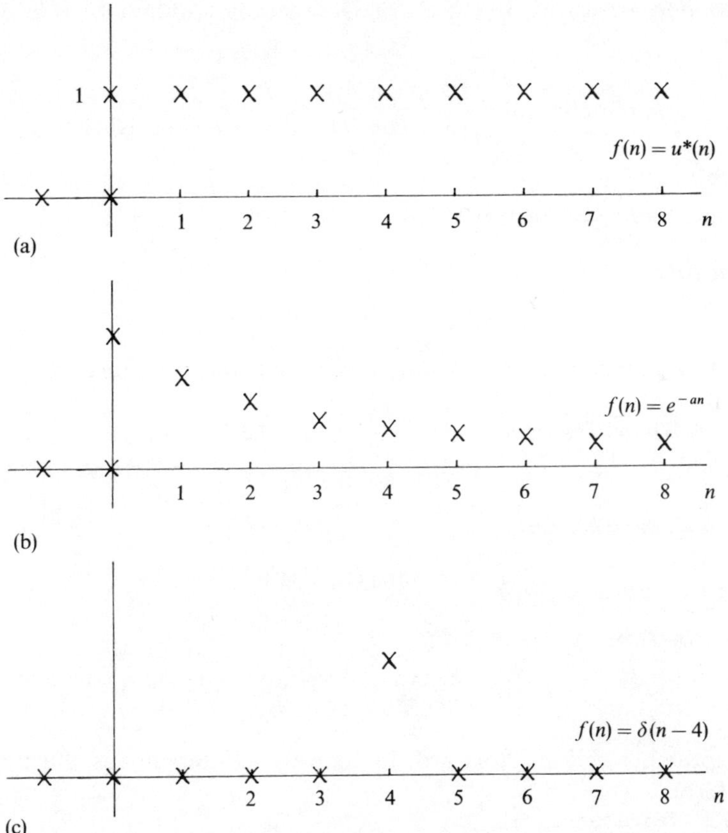

Fig. 1.11 Some simple functions. (a) Unit step function. (b) Exponential decay. (c) Delta function (App. C.2).

This can be written in a convenient closed form (using the binomial expansion):

$$\bar{F}(z) = \frac{1}{1 - z^{-1}}.$$

(b) Consider the exponential decay function (Fig. 1.11(b)):

$$f(n) = \begin{cases} 0, & -\infty \leqslant n < 0 \\ e^{-an}, & 0 \leqslant n \leqslant +\infty \end{cases}$$

$$\bar{F}(z) = \sum_{n=0}^{\infty} e^{-an} z^{-n}$$

$$= \frac{1}{1 - z^{-1} e^{-a}}.$$

(c) Consider the unit impulse, or Kronecker δ-function (Fig. 1.11(c)) (App. C.2):

$$f(n) = \delta(n-a) = \begin{cases} 0, & n \neq a \\ 1, & n = a \quad (a = 4 \text{ in Fig. 1.11(c)}) \end{cases}$$

$$\bar{F}(z) = z^{-a}.$$

Some important properties of z transforms are:

(a) Linearity

$$Z\{a_1 f_1(n) + a_2 f_2(n)\} = a_1 \bar{F}_1(z) + a_2 \bar{F}_2(z)$$

where a_1, a_2 are constants, $Z\{f_1(n)\} = \bar{F}_1(z), Z\{f_2(n)\} = \bar{F}_2(z)$.

(b) Initial Value Theorem

$$f(0) = \lim_{z \to \infty} \bar{F}(z). \tag{1.26}$$

(c) Final Value Theorem

$$f(\infty) = \lim_{z \to 1} [(z-1)\bar{F}(z)]. \tag{1.27}$$

(d) Shift in Time

$$Z\{f(n+k)\} = z^k \bar{F}(z) - \sum_{n=0}^{k-1} z^{-n} f(n). \tag{1.28}$$

This represents shifting 'forward' in time by k intervals, as illustrated in Fig. 1.12(b).

$$Z\{f(n-k)\} = z^{-k}\bar{F}(z). \tag{1.29}$$

This represents 'delay' by k intervals, as illustrated in Fig. 1.12(c).

(e) Multiplication by e^{an}

$$Z\{e^{-an} f(n)\} = \bar{F}(ze^a).$$

(ii) The Inverse z Transform

Three methods are in common use for finding the inverse z transform of a function of z,

$$f(n) = Z^{-1}[\bar{F}(z)].$$

(a) The General Formula (refs. 2, 4)

The individual elements, $f(n)$, can be found from the line integral:

$$f(n) = \frac{1}{2\pi j} \oint_C \bar{F}(z) z^{n-1}\, \mathrm{d}z. \tag{1.30}$$

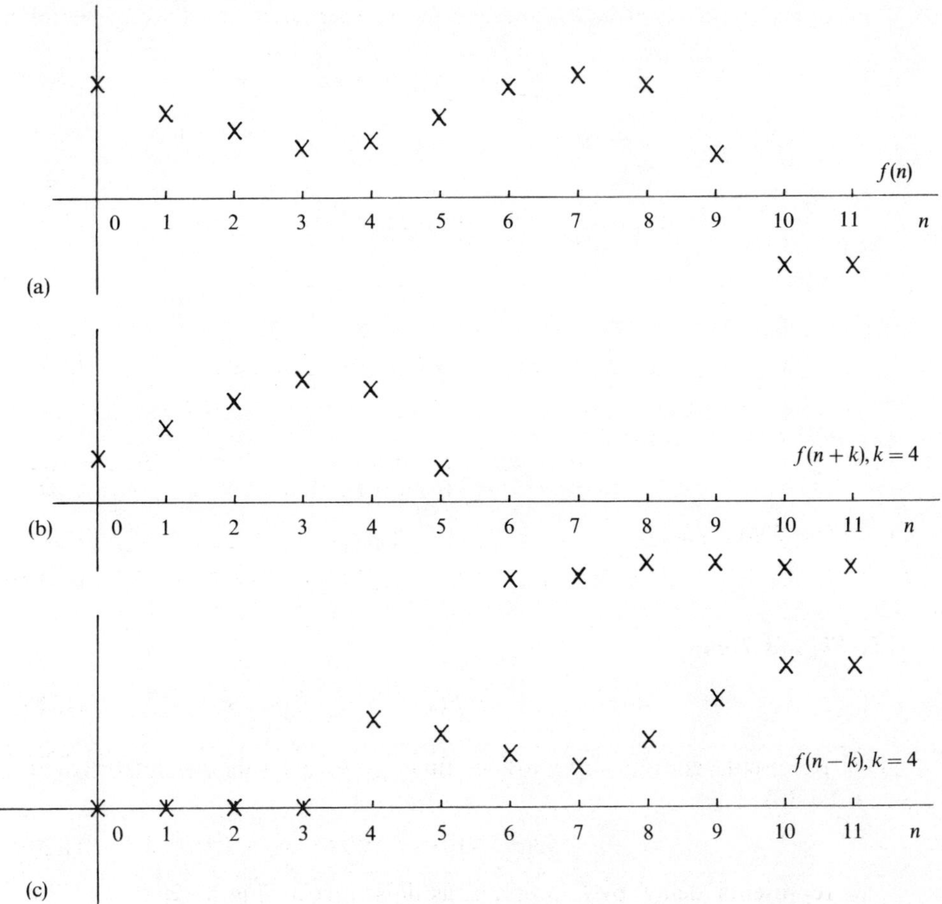

Fig. 1.12 Shifting in time. (a) Discrete signal. (b) Advanced signal. (c) Delayed signal.

C represents a closed path in the z plane which encloses all the singular points of $(\bar{F}(z)z^{n-1})$.

By the residue theorem, the RHS of Eq. (1.30) is equal to the sum of the residues of $[\bar{F}(z)z^{n-1}]$ at the 'poles' of this function (cf. Sec. 1.3(ii)):

$$\begin{aligned} f(n) &= \sum \mathscr{R}[\bar{F}(z)z^{n-1}] \\ &= \sum \lim_{z \to z_0} \left\{ \frac{1}{(r-1)!} \frac{\mathrm{d}^{r-1}}{\mathrm{d}z^{r-1}} [(z - z_0)^r \bar{F}(z) z^{n-1}] \right\} \end{aligned} \tag{1.31}$$

where a pole of $\bar{F}(z)$ (and so of $\bar{F}(z)z^{n-1}$) of order r lies at $z = z_0$, and the sign $\sum$ represents summation over all the poles of $\bar{F}(z)$.

Time Function, $f(t)$	Laplace Transform, $F(s)$	z Transform, $\bar{F}(z)$	Modified z Transform, $\bar{F}(z,m)$
$u^*(t)$	$\frac{1}{s}$	$\frac{1}{1-z^{-1}}$	$\frac{z^{-1}}{1-z^{-1}}$
t	$\frac{1}{s^2}$	$\frac{Tz^{-1}}{(1-z^{-1})^2}$	$\frac{mTz^{-1}}{1-z^{-1}}+\frac{Tz^{-2}}{(1-z^{-1})^2}$
t^2	$\frac{2!}{s^3}$	$\frac{T^2z^{-1}(1+z^{-1})}{(1-z^{-1})^3}$	$T^2\left[\frac{m^2z^{-1}}{1-z^{-1}}+\frac{(2m+1)z^{-2}}{(1-z^{-1})^2}+\frac{2z^{-3}}{(1-z^{-1})^3}\right]$
t^{n-1}	$\frac{(n-1)!}{s^n}$	$\lim_{a\to 0}(-1)^{n-1}\frac{\partial^{n-1}}{\partial a^{n-1}}\left(\frac{1}{1-e^{-aT}z^{-1}}\right)$	$\lim_{a\to 0}(-1)^{n-1}\frac{\partial^{n-1}}{\partial a^{n-1}}\left(\frac{e^{-amT}z^{-1}}{1-e^{-aT}z^{-1}}\right)$
e^{-at}	$\frac{1}{s+a}$	$\frac{1}{1-e^{-aT}z^{-1}}$	$\frac{e^{-amT}z^{-1}}{1-e^{-aT}z^{-1}}$
te^{-at}	$\frac{1}{(s+a)^2}$	$\frac{Te^{-aT}z^{-1}}{(1-e^{-aT}z^{-1})^2}$	$\frac{Te^{-amT}z^{-1}[m+(1-m)e^{-aT}z^{-1}]}{(1-e^{-aT}z^{-1})^2}$
$\frac{1}{b-a}(e^{-at}-e^{-bt})$	$\frac{1}{(s+a)(s+b)}$	$\frac{1}{b-a}\left(\frac{1}{1-e^{-aT}z^{-1}}-\frac{1}{1-e^{-bT}z^{-1}}\right)$	$\frac{z^{-1}}{b-a}\left(\frac{e^{-amT}}{1-e^{-aT}z^{-1}}-\frac{e^{-bmT}}{1-e^{-bT}z^{-1}}\right)$
$\frac{1}{a}(u^*(t)-e^{-at})$	$\frac{1}{s(s+a)}$	$\frac{1}{a}\frac{(1-e^{-aT})z^{-1}}{(1-z^{-1})(1-e^{-aT}z^{-1})}$	$\frac{z^{-1}}{a}\left(\frac{1}{1-z^{-1}}-\frac{e^{-amT}}{1-e^{-aT}z^{-1}}\right)$
$\frac{1}{a}\left(t-\frac{1-e^{-at}}{a}\right)$	$\frac{1}{s^2(s+a)}$	$\frac{1}{a}\left[\frac{Tz^{-1}}{(1-z^{-1})^2}-\frac{(1-e^{-aT})z^{-1}}{a(1-z^{-1})(1-e^{-aT}z^{-1})}\right]$	$\frac{z^{-1}}{a}\left[\frac{T}{(1-z^{-1})^2}+\frac{amT-1}{a(1-z^{-1})}+\frac{e^{-amT}}{a(1-e^{-aT}z^{-1})}\right]$

$\frac{(a-b)}{a^2}u^*(t)+\frac{b}{a}t+\frac{1}{a}\left(\frac{b}{a}-1\right)e^{-at}$	$\frac{s+b}{s^2(s+a)}$	$\frac{z^{-1}}{a}\left[\frac{bT}{(1-z^{-1})^2}+\frac{(a-b)(1-e^{-aT})}{a(1-z^{-1})(1-e^{-aT}z^{-1})}\right]$	$\frac{z^{-1}}{a}\left[\frac{bTz^{-1}}{(1-z^{-1})^2}+\left(bmT+1-\frac{b}{a}\right)\frac{1}{1-z^{-1}}+\frac{b-a}{a}\frac{e^{-amT}}{1-e^{-aT}z^{-1}}\right]$
$\frac{1}{ab}\left(u^*(t)+\frac{b}{a-b}e^{-at}-\frac{a}{a-b}e^{-bt}\right)$	$\frac{1}{s(s+a)(s+b)}$	$\frac{1}{ab}\left[\frac{1}{1-z^{-1}}+\frac{b}{(a-b)(1-e^{-aT}z^{-1})}-\frac{a}{(a-b)(1-e^{-bT}z^{-1})}\right]$	$\frac{z^{-1}}{ab}\left[\frac{1}{1-z^{-1}}+\frac{be^{-amT}}{(a-b)(1-e^{-aT}z^{-1})}-\frac{ae^{-bmT}}{(a-b)(1-e^{-bT}z^{-1})}\right]$
$\sin at$	$\frac{a}{s^2+a^2}$	$\frac{z^{-1}\sin aT}{1-2z^{-1}\cos aT+z^{-2}}$	$\frac{z^{-1}\sin amT+z^{-2}\sin(1-m)aT}{1-2z^{-1}\cos aT+z^{-2}}$
$\frac{1}{b}e^{-at}\sin bt$	$\frac{1}{(s+a)^2+b^2}$	$\frac{1}{b}\left[\frac{z^{-1}e^{-aT}\sin bT}{1-2z^{-1}e^{-aT}\cos bT+e^{-2aT}z^{-2}}\right]$	$\frac{z^{-1}e^{-amT}}{b}\frac{[\sin bmT+z^{-1}e^{-aT}\sin(1-m)bT]}{[1-2z^{-1}e^{-aT}\cos bT+e^{-2aT}z^{-2}]}$
$e^{-at}\cos bt$	$\frac{s+a}{(s+a)^2+b^2}$	$\frac{1-z^{-1}e^{-aT}\cos bT}{1-2z^{-1}e^{-aT}\cos bT+e^{-2aT}z^{-2}}$	$\frac{e^{-amT}z^{-1}[\cos bmT+z^{-1}e^{-aT}\sin(1-m)bT]}{1-2z^{-1}e^{-aT}\cos bT+e^{-2aT}z^{-2}}$
$\cos at$	$\frac{s}{s^2+a^2}$	$\frac{1-z^{-1}\cos aT}{1-2z^{-1}\cos aT+z^{-2}}$	$\frac{z^{-1}\cos amT-z^{-2}\cos(1-m)aT}{1-2z^{-1}\cos aT+z^{-2}}$

Note: $u^*(t)$ denotes a unit step function; $u^*(t)=\begin{cases}0, & -\infty\leqslant t<0\\ 1, & 0\leqslant t\leqslant+\infty\end{cases}$

Fig. 1.13 Common Laplace and z transforms. (Reproduced with permission from Smith, C.L., *Digital Computer Process Control.* Intext, 1972.)

EXAMPLE

Consider:

$$f(n) = Z^{-1}\left\{\frac{z(1-e^{-T})}{(z-1)(z-e^{-T})}\right\}.$$

In this case $r = 1$ for both poles, $z = 1$, $z = e^{-T}$.

$$\begin{aligned} f(n) &= \sum \lim_{z \to z_0} \{(z - z_0)\bar{F}(z)z^{n-1}\} \\ &= \lim_{z \to 1}\left\{\frac{(z-1)z(1-e^{-T})z^{n-1}}{(z-1)(z-e^{-T})}\right\} \\ &\quad + \lim_{z \to e^{-T}}\left\{\frac{(z-e^{-T})z(1-e^{-T})z^{n-1}}{(z-1)(z-e^{-T})}\right\} \\ &= 1 - e^{-nT}. \end{aligned}$$

Hence

$$f(n) = 1 - e^{-nT}, \qquad n = 0, 1, 2, \ldots .$$

(*b*) *Table Look-up*

A table of useful z transforms is in Fig. 1.13 together with related Laplace transforms. The functions $f(t)$ are assumed to be fed to an ADC clocked at intervals T to give the discrete signal $f(n)$ with corresponding z transform $\bar{F}(z)$. A function $f(t)$ thus becomes a series $f(nT)$ with t replaced by (nT). For example,

$$e^{at} \text{ becomes } e^{anT}, \qquad n = 0, 1, \ldots .$$

$Z^{-1}\{\bar{F}(z)\}$ can often be found by manipulating $\bar{F}(z)$ into the sum of expressions listed in Fig. 1.13.

EXAMPLE

Consider:

$$\begin{aligned} f(n) &= Z^{-1}\left\{\frac{z(1-e^{-T})}{(z-1)(z-e^{-T})}\right\} \\ &= Z^{-1}\left\{\frac{z}{z-1} - \frac{z}{z-e^{-T}}\right\} \\ &= u^*(n) - e^{-nT} \qquad \text{(Fig. 1.13)}. \end{aligned}$$

By replacing n with t/T(or nT with t) the function $f(t) = u^*(t) - e^{-t}$ is found, from which $f(n)$ could be derived via an ADC.

(c) *Power Series*

The inverse transform can often be found directly – though not in closed form – simply by dividing out $\bar{F}(z)$.

EXAMPLE

Consider:

$$f(n) = Z^{-1}\left\{\frac{Tz}{(z-1)^2}\right\}$$

$$= Z^{-1}\left\{\frac{Tz}{z^2 - 2z + 1}\right\}.$$

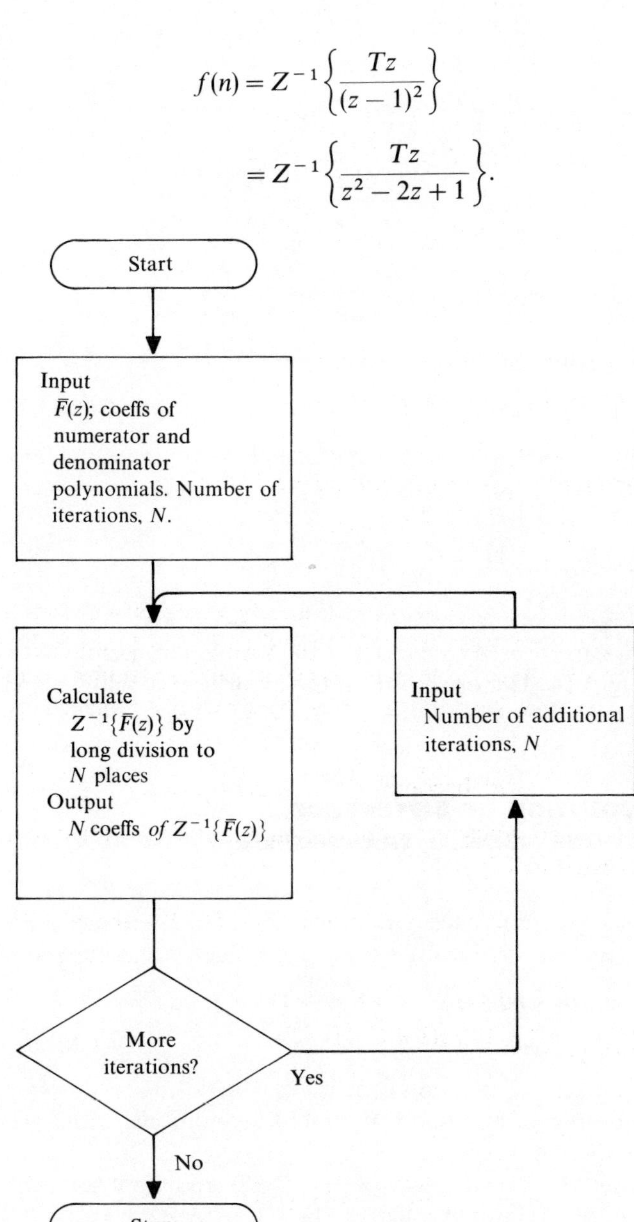

Fig. 1.14 Algorithm for inverse z transforms.

A long division is performed:

$$
\begin{array}{r|l}
 & Tz^{-1}+2Tz^{-2}+3Tz^{-3}+4Tz^{-4}+\cdots \\
\hline
z^2-2z+1 & Tz \\
 & Tz-2T+\ Tz^{-1} \\
 & \quad 2T-\ Tz^{-1} \\
 & \quad 2T-4Tz^{-1}+2Tz^{-2} \\
 & \qquad 3Tz^{-1}-2Tz^{-2} \\
 & \qquad 3Tz^{-1}-6Tz^{-2}+3Tz^{-3} \\
 & \qquad\qquad 4Tz^{-2}-3Tz^{-3}
\end{array}.
$$

Thus

$$\bar{F}(z)=Tz^{-1}+2Tz^{-2}+3Tz^{-3}+\cdots$$

and from the definition of the z transform (Eq. (1.25)):

$$\{f(n)\}=\{0,T,2T,3T,\ldots\}=\{nT\},\qquad \text{or} \qquad f(n)=nT.$$

Equivalently, a continuous time signal giving rise to this sequence is found by replacing (nT) with t (or n with t/T):

$$f(t)=t.$$

CAD Facility

This method can be used to construct the simple but useful CAD algorithm shown in Fig. 1.14. The operation of the algorithm is self-explanatory.

1.9 THE SOLUTION OF DIFFERENCE EQUATIONS USING z TRANSFORMS

A useful class of difference equations (Eq. (1.22)), those with constant coefficients, can be solved using z transforms. Such equations are of the form:

$$
\begin{aligned}
&\alpha_m y(n+m)+\alpha_{m-1}y(n+m-1)+\cdots+\alpha_0 y(n)\\
&\quad=\beta_m u(n+m)+\beta_{m-1}u(n+m-1)+\cdots+\beta_0 u(n)
\end{aligned}
\tag{1.32}
$$

where $u(n)$ is a known function of n.

A solution can be found if 'initial' conditions, $y(0)$, $y(1)$, $y(2),\ldots$, $y(m-1)$ are specified.

By taking the z transform of Eq. (1.32) and applying Eq. (1.28), an algebraic equation is formed which can be solved for $\bar{Y}(z)=Z\{y(n)\}$, given $u(n)$ and the appropriate initial conditions. It then remains to find $Z^{-1}\{\bar{Y}(z)\}$ to give the solution, $y(n)$.

EXAMPLE

Consider the equation:

$$y(n+2) - 1.1y(n+1) + 0.3y(n) = 1$$

with initial conditions:

$$y(0) = y(1) = 0.$$

Taking z transforms:

$$z^2\bar{Y}(z) - 1.1z\bar{Y}(z) + 0.3\bar{Y}(z) = \frac{1}{1-z^{-1}}.$$

Hence

$$\begin{aligned}\bar{Y}(z) &= \frac{z}{(z-1)(z-0.5)(z-0.6)} \\ &= \frac{5}{z-1} + \frac{10}{z-0.5} - \frac{15}{z-0.6}.\end{aligned}$$

From Fig. 1.13, the solution is:

$$y(n) = 5u^*(n-1) + 10(0.5)^{n-1} - 15(0.6)^{n-1}, n \geqslant 1.$$

The solution of equations by this method often depends on an ability to factorize polynomials in z, and the CAD algorithm described in Sec. 1.4 is applicable here.

1.10 SYSTEM TRANSFER FUNCTIONS AND DYNAMIC BEHAVIOR (DISCRETE TIME)

The methods outlined in Secs. 1.5 and 1.6 can readily be applied to discrete linear systems.

A single-input, single-output discrete linear system Σ is illustrated in Fig. (1.15).

If the behavior of the system Σ can be described by a difference equation of the type Eq. (1.32), the system is linear 'time invariant', the term 'time invariant' referring to the constant coefficients $\alpha_0, \alpha_1, \ldots, \alpha_m$, β_0, $\beta_1, \ldots, \beta_m$.

$u(n)$ is the input to the system Σ, $y(n)$ is the output.

Taking z transforms of Eq. (1.32) and assuming zero initial conditions, it is easy to derive:

$$\bar{G}(z) = \frac{\bar{Y}(z)}{\bar{U}(z)} = \frac{\beta_m z^m + \beta_{m-1}z^{m-1} + \cdots + \beta_0}{\alpha_m z^m + \alpha_{m-1}z^{m-1} + \cdots + \alpha_0}. \tag{1.33}$$

$\bar{G}(z)$, the transfer function of the system Σ, is the ratio of the z transform of the output to that of the input, assuming zero initial conditions.

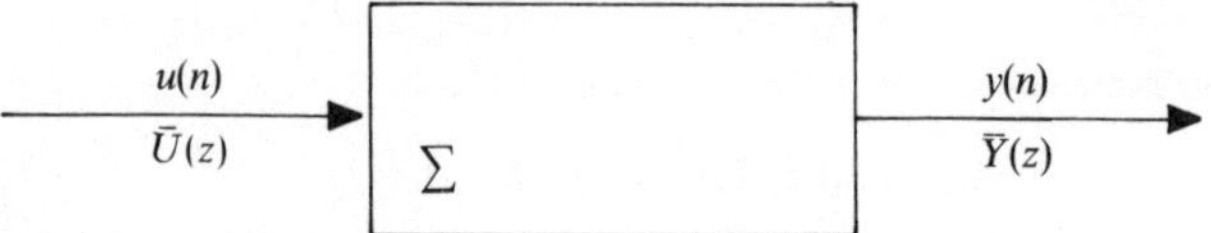

Fig. 1.15 Discrete single-input, single-output system.

$\bar{G}(z)$, which is independent of the input $\bar{U}(z)$ and the output $\bar{Y}(z)$, describes the natural modes, or responses, of the system which occur whenever an input is applied in addition to the response particular to that input.

By factorizing the numerator and denominator of $\bar{G}(z)$, the 'zeros' and 'poles' can be found, and it is instructive to consider cases where these occur in pairs:

$$\bar{G}(z) = \frac{k(z - r_{n1}e^{j\beta_{n1}})(z - r_{n1}e^{-j\beta_{n1}})(z - r_{n2}e^{j\beta_{n2}})\dots}{(z - r_{d1}e^{j\beta_{d1}})(z - r_{d1}e^{-j\beta_{d1}})(z - r_{d2}e^{j\beta_{d2}})\dots}.$$

The zeros are the values of z for which $\bar{G}(z) = 0$:

$$z = r_{n1}e^{j\beta_{n1}}, r_{n1}e^{-j\beta_{n1}}, r_{n2}e^{j\beta_{n2}}, \dots.$$

The poles are the values of z for which $\bar{G}(z) = \infty$:

$$z = r_{d1}e^{j\beta_{d1}}, r_{d1}e^{-j\beta_{d1}}, r_{d2}e^{+j\beta_{d2}}, \dots.$$

The poles are of particular interest since the inverse transform of the system output

$$y(n) = Z^{-1}\{\bar{Y}(z)\} = Z^{-1}\{\bar{G}(z)\bar{U}(z)\}$$

almost invariably involves expressions of the form:

$$\begin{aligned} r_{d1}(e^{j\beta_{d1}n} + e^{-j\beta_{d1}n}) &= 2r_{d1}^n \cos(\beta_{d1}n) \\ r_{d2}(e^{j\beta_{d2}n} + e^{-j\beta_{d2}n}) &= 2r_{d2}^n \cos(\beta_{d2}n). \\ &\vdots \end{aligned}$$

These expressions, and so the poles, describe the natural modes of the system, while the zeros indirectly represent the relative amplitudes of these modes.

The zeros and poles of a transfer function $\bar{G}(z)$ may be represented on a z plane Argand diagram, or 'pole–zero' diagram. The natural modes may be deduced from these using the following simple rules.

1. The argument of a pole pair, β, represents the 'frequency' (radians/sample) of the discrete time mode. Moreover,

$$\beta = \omega T$$

   where ω is the frequency in radians/unit time and T is the sampling period.
2. The modulus of a pole pair (or single real pole), r, represents the behavior of the amplitude of the discrete time mode.

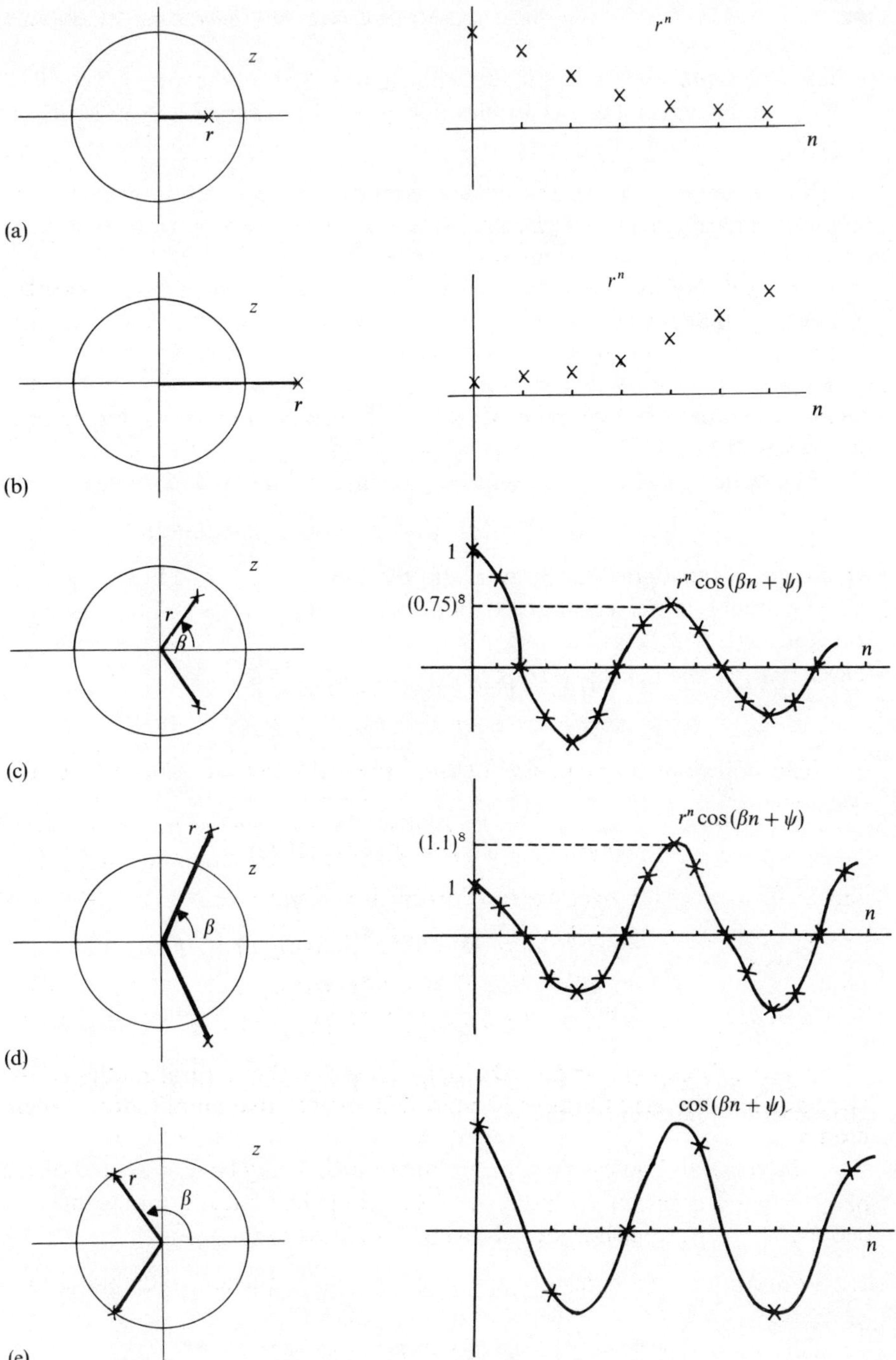

Fig. 1.16 Poles and corresponding time domain modes. (a) Single pole, real axis, inside $|z| = 1$. (b) Single pole, real axis, outside $|z| = 1$. (c) Complex conjugate poles. $\beta = \pi/4, r = 0.75, N = 8$. (d) Complex conjugate poles $\beta = \pi/4, r = 1.1, N = 8$. (e) Complex conjugate poles. $\beta = 3\pi/4, r = 1, N = \frac{8}{3}$.

Thus the signal corresponding to a pole pair $re^{\pm j\beta}$ is of the form $[r^n \cos \beta n]$. Note that the number of samples per cycle of the signal is $N = 2\pi/\beta$.

Some diagrams illustrating these points are in Fig. 1.16.

If a system has any zeros outside the unit circle, $|z| = 1$, it is termed 'non-minimum phase' (cf. See. 1.5). Such a system has a negative gain mode(s) which responds to an input in the sense opposite to that expected, and this feature, if present, naturally increases the difficulty of designing a satisfactory control strategy for that system.

EXAMPLE

Consider the real-time computer algorithm, Σ, executed at regular intervals T, illustrated in Fig. 1.17.

The behavior of the algorithm is governed by the difference equation:

$$y(n) = u(n-1) + 0.5u(n-2) + y(n-1) - 0.5y(n-2)$$

where $y(n)$ is the algorithm output, $u(n)$ the input.

Assuming initial conditions to be zero, $y(0) = y(1) = 0$, and taking z transforms (Eq. (1.29)):

$$\bar{Y}(z)[1 - z^{-1} + 0.5z^{-2}] = \bar{U}(z)[z^{-1} + 0.5z^{-2}]$$

$$\bar{G}(z) = \frac{\bar{Y}(z)}{\bar{U}(z)} = \frac{z^{-1} + 0.5z^{-2}}{1 - z^{-1} + 0.5z^{-2}}$$

$$= \frac{z + 0.5}{z^2 - z + 0.5}$$

$$= \frac{z + 0.5}{(z - (0.5 + j0.5)(z - (0.5 - j0.5))}$$

$$= \frac{z - 0.5e^{j\pi}}{(z - e^{j\pi/4}/\sqrt{2})(z - e^{-j\pi/4}/\sqrt{2})}.$$

The pole–zero diagram for this is in Fig. 1.18(a) together with the corresponding natural mode (b).

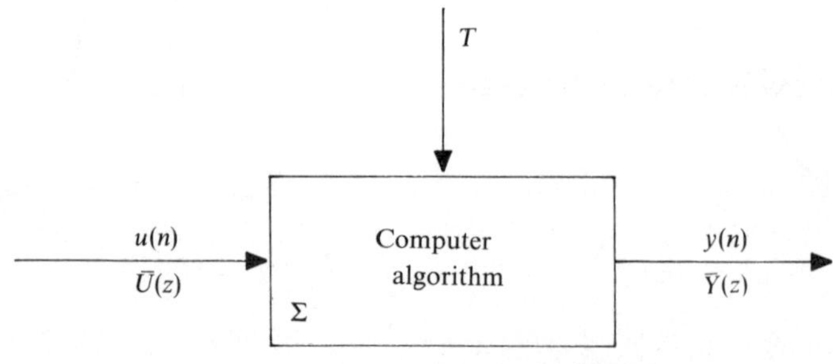

Fig. 1.17 Computer algorithm (example).

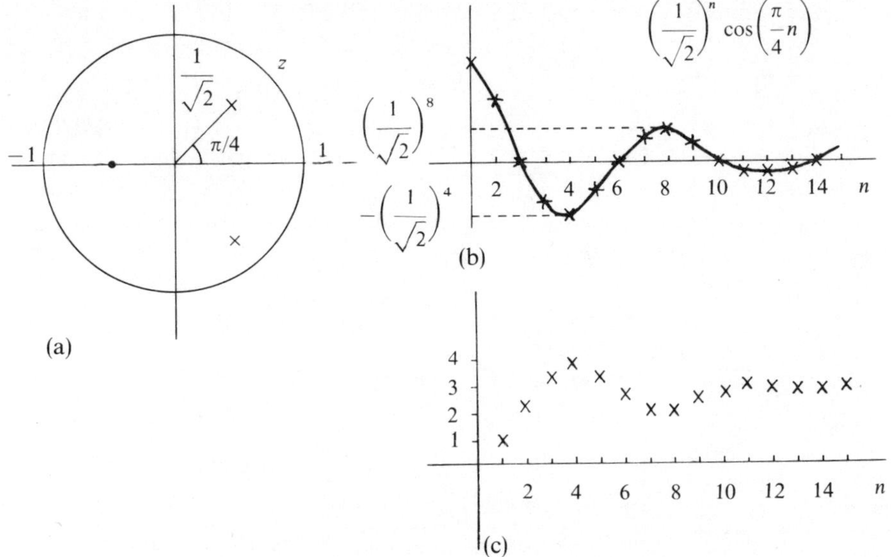

Fig. 1.18 Pole-zero diagram and natural mode (example). (a) Pole-zero diagram. (b) Natural mode (c) Step response

Consider the particular case of a step function input of 1 unit at $n = 0$; the output is readily found:

$$
\begin{aligned}
y(n) &= Z^{-1}\{\bar{G}(z)\bar{U}(z)\} \\
&= Z^{-1}\left\{\frac{z+0.5}{z^2 - z + 0.5}\cdot\frac{z}{z-1}\right\} \\
&= Z^{-1}\left\{\begin{matrix} z^{-1} + 2.5z^{-2} + 3.5z^{-3} \\ + 3.75z^{-4} + 3.5z^{-5} + 3.125z^{-6} + 2.875z^{-7} + \ldots \end{matrix}\right\}
\end{aligned}
$$

This is derived by the method of Sec. 1.8(ii)(c) and is illustrated in Fig. 1.18(c). The presence of the natural mode and that particular to the input are worth noting.

1.11 SAMPLED DATA SYSTEMS

A class of discrete time systems of particular interest concerns continuous time systems 'buffered' by digital to analog (DAC) and analog to digital (ADC) converters. A system of this kind with the continuous time system *S*, transfer function $G(s)$, is shown in Fig. 1.19(a).

Assuming the DAC and ADC are clocked simultaneously at intervals *T*, typical waveforms are as shown in Fig. 1.19(b).

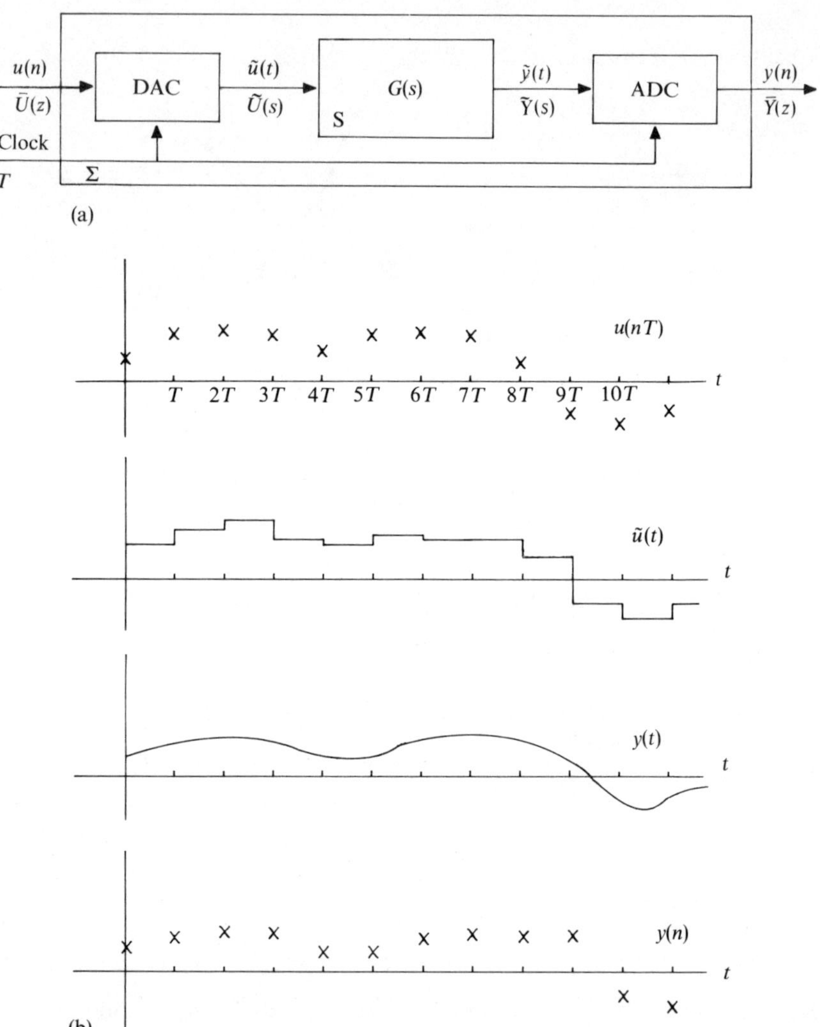

Fig. 1.19 Sampled data system and waveforms. (a) Sampled data system. (b) Waveforms.

The z transfer function of the overall system Σ (ref. 2) is:

$$\overline{GH}(z) = Z\left\{\frac{1-e^{-sT}}{s}G(s)\right\} = (1-z^{-1})Z\left\{\frac{1}{s}G(s)\right\} \tag{1.34}$$

where the RHS signifies, in a convenient though strictly incorrect fashion, the z transform 'equivalent' to the Laplace transform within the brackets.

The table of Fig. 1.13 lists a number of useful 'equivalent' transforms.

EXAMPLE

Consider the z transform analysis of the system shown in Fig. 1.20(a) assuming $T = 50$ ms.

In this case

$$G(s) = \frac{1}{1 + 0.1s}$$

$$\overline{GH}(z) = Z\left\{\frac{1 - e^{-0.05s}}{s} \cdot \frac{1}{1 + 0.1s}\right\} = (1 - z^{-1})Z\left\{\frac{10}{s(s + 10)}\right\}$$

$$= \frac{0.393}{z - 0.607} \qquad \text{(Fig. 1.13).}$$

The response of the system to a unit step is:

$$\bar{Y}(z) = \frac{0.393}{z - 0.607} \cdot \frac{z}{z - 1}.$$

Taking inverse z transforms (Fig. 1.13):

$$y(n) = u^*(n) - e^{-0.5n}.$$

$u(n)$ and $y(n)$ are illustrated in Fig. 1.20(b).

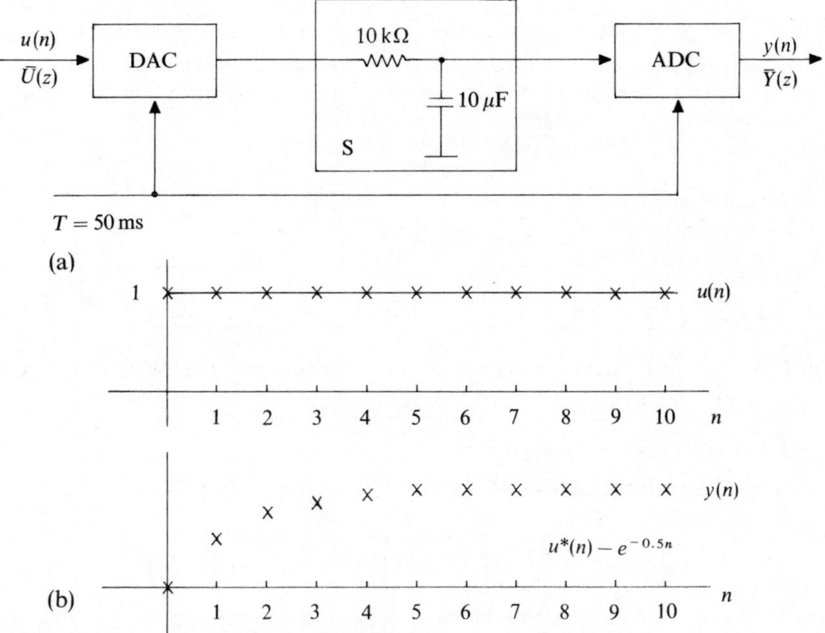

Fig. 1.20 Sampled data system with step function response (example). (a) Sampled data system. (b) Step function response.

1.12 RELATIONSHIP BETWEEN s and z PLANE POLES

The poles of system Σ (Fig. 1.19) are readily found from the poles of S.

Recalling that $s = \sigma + j\omega$ and $z = re^{j\beta}$, the poles are related by:

$$\begin{aligned} z_p &= e^{s_p T} = e^{\sigma_p T} e^{j\omega_p T} \\ &= r_p e^{j\beta_p} \end{aligned} \tag{1.35}$$

where $z_p, s_p, \sigma_p, \omega_p, r_p, \beta_p$ refer to the corresponding pole positions in the s and z planes. This applies to poles in the 'primary strip' of the s plane $|\omega| \leqslant |\omega_s/2|$ where:

$$\omega_s = \frac{2\pi}{T}.$$

ω_s is the sampling frequency (radians/unit time). This is illustrated in Fig. 1.21. The contour ABCDE in the s plane corresponds to $\bar{\text{A}}\bar{\text{B}}\bar{\text{C}}\bar{\text{D}}\bar{\text{E}}$ in the z plane, and stable poles (in both planes) are in the (open) shaded areas, which also correspond.

Poles outside the primary strip correspond to z plane poles at:

$$\begin{aligned} &\exp(\sigma_p T)\exp[j(\omega_p T \pm 2\pi k)], \qquad k = 1, 2, \ldots \\ &= \exp(\sigma_p T)\exp(j\omega_p T). \end{aligned}$$

Thus poles, and so modes, of $G(s)$ which lie outside the primary strip are 'aliased' into poles within the z plane. This can only be avoided by selecting a sampling frequency, ω_s, sufficiently high to ensure that all the poles of $G(s)$ lie within the primary strip (cf. Sec. 1.18), which then corresponds to the whole z plane.

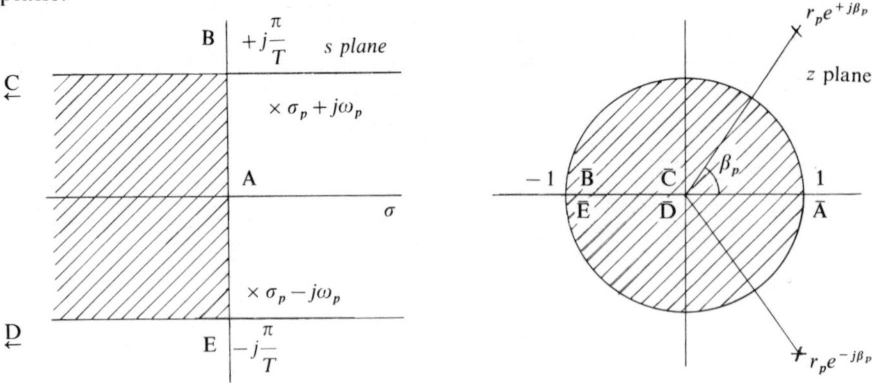

Fig. 1.21 s-Plane primary strip.

EXAMPLE

Consider the continuous time system with transfer function:

$$G(s) = \frac{s+3}{(s+2+3j)(s+2-3j)}$$

If this is incorporated in the system of Fig. 1.19, then to ensure that the poles of $G(s)$ lie in the primary strip, a sampling frequency ω_s must be chosen so that:

$$\frac{\omega_s}{2} \geqslant 3$$

or $\omega_s \geqslant 6$ radians/unit time. Equivalently $2\pi/T \geqslant 6$, or $T \leqslant \pi/3$.

Select $T = 0.5$.

The poles of $\overline{GH}(z)$ are at $e^{(-2+3j)0.5}, e^{(-2-3j)0.5}$ i.e. $0.368e^{1.5j}$, $0.368e^{-1.5j}$ (Eq. (1.35)).

These s and z plane poles are within the shaded areas of Fig. 1.21.

The correspondence between s and z plane poles described by Eq. (1.35) does not apply to zeros. Even the number of finite s and z plane zeros in $G(s)$

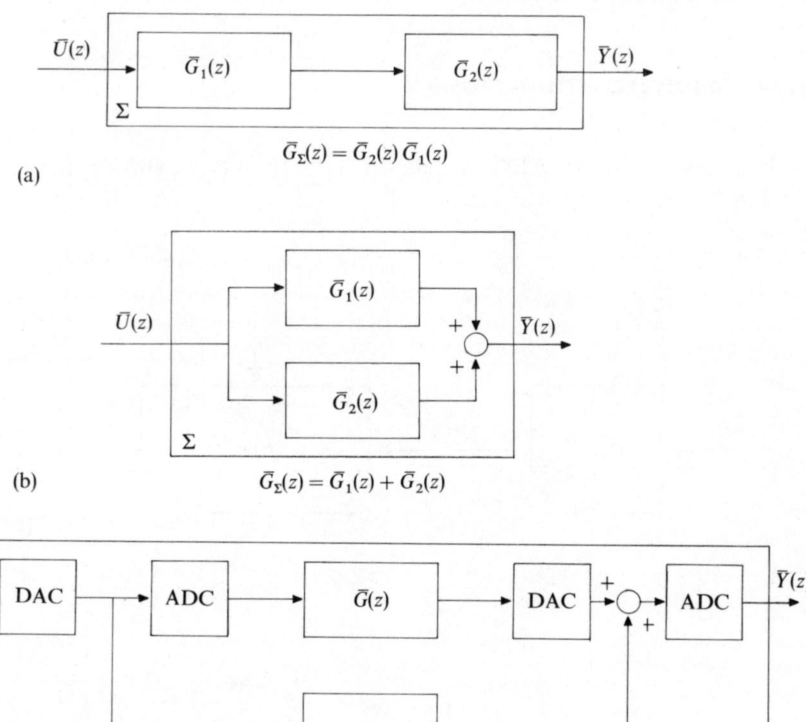

$$\bar{G}_\Sigma(z) = \bar{G}(z) + \overline{FH}(z) \text{ where } \overline{FH}(z) = Z\left\{\frac{1-e^{-sT}}{s}F(s)\right\}$$

Fig. 1.22 Composite systems and their transfer functions. (a) Series connected systems. (b) Parallel connected systems. (c) Combined analog/digital systems.

and $\overline{GH}(z)$ often differ, and furthermore, while $G(s)$ might have left-half s plane zeros, $\overline{GH}(z)$ can have zeros *outside* the unit circle, $|z| = 1$.

The calculation of $\overline{GH}(z)$ from $G(s)$ is thus in no way a conformal mapping from one Argand diagram to another, only the pole positions can be mapped directly.

1.13 BLOCK DIAGRAMS AND COMPOSITE SYSTEMS

The transfer functions of composite continuous and discrete time systems can usually be found quite easily. Some common ones are illustrated in Fig. 1.22.

1.14 MODIFIED z TRANSFORMS

If the input $u(t)$ to an ADC is delayed by a time θ, as shown in Fig. 1.23, modified z transforms are useful.

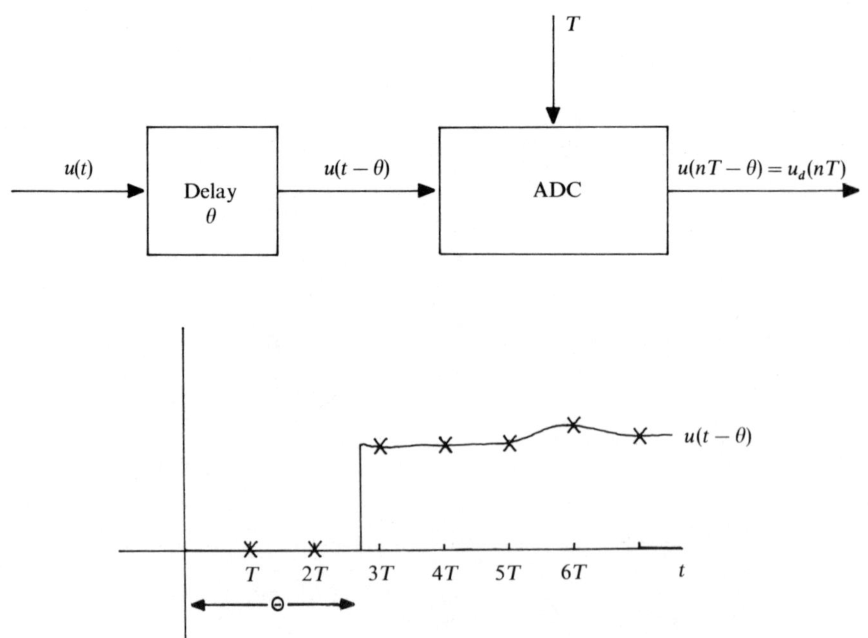

Fig. 1.23 Analog to digital conversion of a delayed signal.

The z transform of the output can be found by sampling the waveform $u(t-\theta)$ and applying Eq. (1.25). The same result can usually be obtained more easily from the 'modified' z transform table of Fig. 1.13.

Suppose
$$\theta = NT + \psi$$
where N is an integer, T is the sampling period of the ADC, ψ is a time less than T.
Then:

$$\begin{aligned} \bar{U}_d(z) &= Z\{u(t-\theta)|_{t=nT}\} \\ &= z^{-N} Z_m\{u(n)\}. \end{aligned} \tag{1.36}$$

where the variable m is defined:

$$m = 1 - \frac{\psi}{T}$$

and Z_m signifies the 'modified' z transform (Fig. 1.13).

EXAMPLE

Referring to Figure 1.24, consider:

$$u(t) = e^{-3t}, \quad \theta = 0.35, \quad T = 0.1, \quad u_d(nT) = e^{-3(nT-0.35)}.$$

Clearly, from Figure 1.24,

$$\{u[n(0.1) - 0.35]\} = \{0\ 0\ 0\ 0 \quad e^{-0.15} \quad e^{-0.45} \quad e^{-0.75} \quad \ldots\}.$$

From Eq. (1.25),

$$\bar{U}(z) = e^{-0.15} z^{-4} + e^{-0.45} z^{-5} + e^{-0.75} z^{-6} + \cdots.$$

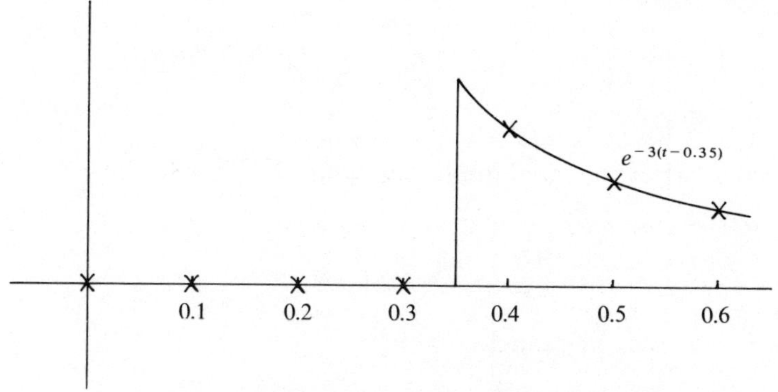

Fig. 1.24 Delayed and sampled signal (example).

Alternatively, using modified z transforms (Fig. 1.13):

$$N = 3$$
$$\psi = 0.05$$
$$m = 1 - \frac{0.05}{0.1} = 0.5$$
$$\bar{U}_d(z) = z^{-3} Z_{0.5}\{e^{-0.3n}\}$$
$$= z^{-3}\left\{\frac{e^{-0.15}z^{-1}}{1 - e^{-0.3}z^{-1}}\right\}$$
$$= e^{-0.15}z^{-4} + e^{-0.45}z^{-5} + e^{-0.75}z^{-6} + \cdots.$$

1.15 THE FOURIER TRANSFORM (CONTINUOUS FUNCTIONS OF TIME) (REF. 6)

The frequency characteristics – or *content* – of a function of time, $f(t)$, can be studied using its Fourier transform, defined:

$$\mathscr{F}\{f(t)\} = F(j\omega) = \int_{-\infty}^{+\infty} f(t)e^{-j\omega t}\,dt \tag{1.37}$$

where ω is frequency in radians/unit time.

The corresponding inverse transform, which is often difficult to evaluate, and is usually of less interest in this context, is:

$$\mathscr{F}^{-1}\{F(j\omega)\} = f(t) = \frac{1}{2\pi}\int_{-\infty}^{+\infty} F(j\omega)e^{j\omega t}\,d\omega. \tag{1.38}$$

EXAMPLES

(a) Consider the exponential decay function:

$$f(t) = \begin{cases} 0, & t < 0 \\ e^{-at}, & t \geqslant 0 \end{cases}$$

$$F(j\omega) = \int_{0}^{+\infty} e^{-at}e^{-j\omega t}\,dt = \frac{1}{a + j\omega}$$

$$f(t) = \frac{1}{2\pi} \int_{-\infty}^{+\infty} \frac{1}{a + j\omega} e^{j\omega t} \mathrm{d}\omega$$

$$= \begin{cases} 0\ , & t < 0 \\ e^{-at}, & t \geqslant 0 . \end{cases}$$

(b) Consider the sinusoidal signal:

$$f(t) = \cos \omega_0 t$$

$$F(j\omega) = \int_{-\infty}^{\infty} \cos \omega_0 t \, e^{-j\omega t} \, \mathrm{d}t$$

$$= \pi\delta(\omega - \omega_0) + \pi\delta(\omega + \omega_0).$$

Some properties of Fourier transforms are:

(a) Linearity

$$\mathscr{F}\{a_1 f_1(t) + a_2 f_2(t)\} = a_1 F_1(j\omega) + a_2 F_2(j\omega). \tag{1.39}$$

(b) Shifting in Time

$$\mathscr{F}\{f(t \pm a)\} = e^{\pm j\omega a} F(j\omega). \tag{1.40}$$

(c) Shifting in Frequency

$$\mathscr{F}^{-1}\{F(j(\omega \pm \omega_0))\} = e^{\mp \omega_0 t} f(t). \tag{1.41}$$

(d) Differentiation of f(t) with respect to t

$$\mathscr{F}\left\{\frac{\mathrm{d}}{\mathrm{d}t} f(t)\right\} = j\omega F(j\omega)$$

$$\mathscr{F}\left\{\frac{\mathrm{d}^n}{\mathrm{d}t^n} f(t)\right\} = (j\omega)^n F(j\omega). \tag{1.42}$$

(e) Differentiation of F(jω) with respect to ω

$$\mathscr{F}^{-1}\left\{\frac{\mathrm{d}^n}{\mathrm{d}\omega^n} F(j\omega)\right\} = (-jt)^n f(t). \tag{1.43}$$

(f) Real and Imaginary Parts of F(jω)

$$F(j\omega) = \int_{-\infty}^{+\infty} f(t) (\cos \omega t - j \sin \omega t) \, \mathrm{d}t$$

$$= R(\omega) - jX(\omega) \tag{1.44}$$

where

$$R(\omega) = \int_{-\infty}^{\infty} f(t) \cos \omega t \, \mathrm{d}t$$

$$X(\omega) = \int_{\infty}^{\infty} f(t) \sin \omega t \, \mathrm{d}t.$$

If $f(t)$ is real, $R(\omega)$ is even, and $X(\omega)$ is odd.
If $f(t)$ is even, $X(\omega) = 0$
If $f(t)$ is odd, $R(\omega) = 0$

(*g*) *Amplitude and Phase Characteristics*

From Eqs. (1.38) and (1.44), $f(t)$ can be regarded as consisting of the sum of an infinite number of components of amplitude:

$$\frac{1}{2\pi} |F(j\omega)| \, \mathrm{d}\omega = \frac{1}{2\pi} \sqrt{R^2(\omega) + X^2(\omega)} \, \mathrm{d}\omega$$

and angle of phase lag:

$$\angle F(j\omega) = \tan^{-1} \frac{X(\omega)}{R(\omega)}.$$

The frequency characteristics of relative amplitudes and phase angles are found by evaluating the Fourier spectrum, $F(j\omega)$, $-\infty \leqslant \omega \leqslant +\infty$.

EXAMPLE

Consider the function:

$$f(t) = \begin{cases} 0 \quad , & t < 0 \\ e^{-at}, & t \geqslant 0 \end{cases}$$

$$F(j\omega) = \frac{1}{a + j\omega}$$

$$= \frac{a}{a^2 + \omega^2} - \frac{j\omega}{a^2 + \omega^2}$$

$$= R(\omega) - jX(\omega).$$

Hence

$$|F(j\omega)| = \frac{1}{a^2 + \omega^2}$$

$$\angle F(j\omega) = \tan^{-1} \frac{\omega}{a}.$$

This is illustrated in Fig. 1.25.

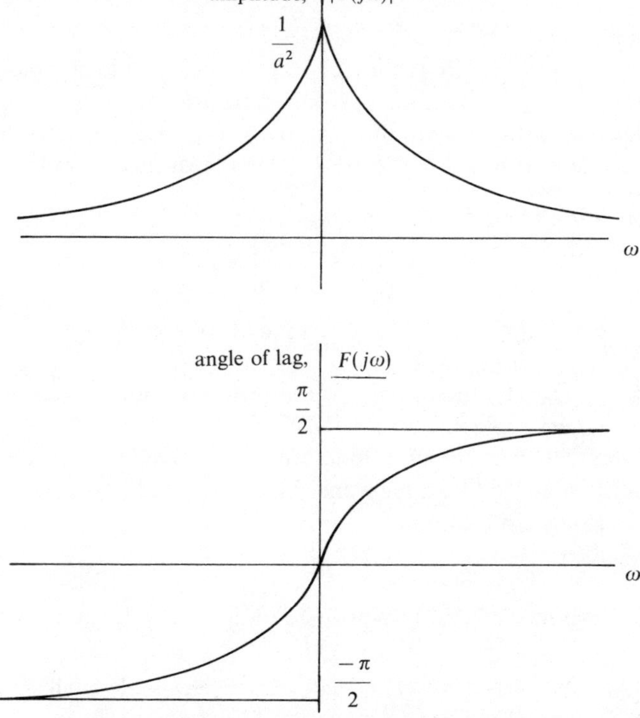

Fig. 1.25 Frequency characteristics (example).

(*h*) *Common Fourier Transforms*

A table of common Fourier transforms is in Fig. 1.26.

Time Function $f(t)$	*Fourier Transform* $F(j\omega)$	*Time Function* $f(t)$	*Fourier Transform* $F(j\omega)$
1	$2\pi\delta(\omega)$	$\frac{a}{a^2+t^2}$, $a>0,\ t\geqslant 0$	$\pi e^{-a\lvert\omega\rvert}$
e^{-at}, $a>0,\ t\geqslant 0$	$\frac{1}{a+j\omega}$	$\cos at$	$\pi\delta(\omega-a)+\pi\delta(\omega+a)$
te^{-at}, $a>0,\ t\geqslant 0$	$\frac{1}{(a+j\omega)^2}$	$\sin at$	$-\pi j\delta(\omega-a)+\pi j\delta(\omega+a)$
		$\exp(-a\lvert t\rvert)$,	$\frac{2a}{a^2+\omega^2}$, $a>0$
$t^n e^{-at}$, $a>0,\ t\geqslant 0$	$\frac{n!}{(a+j\omega)^{n+1}}$	$u^*(t)$	$\pi\delta(\omega)+\frac{1}{j\omega}$

Note: $\delta(\omega)$ is the Dirac delta function (App. C.1)

$u^*(t)$ denotes a unit step function; $u^*(t)=\begin{cases}0, & -\infty\leqslant t<0\\ 1, & 0\leqslant t\leqslant+\infty\end{cases}$

Fig. 1.26 Common Fourier transforms.

1.16 SYSTEM FREQUENCY BEHAVIOR

The behavior, in terms of frequency characteristics, of linear time-invariant (LTI) systems may be studied using Fourier transforms (refs. 2, 6).

Consider the LTI system illustrated in Fig. 1.27, where $u(t)$ is the input to the system S, $U(j\omega) = \mathscr{F}\{u(t)\}$, $y(t)$ is the output from the system S, $Y(j\omega) = \mathscr{F}\{y(t)\}$.

$$G(j\omega) = \frac{Y(j\omega)}{U(j\omega)}. \tag{1.45}$$

$G(j\omega)$, the Fourier transfer function of S, is the ratio of the Fourier transform of the output to that of the input.

The Fourier transfer functions of some common simple circuits are easily found in Fig. 1.7 by replacing s with $j\omega$.

Given the Fourier transfer function of a system, and the Fourier transform of its input, the Fourier transform and so the frequency characteristics of the output can be found.

EXAMPLE

Consider the response of the network shown in Fig. 1.28 to an input of the form:

$$u(t) = \begin{cases} 0 \quad , & t < 0 \\ 60e^{-30t}, & t \geqslant 0 \end{cases}$$

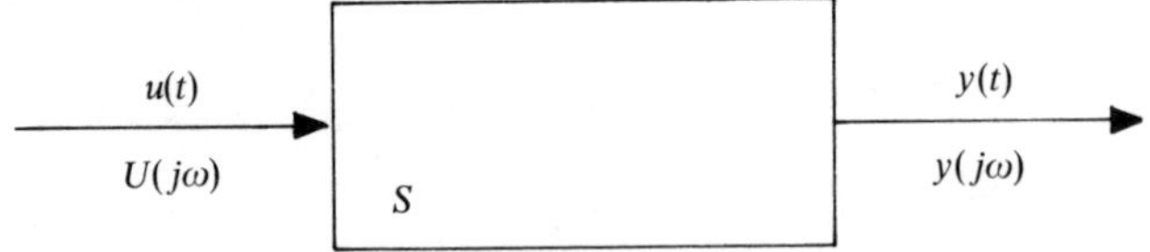

Fig. 1.27 LTI system.

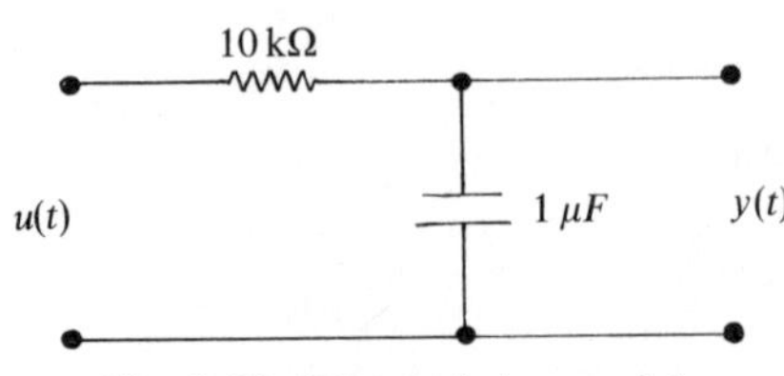

Fig. 1.28 Network (example).

$$G(j\omega) = \frac{1/(j\omega 10^{-6})}{10^4 + 1/(j\omega 10^{-6})}$$

$$= \frac{100}{100 + j\omega}$$

$$U(j\omega) = \frac{60}{30 + j\omega} \qquad \text{(Fig. 1.26)}$$

$$Y(j\omega) = \frac{100}{100 + j\omega} \cdot \frac{60}{30 + j\omega}$$

$$= \frac{6000}{3000 + 130 j\omega - \omega^2}.$$

The Fourier spectra of input and output, $|U(j\omega)|$ and $|Y(j\omega)|$ respectively, are shown in Fig. 1.29.

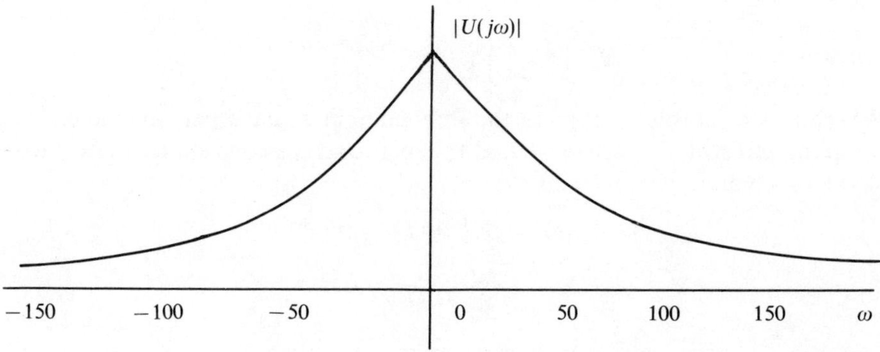

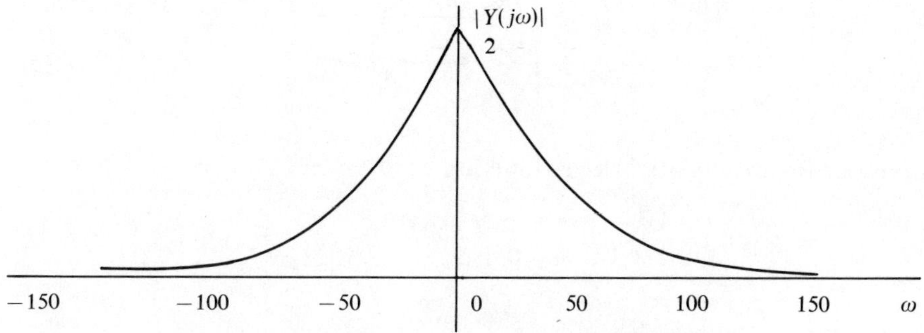

Fig. 1.29 Fourier spectra (example).

1.17 THE FOURIER TRANSFORM (DISCRETE FUNCTIONS OF TIME) (REF. 6)

The frequency characteristics – or *content* – of a discrete function of time, $f(n)$, can be studied using the Fourier transform of $f(n)$, defined:

$$\mathscr{F}\{f(n)\} = \bar{F}(e^{j\beta})$$
$$= \sum_{-\infty}^{+\infty} f(n)e^{-jn\beta} \tag{1.46}$$

where β is 'frequency' in radians per sample (Sec. 1.10).

$\bar{F}(e^{j\beta})$ is a continuous complex function of frequency, $-\infty \leqslant \beta \leqslant +\infty$, which is periodic with period 2π. Any interval of length 2π is sufficient to describe $\bar{F}(e^{j\beta})$ and generally $-\pi \leqslant \beta \leqslant \pi$ is used.

The corresponding inverse transform, which is often very difficult to evaluate and is usually of less interest in this context, is:

$$\mathscr{F}^{-1}\{\bar{F}(e^{j\beta})\} = f(n)$$

where

$$f(n) = \frac{1}{2\pi}\int_{-\pi}^{+\pi} \bar{F}(e^{j\beta})e^{j\beta n}\,\mathrm{d}\beta. \tag{1.47}$$

Where it is desirable to express the frequency in radians per unit time, the sampling interval T is involved, and the relationships corresponding to Eqs. (1.46), (1.47) are:

$$\mathscr{F}\{f(n)\} = \mathscr{F}\{f(nT)\} = \bar{F}(e^{j\omega T})$$
$$= \sum_{-\infty}^{+\infty} f(nT)e^{-j\omega nT} \tag{1.48}$$

where ω is frequency in radians per unit time, and T is the sampling period. $\bar{F}(e^{j\omega T})$ is periodic with period $2\pi/T$.

The inverse relationship is:

$$\mathscr{F}^{-1}\{\bar{F}(e^{j\omega T})\} = f(nT) = f(n)$$
$$= \frac{T}{2\pi}\int_{-\pi/T}^{\pi/T} \bar{F}(e^{j\omega T})e^{j\omega nT}\,\mathrm{d}\omega. \tag{1.49}$$

EXAMPLE

Consider the exponential decay function:

$$f(n) = \begin{cases} 0, & n = -\infty, \dots, -1 \\ a^n, & n = 0, 1, 2, \dots, \infty, |a| < | \end{cases}$$

$$\bar{F}(e^{j\beta}) = \sum_{0}^{+\infty} a^n e^{-j\beta n}$$
$$= \frac{1}{1 - ae^{-j\beta}}.$$

The inverse relationship can be evaluated, with some difficulty, using Eqs. (1.47), (1.49).

Some properties of Fourier Transforms of discrete functions are:

(a) Linearity

$$\mathscr{F}\{a_1 f_1(n) + a_2 f_2(n)\} = a_1 \bar{F}_1(e^{j\beta}) + a_2 \bar{F}_2(e^{j\beta}).$$

(b) Shifting in Time

$$\mathscr{F}\{f(n \pm k)\} = e^{\pm j\beta k} \bar{F}(e^{j\beta}).$$

(c) Shifting in Frequency

$$\mathscr{F}^{-1}\{\bar{F}(e^{j(\beta \pm \beta_0)})\} = \{\mp e^{j\beta_0} f(n)\}.$$

(d) Amplitude and Phase Characteristics

From Eq. (1.47), $f(n)$ can be regarded as consisting of the sum of an infinite number of components of frequency dependent amplitude and phase lag (cf. Sec. 1.15(f), (g)):

$$\frac{1}{2\pi}|\bar{F}(e^{j\beta})|, \quad -\underline{|\bar{F}(e^{j\beta})}.$$

The frequency characteristics of relative amplitude and phase angle are found by evaluating these in the interval $-\pi \leqslant \beta \leqslant \pm\pi$.

For a real function of time, $f(n)$,

$|\bar{F}(e^{j\beta})|$ is symmetric and $-\underline{|\bar{F}(e^{j\beta})}$ is antisymmetric about $\beta = 0$ over the interval $-\pi \leqslant \beta \leqslant \pi$.

EXAMPLE

Consider the function:

$$f(n) = \begin{cases} 0 & n < 0 \\ a^{-n}, & n = 0, 1, 2, \ldots, \infty, |a| < | \end{cases}$$

$$|\bar{F}(e^{j\beta})| = \left|\frac{1}{1 - ae^{-j\beta}}\right| = \left[\frac{1}{(1 - a\cos\beta)^2 + (a\sin\beta)^2}\right]^{1/2}$$

$$\underline{|\bar{F}(e^{j\beta})} = \underline{\left|\frac{1}{1 - ae^{-j\beta}}\right.} = \tan^{-1}\left[\frac{-a\sin\beta}{1 - a\cos\beta}\right], \qquad -\pi \leqslant \beta \leqslant +\pi.$$

(e) Common Fourier Transforms

Fourier transforms of some common functions with $f(n) = 0$, $n < 0$, can be found from the z transform table (Fig. 1.13) by substituting $z = e^{j\beta}$.

1.18 SHANNON'S SAMPLING THEOREM (CF. SEC.1.12)

The frequency content of a discrete signal derived via an ADC from a continuous one, as illustrated in Fig. 1.30, is of interest.

It can be shown (ref. 6) that $\mathscr{F}\{x(t)\}$ and $\mathscr{F}\{x(n)\}$ are related as follows:

$$\bar{X}(e^{j\omega T}) = \frac{1}{T}\sum_{m=-\infty}^{+\infty} X\left(\omega + \frac{2\pi}{T}m\right), \quad m = 0, 1, 2 \ldots \tag{1.50}$$

where

$$\mathscr{F}\{x(t)\} = X(j\omega) \quad \text{and} \quad \mathscr{F}\{x(n)\} = \bar{X}(e^{j\omega T}) = \bar{X}(e^{j\beta}),$$

according to the definitions of Eqs. (1.37) and (1.48).

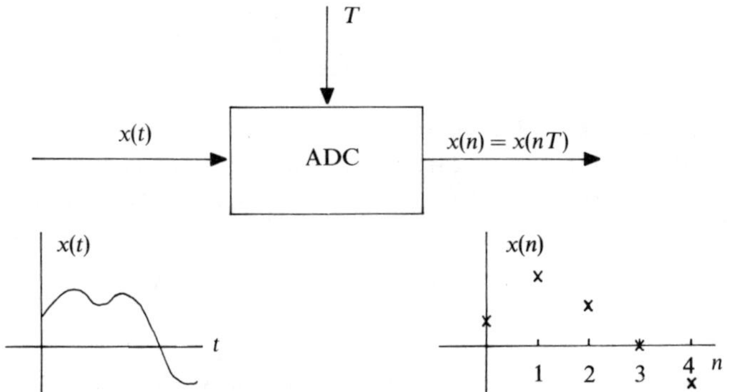

Fig. 1.30 Continuous to discrete signal conversion.

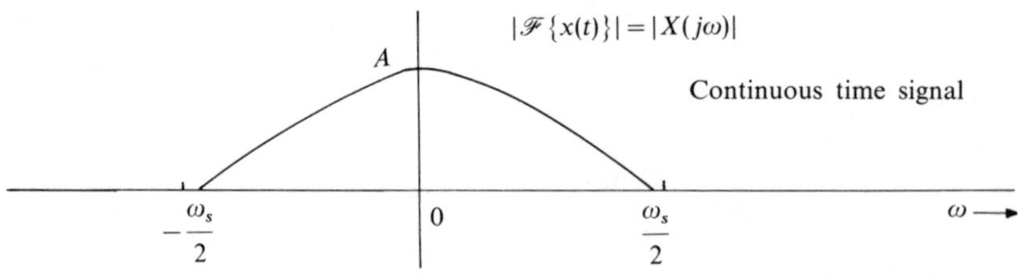

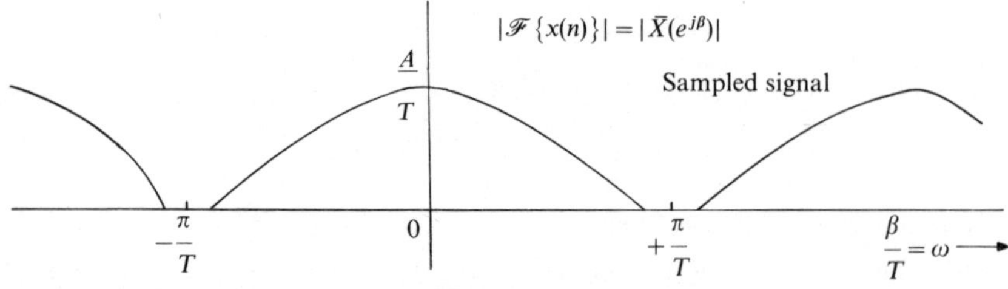

Fig. 1.31 Frequency response for bandlimited sampled signal.

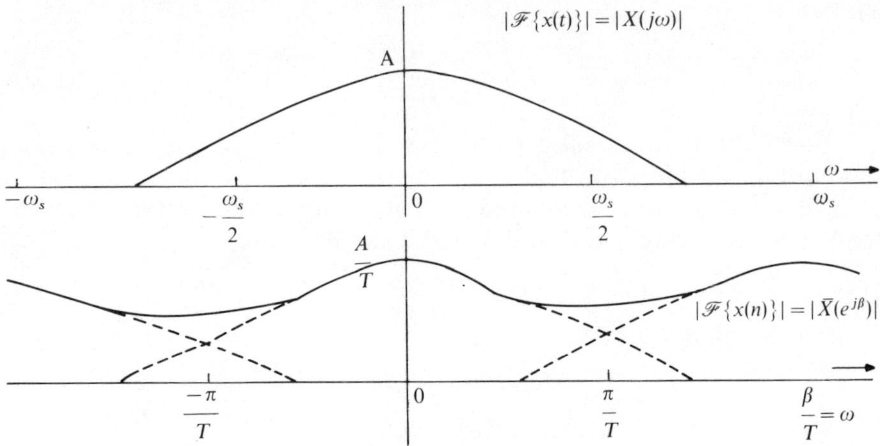

Fig. 1.32 Aliasing of frequencies.

This relationship is known as *Shannon's Sampling Theorem.* Where the frequency content of $x(t)$ is limited to half the sampling frequency (or less), i.e. $|X(j\omega)| = 0, |\omega| > \pi/T$, the frequency response of $x(n)$ is as shown in Fig. 1.31.

However, if $x(t)$ is not bandlimited in this way, Eq. (1.50) shows that signal components with higher than half the sampling frequency, $(|\omega| > \pi/T)$ are aliased, or reflected, into components of lower frequencies, $(|\omega| < \pi/T)$ as illustrated in Fig. 1.32.

To avoid aliasing altogether, therefore, the sampling frequency ($1/T$ samples/unit time) must be at least twice the frequency (cycles/unit time) of the highest frequency component of $x(t)$. In practice, of course, if high frequency components of $x(t)$ are of low relative amplitude, some aliasing may be tolerable.

1.19 DISCRETE TIME SYSTEM FREQUENCY BEHAVIOR ANALYSIS

The behavior – in terms of frequency characteristics – of LTI discrete systems may be studied using Fourier transforms.

Consider the LTI discrete system illustrated in Fig. 1.33, where $u(n)$ is the input to the system Σ, $\bar{U}(e^{j\beta}) = \mathcal{F}\{u(n)\}$, $y(n)$ is the output,

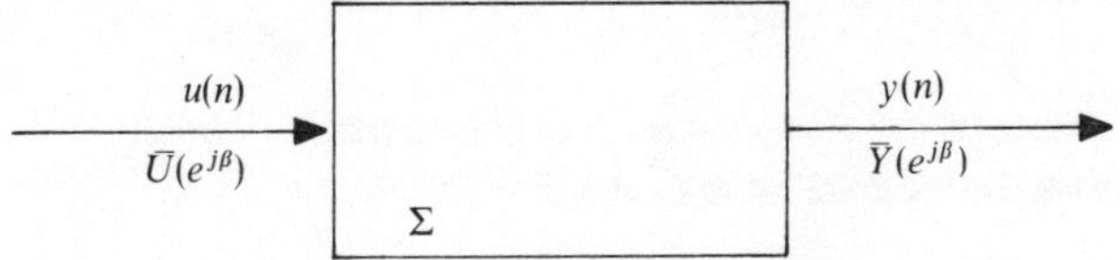

Fig. 1.33 LTI discrete system.

$\bar{Y}(e^{j\beta}) = \mathscr{F}\{y(n)\}$

$$\bar{G}(e^{j\beta}) = \frac{\bar{Y}(e^{j\beta})}{\bar{U}(e^{j\beta})}. \tag{1.51}$$

$\bar{G}(e^{j\beta})$, the Fourier transfer function of Σ, is the ratio of the Fourier transform of the output to that of the input.

Thus, if the Fourier spectrum $\bar{U}(e^{j\beta})$ of the input and $\bar{G}(e^{j\beta})$ are known, the Fourier spectrum $\bar{Y}(e^{j\beta})$ of the output can be found.

EXAMPLE

Consider an algorithm:

$$y(n) = u(n) + by(n-1).$$

This has z transfer function (Sec. 1.9):

$$\frac{\bar{Y}(z)}{\bar{U}(z)} = \frac{1}{1 - bz^{-1}}$$

and Fourier transfer function (Sec. 1.17):

$$\frac{\bar{Y}(e^{j\beta})}{\bar{U}(e^{j\beta})} = \frac{1}{1 - be^{-j\beta}}.$$

Consider the input:

$$u(n) = \begin{cases} 0, & n < 0 \\ a^n, & n \geqslant 0, |a| < |. \end{cases}$$

Then

$$\mathscr{F}\{u(n)\} = \frac{1}{1 - ae^{-j\beta}}$$

and

$$\mathscr{F}\{y(n)\} = \frac{1}{(1 - ae^{-j\beta})} \frac{1}{(1 - be^{-j\beta})}$$

$$= \frac{1}{(1 - (a+b)e^{-j\beta} + abe^{-2j\beta}}.$$

The amplitude and phase lag frequency characteristics of $y(n)$ are easily found from this for the frequency range of interest, $-\pi \leqslant \beta \leqslant \pi$. Beyond this range, the characteristics are periodic (Figs. 1.31, 1.32).

1.20 THE DISCRETE FOURIER TRANSFORM AND ITS INVERSE (REF. 6)

When Fourier transforms (Eq. (1.37)) are to be evaluated numerically, $F(j\omega)$ is obtained at a number of discrete frequencies $k\Delta\omega$ to give $\bar{F}(k)$, using a number of discrete values $f(n)$ typically sampled at intervals T from the time domain function $f(t)$.

In this case the Fourier transform is:

$$\bar{F}(k) = \mathscr{F}\{f(n)\}$$
$$= \sum_{n=0}^{N-1} f(n) e^{-j(2\pi/N)kn}. \tag{1.52}$$

Typically, $f(t)$ is sampled at nT, $n = 0, 1, \ldots, N-1$ (i.e. N points), and $F(j\omega)$ is evaluated at $k\Delta\omega$, $k = 0, 1, \ldots, N-1$ (i.e. N frequencies).

$\bar{F}$ (k) is actually symmetrical about $k = N/2$, $N/2 - 1$, and Eq. (1.52) assumes periodicity of $f(n)$ with period N:

$$\Delta\omega = \frac{2\pi}{NT}.$$

The corresponding inverse transform is:

$$f(n) = \mathscr{F}^{-1}\{\bar{F}(k)\}$$
$$= \frac{1}{N} \sum_{k=0}^{N-1} \bar{F}(k) e^{j(2\pi/N)kn}. \tag{1.53}$$

(i) Calculation of the Discrete Fourier Transform

From Eq. (1.52), using N values of $f(n)$, N discrete values of $\bar{F}(k)$ can be calculated by means of approximately N^2 complex multiplications and $N(N-1)$ complex additions. Since N might typically be 1000, this can amount to a great deal of computation.

A very significant amelioration can be achieved using so-called 'fast Fourier transform' (FFT) techniques, one of the simplest of which (ref. 6) depends on N being chosen as a power of 2, $N = 2^{\gamma}$.

The so called 'decimation in time' algorithm is based on the idea that $\{f(n)\}$ can be expressed as two series, one of even numbered terms $f(2n)$ and the other of odd, $f(2n+1)$, $n = 0, 1, \ldots, N/2$, and that as a result $\bar{F}(k)$ can be expressed as the sum of two separate Fourier transforms:

$$\bar{F}(k) = \sum_{n=0}^{N/2-1} f(2n) e^{-j(2\pi/N)2kn} + \sum_{n=0}^{N/2-1} f(2n+1) e^{-j(2\pi/N)k(2n+1)}$$
$$= \begin{cases} \bar{F}_1(k) + W_N^k \cdot \bar{F}_2(k), & 0 \leqslant k \leqslant N/2 - 1 \\ \bar{F}_1(k - N/2) + W_N^k \cdot \bar{F}_2(k - N/2) & N/2 \leqslant k \leqslant N - 1 \end{cases} \tag{1.54}$$

where $W_N = e^{-j(2\pi/N)}$

$\bar{F}_1(k)$ is the Fourier transform of the $N/2$ term sequence $\{f(2n)\}$, $n = 0, 1, \ldots, (N/2–1)$

$\bar{F}_2(k)$ is the Fourier transform of the $N/2$ term sequence $\{f(2n+1)\}$ $n = 0, 1, \ldots, (N/2 - 1)$.

The same scheme may be used to evaluate $\bar{F}_1(k)$, $\bar{F}_2(k)$ according to the rules:

$$\bar{F}_1(k) = \begin{cases} \bar{F}_{11}(k) + W_N^{2k}\bar{F}_{12}(k) & 0 \leqslant k \leqslant N/4 - 1 \\ \bar{F}_{11}(k - N/4) + W_N^{2k}\bar{F}_{12}(k - N/4), & N/4 \leqslant k \leqslant N/2 \end{cases} \quad (1.55)$$

$$\bar{F}_2(k) = \begin{cases} \bar{F}_{21}(k) + W_N^{2k}\bar{F}_{22}(k), & 0 \leqslant k \leqslant N/4 - 1 \\ \bar{F}_{21}(k - N/4) + W_N^{2k}\bar{F}_{22}(k - N/4), & N/4 \leqslant k \leqslant N/2. \end{cases}$$

[Note: $W_{N/2} = e^{-j(2\pi/N)2} = W_N^2$.]

Again, the same scheme can be used to evaluate $\bar{F}_{11}$ in terms of $\bar{F}_{111}$, $\bar{F}_{112}$:

$$\bar{F}_{11}(k) = \begin{cases} \bar{F}_{111}(k) + W_N^{4k}\bar{F}_{112}(k), & 0 \leqslant k \leqslant N/8 - 1 \\ \bar{F}_{111}(k - N/8) + W_N^{4k}\bar{F}_{112}(k - N/8) & N/8 \leqslant k \leqslant N/4 \end{cases}$$

$$\bar{F}_{12}(k) = \begin{cases} \bar{F}_{121}(k) + W_N^{4k}\bar{F}_{122}(k), & 0 \leqslant k \leqslant N/8 - 1 \\ \bar{F}_{121}(k - N/8) + W_N^{2k}\bar{F}_{122}(k - N/8), & N/8 \leqslant k \leqslant N/4. \end{cases} \quad (1.56)$$

Similarly, $\bar{F}_{21}(k)$, $\bar{F}_{22}(k)$ can be expressed in terms of $\bar{F}_{211}(k)$, $\bar{F}_{212}(k)$, $\bar{F}_{221}(k)$, $\bar{F}_{222}(k)$, etc.

The process can be repeated until $2^{\gamma-1}$ sets of 2-point Fourier transforms are found according to rules like:

$$\bar{F}_{11\ldots1}(0) = f_s(0) + f_s(1)$$

$$\bar{F}_{11\ldots1}(1) = f_s(0) - f_s(1)$$

in which $f_s(0)$, $f_s(1)$, $f_s(2)$. and $f_s(3)$, are terms taken from appropriate positions in the original $\{f(n)\}$ sequence.

For the output sequence $\{\bar{F}(k)\}$ to be in the correct order, the input sequence $\{f(n)\}$ must be 'shuffled' so that the indices of the elements in the sequence $\{f_s(n)\}$ are in a so-called binary 'bit-reversed' order compared with $\{f(n)\}$. This is achieved by reversing the order of the binary indices as illustrated for an 8-bit sequence in Fig. 1.34.

Index in $\{f(n)\}$	*Binary Representation*	*Bit-reversed Representation*	*Index in* $\{f_s(n)\}$
0	000	000	0
1	001	100	4
2	010	010	2
3	011	110	6
4	100	001	1
5	101	101	5
6	110	011	3
7	111	111	7

Fig. 1.34 Bit-reversed indices for an 8-bit sequence.

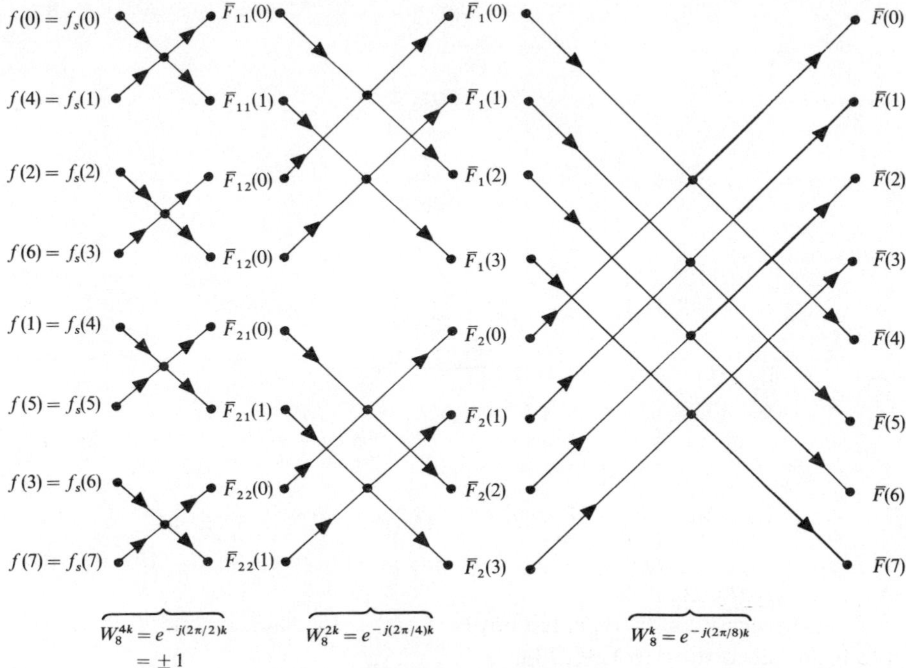

Fig. 1.35 Fast Fourier transform for an 8-point sequence.

It can be seen from Fig. 1.35 that each value is calculated from two values in the previous stage in the sequence according to a formula of the general form:

$$X = A + W_N^{rk} B$$

$$Y = A - W_N^{rk} B, \qquad r = \frac{N}{2}, \frac{N}{4}, \ldots, 1.$$

The detailed forms are specified by Eq. (1.54,), (1.55), (1.56), etc. Conveniently, when the new values X, Y have been calculated, the old values A, B can be discarded, or in practice, overwritten by X, Y.

CAD Facility

The algorithm of Fig. 1.36 utilizes the above FFT technique to generate N-point Fourier transforms ($N = 2^\gamma$) from N-point time sequences $\{f(n)\}$, using about $(N/2)\log_2 N$ complex multiplications.

The operation of the algorithm is as follows.

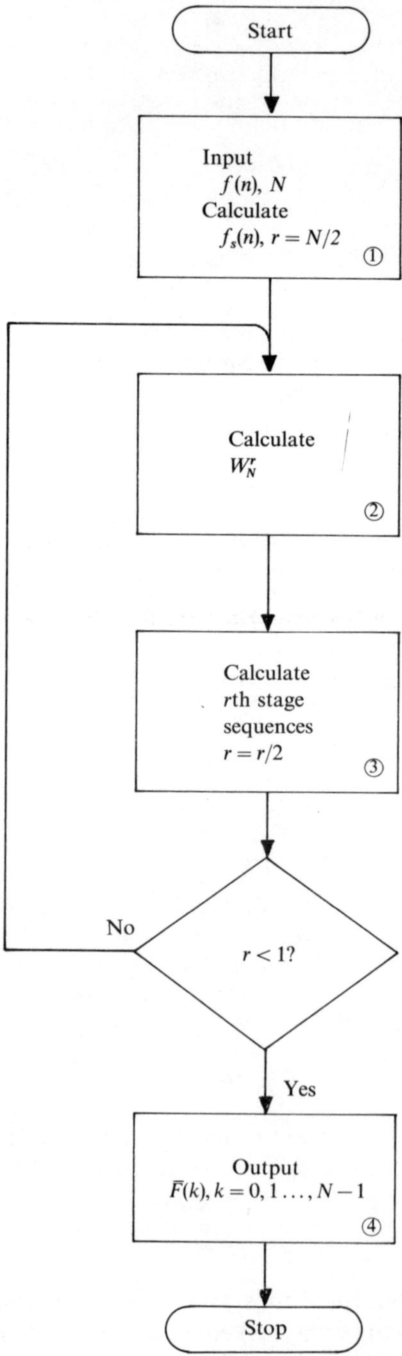

Fig. 1.36 FFT using a decimation-in-time algorithm.

BLOCK 1

The sequence $\{f(n)\}$ and its length, N, are input and $f(n)$ is shuffled according to the index bit-reversal technique described to give a 'shuffled' sequence $\{f_s(n)\}$.

$$r \text{ is set to } r = N/2.$$

BLOCK 2

$W_N^r = e^{-j(2\pi/N)r}$ is calculated.

BLOCK 3

The rth stage sequence is calculated using the appropriate formulas of the form of Eqs. (1.54), (1.55), and (1.56).

r is then halved and the process repeated until $r < 1$.

BLOCK 4

The Fourier transform is output (to a file or display).

(ii) Calculation of the Inverse Discrete Fourier Transform (ref. 6)

The inverse transform is (Eq. (1.53)):

$$f(n) = \frac{1}{N} \sum_{k=0}^{N-1} \bar{F}(k) e^{j(2\pi/N)kn}.$$

The complex conjugate of this is:

$$f^*(n) = \frac{1}{N} \sum_{k=0}^{N-1} \bar{F}^*(k) e^{-j(2\pi/N)kn}$$

where $\bar{F}^*(k)$ is the complex conjugate of $\bar{F}(k)$.

From Eq. (1.52)

$$f^*(n) = \frac{1}{N}[\mathscr{F}_{nk}(\bar{F}^*(k))]$$

where $\mathscr{F}_{nk}$ represents the Fourier transform with n, k interchanged. Taking the complex conjugate of this:

$$f(n) = \frac{1}{N}[\mathscr{F}_{nk}(\bar{F}^*(k))]^*.$$

The inverse transform is thus found simply by calculating the Fourier transform of the (usually complex) complex conjugate of the sequence $\bar{F}(k)$. The algorithm of subsection (i) is of course directly suited to this. Since $f(n)$ is usually real, it should normally be found that $f(n) = f^*(n)$.

1.21 SOME PROPERTIES OF CONTINUOUS TIME STOCHASTIC PROCESSES (REF. 2)

A *stochastic*, or *random, process*, $x(t)$ is defined as a real function of time such that at any time, t, $x(t)$ is a random number. In practice this means that $x(t)$ represents any one of an infinitely large family or *ensemble* of sample functions, $\xi_1(t)$, $\xi_2(t), \ldots$, as illustrated in Fig. 1.37.

At time t the numbers $\xi_1(t)$, $\xi_2(t), \ldots$ make up a population with a probability density function (App. B) $p(x, t)$, which, in principle at least, depends on t.

$p(x, t)$ is a measure of the amplitude of $x(t)$ and, if known, allows the calculation of the mean value of $x(t)$ in the ensemble sense:

$$m_x(t) = \mathscr{E}(x(t)) = \int_{-\infty}^{+\infty} x p(x, t)\,\mathrm{d}x. \tag{1.57}$$

The *variance*, or spread, can also be calculated:

$$\begin{aligned} \sigma_x^2(t) &= \mathscr{E}[x(t) - \mathscr{E}(x(t))]^2 \\ &= \int_{-\infty}^{+\infty} [x(t) - \mathscr{E}(x(t))]^2 p(x, t)\,\mathrm{d}x. \end{aligned} \tag{1.58}$$

If $p(x, t)$ is not a function of time, i.e. $p(x,t) = p(x)$, the process is *stationary*, and

$$m_x(t) = m_x; \qquad \sigma_x(t) = \sigma_x.$$

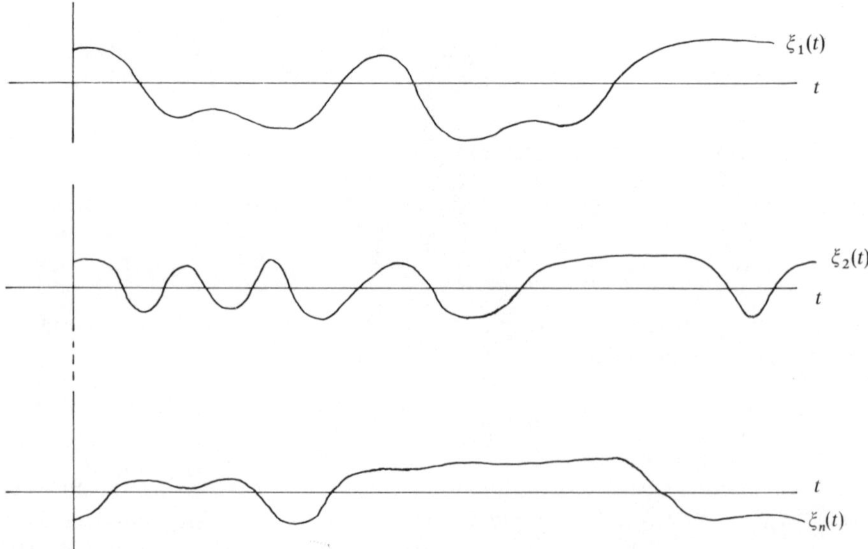

Fig. 1.37 An ensemble of sample functions.

A measure of the rate of change of a stationary process $x(t)$ is given by the *covariance* of $x(t)$:

$$\begin{aligned}\operatorname{cov}(x(t)) &= r_x(\tau) \\ &= \mathscr{E}[(x(t)-m_x)(x(t+\tau)-m_x)].\end{aligned} \tag{1.59}$$

A measure of the relationship, if any, between two stationary processes $x(t)$ and $y(t)$ is the *cross-covariance*:

$$r_{xy}(\tau) = \mathscr{E}[(x(t)-m_x)(y(t+\tau)-m_y)]. \tag{1.60}$$

These functions may be compared with the corresponding *time-averaged* functions. In this case, the mean value of $x(t)$ is:

$$m_t = \lim_{T\to\infty}\frac{1}{2T}\int_{-T}^{+T} x(t)\,\mathrm{d}t \tag{1.61}$$

where $x(t)$ is any one of the sample functions $\xi_1(t)$, $\xi_2(t), \ldots$.

A time-averaged measure of the rate of change of $x(t)$ is the *autocorrelation function*:

$$R_x(\tau) = \lim_{T\to\infty}\frac{1}{2T}\int_{-T}^{+T} x(t)x(t+\tau)\,\mathrm{d}t. \tag{1.62}$$

If the time-averaged functions are equal to the corresponding ensemble functions, i.e.:

$$m_x = m_t$$

and

$$\begin{aligned}R_x(\tau) &= \mathscr{E}[x(t)x(t+\tau)] \\ &= r_x(\tau) + m_x^2\end{aligned} \tag{1.63}$$

then $x(t)$ is *ergodic*, which in practice means that data collected over time from one sample function agree statistically with data collected at one instant from the whole ensemble.

A time-averaged measure of the relationship between two processes $x(t)$, $y(t)$ is given by the *cross-correlation function*:

$$R_{xy}(\tau) = \lim_{T\to\infty}\frac{1}{2T}\int_{-T}^{+T} x(t)y(t+\tau)\,\mathrm{d}t. \tag{1.64}$$

EXAMPLES

(a) Consider the process:

$$x(t) = a + bt$$

where a and b are random numbers with probability density functions $p(a)$, $p(b)$. Some sample functions of $x(t)$ are shown in Fig. 1.38.

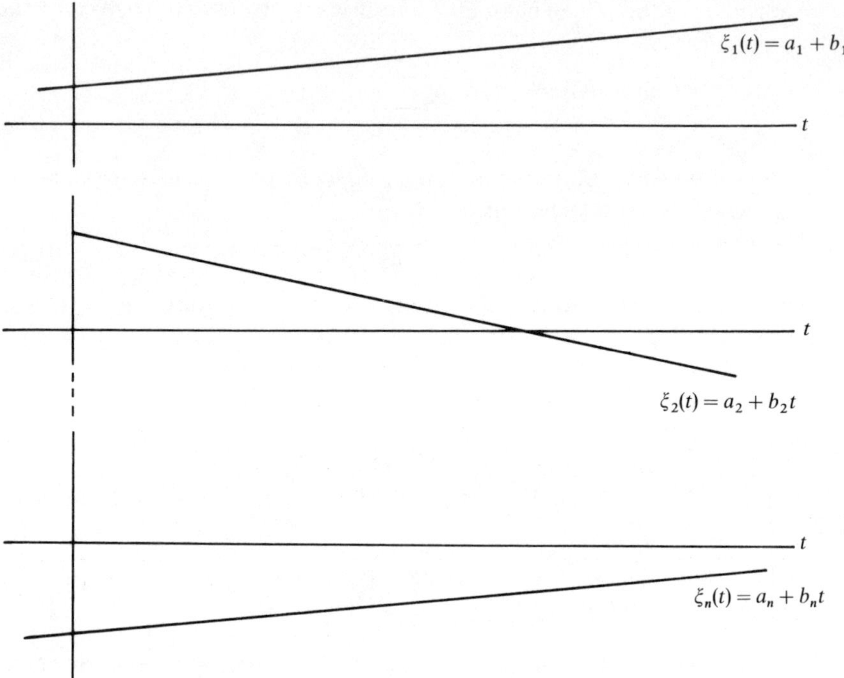

Fig. 1.38 Sample function (example).

$$\begin{aligned} m_x(t) &= \mathscr{E}(x(t)) \\ &= \mathscr{E}(a + bt) \\ &= \mathscr{E}(a) + \mathscr{E}(b)t \end{aligned}$$

$$\begin{aligned} \sigma_x^2(t) &= \mathscr{E}[a + bt - \mathscr{E}(a) - \mathscr{E}(b)t]^2 \\ &= \mathscr{E}[(a - \mathscr{E}(a)) + t(b - \mathscr{E}(b))]^2 \\ &= \sigma_a^2 + 2\mathscr{E}[(a - \mathscr{E}(a))(b - \mathscr{E}(b))]t + t^2\sigma_b^2 \end{aligned}$$

$m_x(t)$ and $\sigma_x^2(t)$ depend on t, indicating a nonstationary process. The value of $\mathscr{E}[(a - \mathscr{E}(a))(b - \mathscr{E}(b))]$ is a measure of any relationship between the random numbers a, b. If a, b are independent, $\mathscr{E}[(a - \mathscr{E}(a))(b - \mathscr{E}(b))] = 0$.

(b) Consider the output $x(t)$ of a DAC (digital to analog converter), clocked at intervals T and fed by a computer algorithm generating random numbers equally likely to be anywhere in the range $1 < x(n) \leqslant 3$.

The system and a sample function are illustrated in Fig. 1.39(a), (b).

In this case (see Appendix B):

$$p(x, t) = p(x) = \begin{cases} 0, & x \leqslant 1 \\ 0.5, & 1 < x \leqslant 3 \\ 0, & 3 < x. \end{cases}$$

The process is stationary since $p(x, t) = p(x)$.

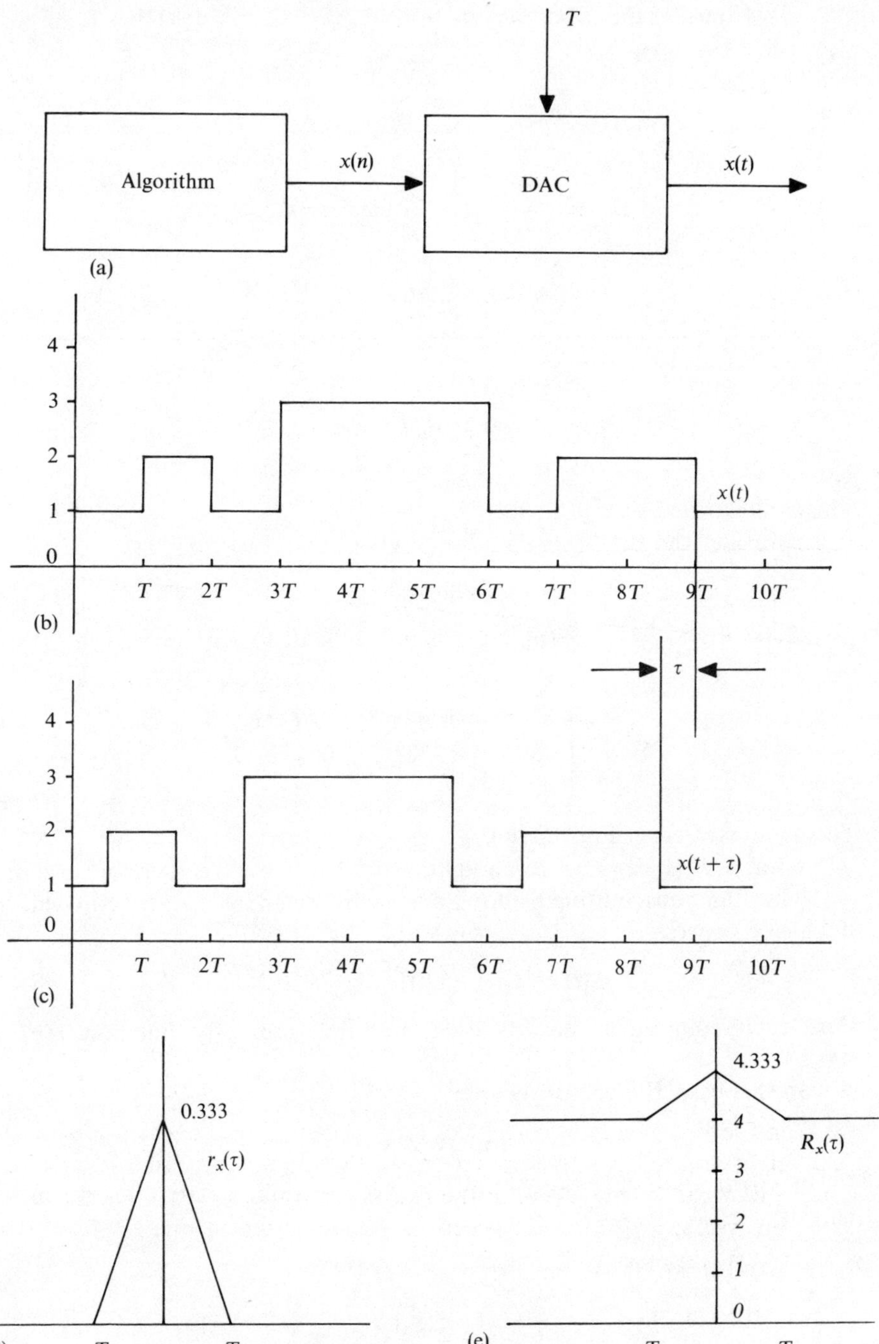

Fig. 1.39 System and sample function (example). (a) System. (b) Sample function. (c) Time shifted sample function. (d) Covariance function. (e) Autocorrelation function.

Consider the ensemble functions $m_x(t)$(Eq. 1.57), $\sigma_x^2(t)$ (Eq. 1.58), $r_x(\tau)$ (Eq. 1.59):

$$m_x(t) = m_x = \int_1^3 x0.5\,\mathrm{d}x = 2$$

$$\sigma_x^2(t) = \sigma_x^2 = \int_1^3 (x-2)^2 0.5\,\mathrm{d}x = 0.333$$

$$r_x(\tau) = \int_1^3 (x(t)-2)(x(t+\tau)-2)0.5\,\mathrm{d}x$$

$$= \begin{cases} 0, & -\infty < \tau \leqslant -T \\ 0.333 + 0.333\tau/T, & -T < \tau \leqslant 0 \\ 0.333 - 0.333\tau/T, & 0 \leqslant \tau \leqslant T \\ 0, & T < \tau \leqslant +\infty. \end{cases}$$

This is illustrated in Fig. 1.39(d).

Consider the time-averaged functions m_t (Eq. 1.61), $R_x(\tau)$ (Eq. 1.62):

$$m_t = 2.$$

Referring to Fig. 1.39(b), (c), it is not difficult to see that:

$$R_x(\tau) = \begin{cases} 4, & -\infty < \tau \leqslant -T \\ 4.333 + 0.333\tau/T, & -T < \tau \leqslant 0 \\ 4.333 - 0.333\tau/T, & 0 \leqslant \tau \leqslant T \\ 4, & T < \tau \leqslant \infty. \end{cases}$$

This is illustrated in Fig. 1.39(e).

Clearly, $R_x(\tau) = r_x(\tau) + m_x^2$, and $m_x = m_t$. The process is therefore ergodic.

The functions outlined above are easily extended to a vector-valued stochastic process:

$$\mathbf{x}(t) = (x_1(t) \quad x_2(t) \quad \cdots \quad x_m(t))^{\mathrm{T}}.$$

Here each component has its own probability density function $p(x_1\ t)$, $p(x_2, t), \ldots.$

In this case, the vector mean:

$$\mathbf{m}_x(t) = \mathscr{E}(\mathbf{x}(t)).$$

If $\mathbf{x}(t)$ is stationary, $\mathbf{m}_x(t) = \mathbf{m}_x$, and a covariance matrix $\bar{\mathbf{R}}_x(\tau)$ can be defined:

$$\bar{\mathbf{R}}_x(\tau) = \operatorname{cov}(\mathbf{x}) = \mathscr{E}[(\mathbf{x}(t) - \mathbf{m}_x)(\mathbf{x}(t+\tau) - \mathbf{m}_x)^{\mathrm{T}}]$$

$$= \begin{bmatrix} \mathscr{E}(x_1(t) - m_{x_1})(x_1(t+\tau) - m_{x_1}) & \mathscr{E}(x_1(t) - m_{x_1})(x_2 t + \tau) - m_{x_2}) & \cdots \\ \mathscr{E}(x_2(t) - m_{x_2})(x_1(t+\tau) - m_{x_1}) & \mathscr{E}(x_2(t) - m_{x_2})(x_2(t+\tau) - m_{x_2}) & \cdots \\ \vdots & \vdots & \cdots \\ \mathscr{E}(x_m(t) - m_{x_m})(x_1(t+\tau) - m_{x_1}) & \cdots & \cdots \end{bmatrix}. \quad (1.65)$$

The diagonal terms are the covariance functions of the vector components, while the off-diagonal elements are the cross-covariances of pairs of components. For an ergodic process $\mathbf{x}(t)$, these terms are related to the corresponding time-averaged functions, $R_{x_1 x_2}(\tau)$, etc. The covariance matrix is symmetric for $\tau = 0$.

A measure of the frequency content of a stationary process is given by the power per unit bandwidth (power/radian/unit time), or *power spectral density* (PSD), $S(\omega)$. It can be shown (Wiener Kinchine Theorem, ref. 2) that $S(\omega)$ is the Fourier transform (w.r.t. the time shift variable τ) of the autocorrelation function, $R_x(\tau)$(Eq. (1.62)):

$$\begin{aligned} S(\omega) &= \mathscr{F}\{R_x(\tau)\} \\ &= \int_{-\infty}^{+\infty} R_x(\tau) e^{-j\omega\tau}\, d\tau. \end{aligned} \tag{1.66}$$

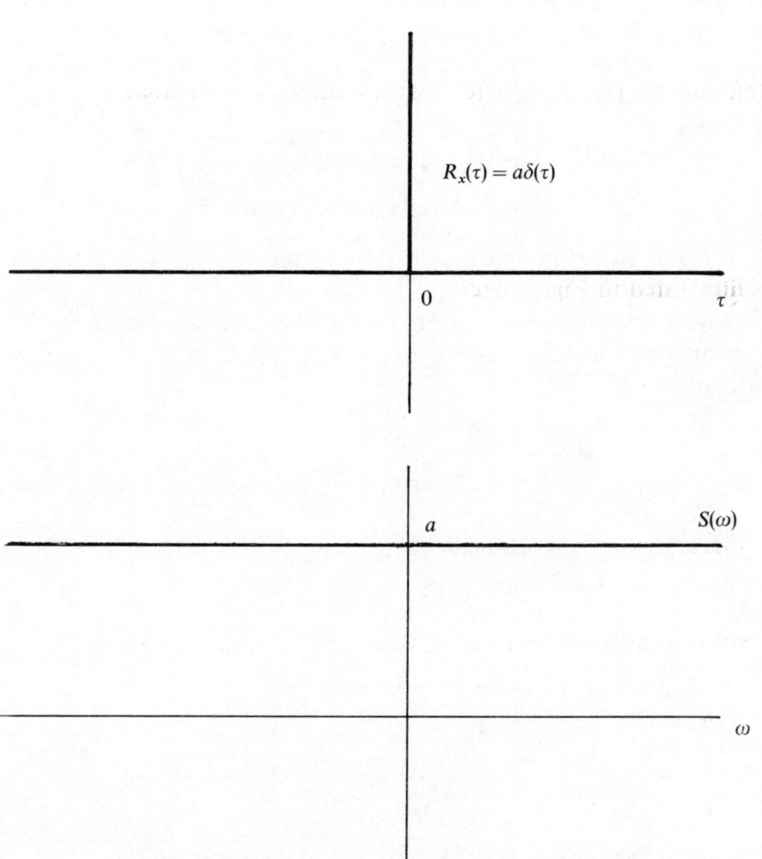

Fig. 1.40 White-noise autocorrelation and power spectral density.

Inversely:

$$R_x(\tau) = \frac{1}{2\pi}\int_{-\infty}^{+\infty} S(\omega)e^{j\omega\tau}\,d\omega. \tag{1.67}$$

Since $R_x(\tau)$ is real and even, $S(\omega)$ is even and real (Sec. 1.15(f)).

EXAMPLE

White noise (App. B.5) $x(t)$, is defined as a signal with constant power spectral density for all frequencies (an impossibility in practice, since this implies infinite power):

$$\begin{aligned} S(\omega) &= a \\ R_x(\tau) &= \mathscr{F}^{-1}\{S(\omega)\} \\ &= a\delta(\tau). \end{aligned}$$

$\delta(\tau)$ is a Dirac delta function (App. C.1). Thus the autocorrelation function of white noise is ∞, $\tau = 0$, and 0, $\tau \neq 0$. This means that the value of $x(t_1)$ is independent of $x(t_2)$, $t_1 \neq t_2$.

$S(\omega)$ and $R_x(\tau)$, $x(t)$ white noise, are illustrated in Fig. 1.40.

The response of an LTI system, Fourier transfer function $G(j\omega)$, to a stochastic input $u(t)$ can be examined using the PSD of $u(t)$. Consider the system in Fig. 1.41.

If the input $u(t)$ has PSD $S_u(\omega)$, the output $y(t)$ has PSD

$$S_y(\omega) = G(j\omega)G(-j\omega)S_u(\omega). \tag{1.68}$$

From this $R_y(\tau)$ can be found (Eq. (1.67)) by taking inverse Fourier transforms (Sec. 1.20).

Consider a resistance–capacitance network (Fig. 1.7(a)) to which a white noise signal is input.

$$G(s) = \frac{1}{1 + sCR}$$

$$G(j\omega) = \frac{1}{1 + j\omega CR}$$

$$S_u(\omega) = a, \quad R_u(\tau) = a\delta(\tau)$$

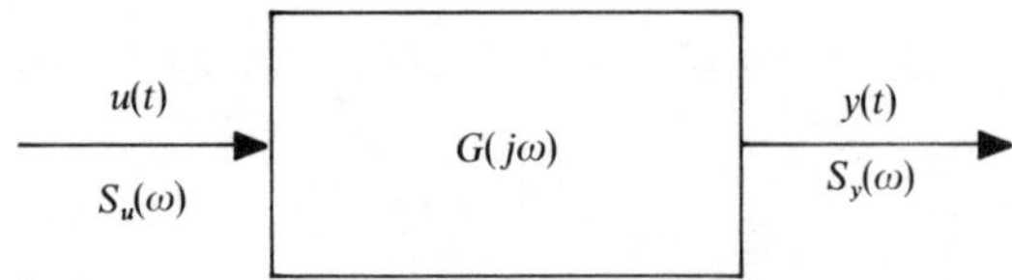

Fig. 1.41 An LTI system with stochastic input.

$$S_y(\omega) = \frac{a}{(1 + j\omega CR)(1 - j\omega CR)} = \frac{a}{1 + \omega^2 C^2 R^2}$$

$$R_y(\tau) = \frac{a}{2CR} \exp\left[-\frac{|\tau|}{CR} \right] \qquad \text{(Fig. 1.26).}$$

1.22 SOME PROPERTIES OF DISCRETE TIME STOCHASTIC PROCESSES (refs. 2, 6)

A discrete stochastic, or random, process $x(n)$ is defined as a real function of discrete time such that at any time n, $x(n)$ is a random number.

In practice (cf. Sec. 1.21) this means that $x(n)$ represents any one of an infinitely large ensemble of discrete sample functions $\xi_1(n)$, $\xi_2(n), \ldots,$ as illustrated in Fig. 1.42.

At time n, the numbers $\xi_1(n)$, $\xi_2(n), \ldots,$ make up a population with probability density function $p(x, n)$ (App. B) which, in principle at least, depends on n.

$p(x, n)$ is a measure of the amplitude of $x(n)$ and, if known, allows the

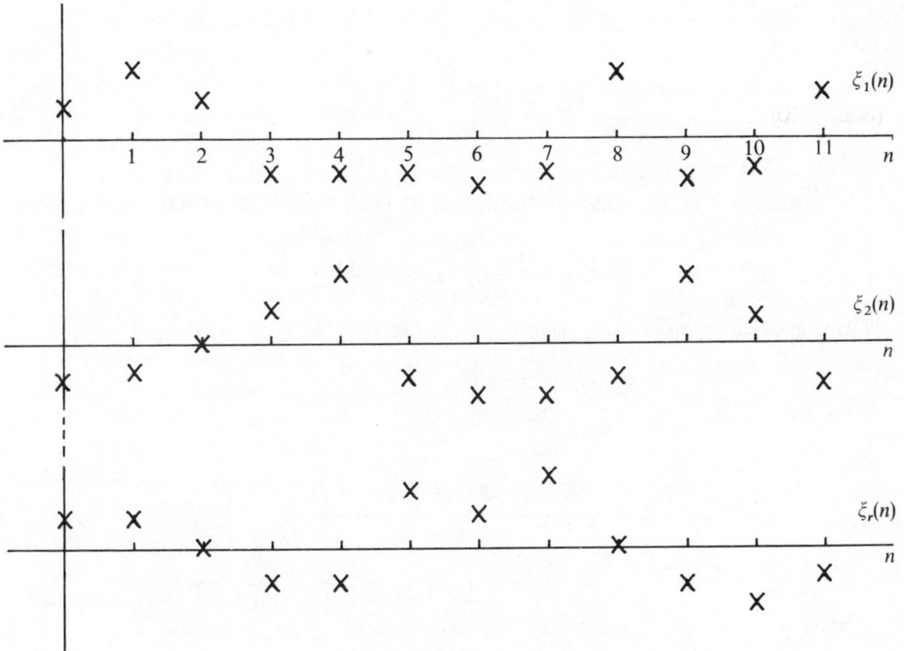

Fig. 1.42 An ensemble of discrete sample functions.

calculation of the mean of $x(n)$ in the ensemble sense.

$$m_x(n) = \mathscr{E}(x(n)) = \int_{-\infty}^{+\infty} xp(x,n)\,\mathrm{d}x. \tag{1.69}$$

The variance of $x(n)$ may also be calculated:

$$\begin{aligned}\sigma_x^2(n) &= \mathscr{E}(x(n) - m_x(n))^2 \\ &= \int_{-\infty}^{+\infty} (x(n) - m_x(n))^2 p(x,n)\,\mathrm{d}x.\end{aligned} \tag{1.70}$$

If $p(x,n)$ is not a function of n, i.e. $p(x,n) = p(x)$, the process $x(n)$ is stationary, and

$$m_x(n) = m_x; \qquad \sigma_x(n) = \sigma_x.$$

A measure of the rate of change of a stationary process, $x(n)$, is given by the covariance of $x(n)$:

$$\begin{aligned}\operatorname{cov}\{x(n)\} &= r_x(k),\ k = -\infty, \ldots -1, 0, 1, 2, \ldots \infty \\ &= \mathscr{E}[(x(n) - m_x)(x(n+k) - m_x)]\end{aligned} \tag{1.71}$$

and a measure of the relationship, if any, between two stationary processes $x(n)$, $y(n)$ is given by the cross-covariance:

$$r_{xy}(k) = \mathscr{E}[(x(n) - m_x)(y(n+k) - m_y)]. \tag{1.72}$$

These functions may be compared with the corresponding time-averaged functions:

$$m_n = \lim_{r\to\infty} \frac{1}{2r} \sum_{n=-r}^{+r} x(n) \tag{1.73}$$

where $x(n)$ is any one of the sample functions $\xi_1(n)$, $\xi_2(n)\ldots$.

A measure of the rate of change of $x(n)$ is the autocorrelation function:

$$R_x(k) = \lim_{r\to\infty} \frac{1}{2r} \sum_{n=-r}^{+r} x(n)\,x(n+k). \tag{1.74}$$

If the time-averaged functions are equal to the corresponding ensemble functions, i.e.:

$$m_n = m_x$$

and

$$\begin{aligned}R_x(k) &= \mathscr{E}[(x(n)x(n+k)] \\ &= r_x(k) + m_x^2\end{aligned}$$

then $x(n)$ is ergodic (cf. Sec. 1.21).

A time-averaged measure of the relationship between two stationary processes, $x(n)$, $y(n)$, is given by the cross-correlation function:

$$R_{xy}(k) = \lim_{r\to\infty} \frac{1}{2r} \sum_{n=-r}^{+r} x(n)y(n+k). \tag{1.75}$$

EXAMPLES

(a) Consider the discrete process defined by sampling the process in the Example (a) of Sec. 1.21 (Fig. 1.38), $T = 0.5$:

$$x(n) = a + bn\,0.5$$

where a, b are random numbers with probability density functions $p(a)$, $p(b)$.

$$m_x(n) = \mathscr{E}(a) + 0.5n\mathscr{E}(b)$$

$$\sigma_x^2(n) = \sigma_a^2 + \mathscr{E}[(a - \mathscr{E}(a))(b - \mathscr{E}(b))]n + 0.25n^2\sigma_b^2.$$

Since $m_x(n)$, $\sigma_x^2(n)$ depend on n, the process is not stationary.

(b) Consider the algorithm in the Example (b) of Sec. 1.21 (Fig. 1.39(a)).

$$m_x(n) = m_x = \int_1^3 x0.5\,\mathrm{d}x = 2$$

$$\sigma_x^2(n) = \sigma_x^2 = \int_1^3 (x-2)^2 0{\cdot}5\,\mathrm{d}x = 0.333$$

$$r_x(k) = \int_1^3 (x(n) - 2)(x(n+k) - 2)0.5\,\mathrm{d}x$$

$$= \begin{cases} 0, & k \neq 0 \\ 0.333, & k = 0. \end{cases}$$

Consider the time-averaged functions:

$$m_n = 2$$

$$R_x(k) = \begin{cases} 4, & k \neq 0 \\ 4.333, & k = 0 \end{cases}$$

$$= r_x(k) + m_x^2.$$

Since $R_x(k) = r_x(k) + m_x^2$ and $m_x = m_n$, the process is ergodic.

The functions outlined above are easily extended to a vector-valued stochastic process $\mathbf{x}(n) = (x_1(n)\ x_2(n)\ \ldots\ x_m(n))^{\mathrm{T}}$, where each component has its own probability density function $p(x_1, n)$, $p(x_2, n), \ldots$

In this case the vector mean is:

$$\mathbf{m}_x(n) = \mathscr{E}(\mathbf{x}(n)).$$

If $\mathbf{x}(n)$ is stationary, $\mathbf{m}_x(n) = \mathbf{m}_x$ and a covariance matrix $\bar{\mathbf{R}}_x(k)$ can be defined:

$$\bar{\mathbf{R}}_x(k) = \operatorname{cov}(\mathbf{x}(n)) = \mathscr{E}[(\mathbf{x}(n) - \mathbf{m}_x)(\mathbf{x}(n+k) - \mathbf{m}_x)^{\mathrm{T}}]. \tag{1.76}$$

The detail of this is similar to Eq (1.65), and similar remarks apply to the elements of this matrix, which is symmetric for $k = 0$.

A measure of the frequency (radians/sample) content of a stationary process – up to a maximum frequency of π radians/sample (half the sampling frequency) – is given by the power spectral density (PSD), $\bar{S}(e^{j\beta})$.

As in the continuous-time case, it can be shown (ref. 6) that $\bar{S}(e^{j\beta})$ is the Fourier transform of the discrete autocorrelation function (Eq. (1.74)):

$$\bar{S}(e^{j\beta}) = \mathscr{F}\{R_x(k)\}$$
$$= \sum_{k=-\infty}^{+\infty} R_x(k) e^{-jk\beta}. \tag{1.77}$$

This is a continuous function in β, $-\pi \leqslant \beta \leqslant +\pi$.

Inversely:

$$R_x(k) = \frac{1}{2\pi} \int_{-\pi}^{\pi} \bar{S}(e^{j\beta}) \, d\beta. \tag{1.78}$$

Since $R_x(k)$ is real and even, $\bar{S}(e^{j\beta})$ is even and real.

EXAMPLE

Consider discrete white noise, $x(n)$, defined as a signal with constant power spectral density, $-\pi \leqslant \beta \leqslant \pi$.

$$\bar{S}(e^{j\beta}) = \begin{cases} a, & -\pi \leqslant \beta \leqslant \pi \\ 0, & |\beta| > \pi \end{cases}$$

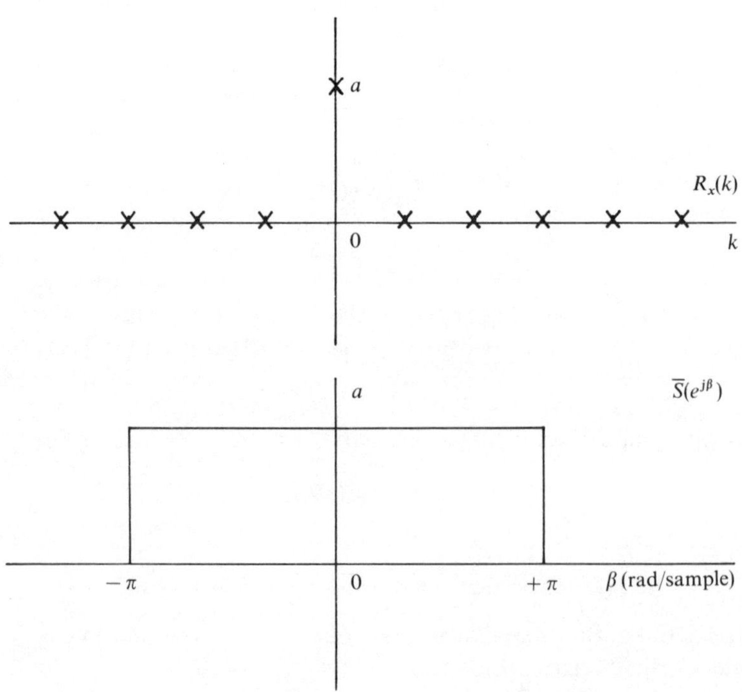

Fig. 1.43 Discrete white noise autocorrelation and PSD.

$$R_x(k) = \frac{1}{2\pi}\int_{-\pi}^{+\pi} a e^{jk\beta}\, d\beta$$
$$= a\delta(k)$$

where $\delta(k)$ is a Kronecker delta function (App. C.2).

Thus the autocorrelation function of discrete white noise has a value for $k = 0$ and is zero otherwise, meaning that $x(n_1)$ is uncorrelated with $x(n_2)$, $n_1 \neq n_2$.

$\bar{S}(e^{j\beta})$ and $R_x(k)$ are illustrated in Fig. 1.43.

Unlike continuous time white noise, discrete white noise is a practical signal and its PSD, which is constant over a finite frequency range, carries no implication of infinite power.

The response of a discrete LTI system, transfer function $\bar{G}(e^{j\beta})$, to a stochastic input $u(n)$ can be found using the PSD of $u(n)$. Consider the system in Fig. 1.44.

If the input $u(n)$ has PSD, $\bar{S}_u(e^{j\beta})$, the output $y(n)$ has PSD

$$\bar{S}_y(e^{j\beta}) = \bar{G}(e^{j\beta})\bar{G}(e^{-j\beta})\bar{S}_u(e^{j\beta}).$$

From this $R_y(k)$ (Eq. (1.78)) can be found.

EXAMPLE

Suppose a white noise signal $u(n)$ of 'strength' a is fed to the system shown in Fig. 1.20(a).

$$\bar{G}(z) = \frac{0.393}{z - 0.607}$$

$$\bar{G}(e^{j\beta}) = \frac{0.393}{e^{j\beta} - 0.607}$$

$$\bar{S}_u(e^{j\beta}) = a$$

$$\bar{S}_y(e^{j\beta}) = \frac{0.154a}{(e^{j\beta} - 0.607)(e^{-j\beta} - 0.607)}$$
$$= \frac{0.154a}{1.368 - 1.214\cos\beta}.$$

$R_y(k)$ may be found, numerically at least, by the method outlined in Sec. 1.20(ii).

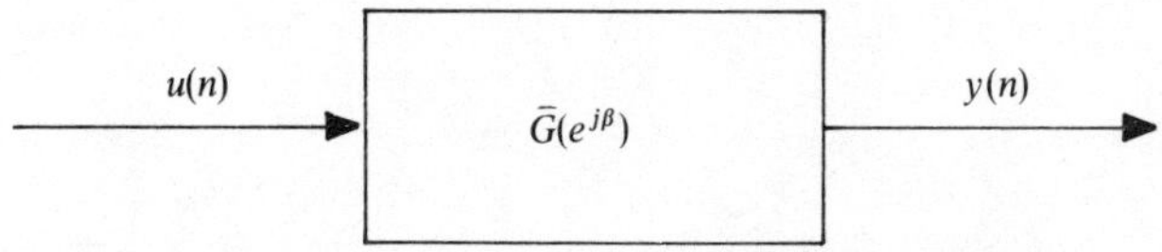

Fig. 1.44 Discrete LTI system with stochastic input.

REFERENCES

1. Ayres, F., *Differential Equations.* Schaum Outline Series, 1952.
2. Chirlian, P.M., *Signals Systems and the Computer.* Intertext, 1973.
3. Di Stefano, J.J., Stubberud, A.R., and Williams, I.J., *Feedback Control Systems.* Schaum Outline Series, 1967.
4. Jury, E.I., *Theory and Application of the z-Transform Method.* Wiley, 1964.
5. Pennington, R.H., ed., *Introductory Computer Methods and Numerical Analysis.* Macmillan, 1971.
6. Rabiner, L.R., and Gold, C., *Theory and Application of Digital Signal Processing.* Prentice-Hall, 1975.
7. Ralston, A., and Rabinowitz, P., *A First Course in Numerical Analysis.* McGraw-Hill, 1978.
8. Ralston, A., and Wilf, H.S., *Mathematical Methods for Digital Computers.* Wiley, 1966.
9. Spiegel, M.R., *Laplace Transforms.* Schaum Outline Series, 1965.

2 CLASSICAL TECHNIQUES FOR CONTINUOUS AND DISCRETE TIME SYSTEMS

2.1 INTRODUCTION

The classical techniques of graphical analysis and design for continuous and discrete time systems are outlined in this chapter. These include Bode plots, root locus and Nyquist diagrams, Nichols charts, and inverse Nyquist diagrams. Detailed suggestions are made for the design of a combined CAD facility which greatly enhances the usefulness of these techniques.

2.2 BODE PLOTS (CONTINUOUS TIME) (REFS. 1, 2, 7)

The frequency response $G(j\omega)$, $0 < \omega \leqslant +\infty$, of a system may be found by experiment on the system, or from a knowledge of the system transfer function $G(s)$, and plotted as a 'Bode plot.'

Factorized into first- and, where necessary, second-order terms:

$$G(j\omega) = G(s)|_{s=j\omega}$$
$$= \frac{K(1+j\omega\tau_{n1})(1+j\omega\tau_{n2})\cdots(1+(j\omega)a_{n_1}+(j\omega)^2 b_{n1})(1+(j\omega)a_{n2}+(j\omega)^2 b_{n2})\cdots}{(j\omega)^N(1+j\omega\tau_{d1})(1+j\omega\tau_{d2})\cdots(1+(j\omega)a_{d1}+(j\omega)^2 b_{d1})(1+(j\omega)a_{d2}+(j\omega)^2 b_{d2})\cdots}. \tag{2.1}$$

The gain magnitude, expressed in decibels, is:

$$\begin{aligned} M &= 20\log_{10}|G(j\omega)| \\ &= 20\log_{10}K + 20\log_{10}|1+j\omega\tau_{n1}| + 20\log_{10}|1+j\omega\tau_{n2}| + \cdots \\ &\quad + 20\log_{10}|1+(j\omega)a_{n1}+(j\omega)^2 a_{n2}| + \cdots \\ &\quad - 20N\log_{10}|j\omega| - 20\log_{10}|1+j\omega\tau_{d1}| - 20\log_{10}|1+j\omega\tau_{d2}|\cdots \\ &\quad - 20\log_{10}|1+j\omega a_{d1}+(j\omega)^2 a_{d2}| - \cdots. \end{aligned}$$

The phase shift, expressed in degrees, is:

$$\begin{aligned}\phi &= \arg(G(j\omega)) = \underline{|G(j\omega)} \\ &= \tan^{-1}\omega\tau_{n1} + \tan^{-1}\omega\tau_{n2} + \cdots + \tan^{-1}\left[\frac{\omega a_{n1}}{1-\omega^2 b_{n1}}\right] + \cdots \\ &\quad - N(90) - \tan^{-1}\omega\tau_{d_1} - \tan^{-1}\omega\tau_{d_2} - \cdots \\ &\quad - \tan^{-1}\left[\frac{\omega a_{d1}}{1-\omega^2 b_{d1}}\right] - \cdots .\end{aligned}$$

The Bode plot for $G(j\omega)$ is found by summing the plots for its individual elements. These are of four types:

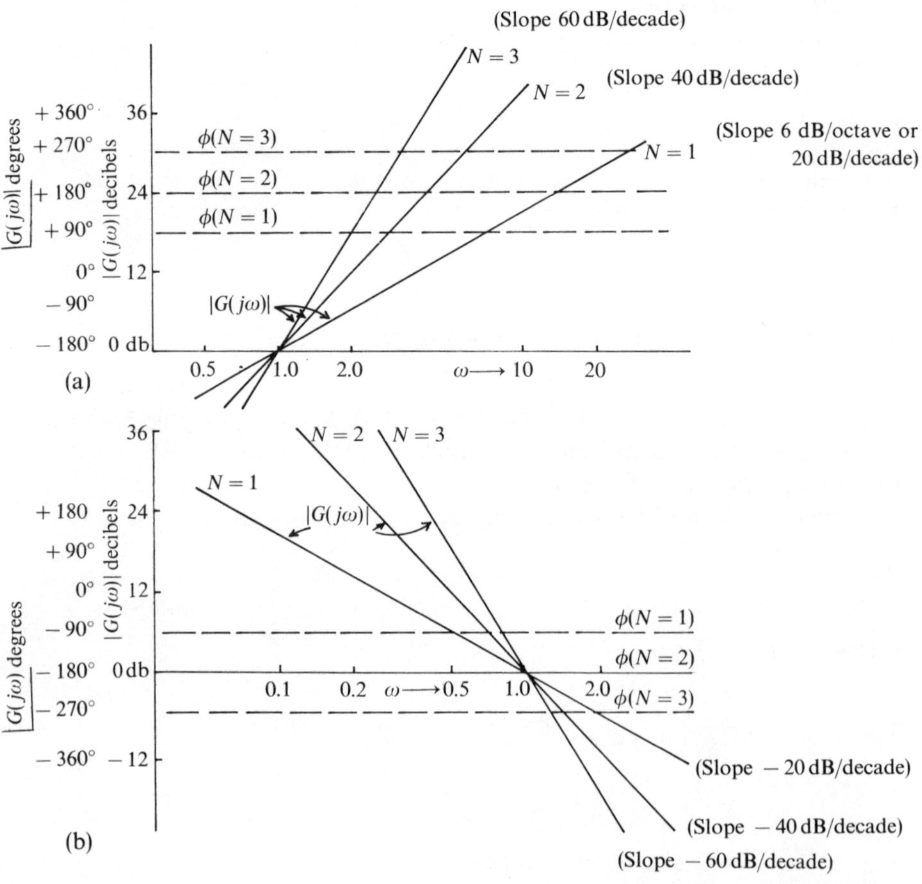

Fig. 2.1 Bode plots for $(j\omega)^{\pm N}$. (a) Bode diagrams for $G(j\omega) = (j\omega)^N$. (b) Bode diagrams for $G(j\omega) = 1/(j\omega)^N$. (Reproduced with permission from Thaler, G.J. and Brown, R.G., *Analysis and Design of Feedback Control Systems.* McGraw-Hill, 1960.)

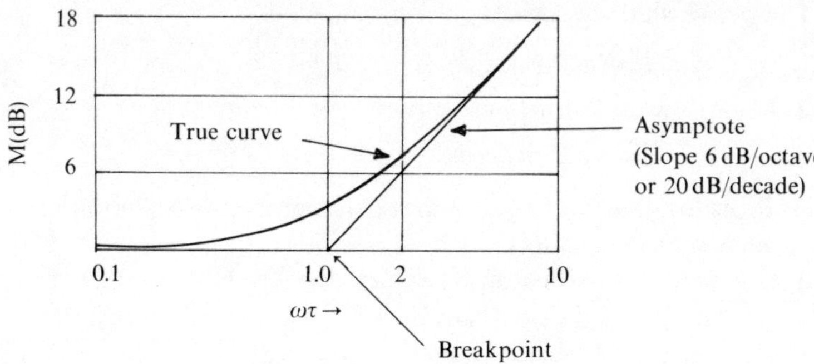

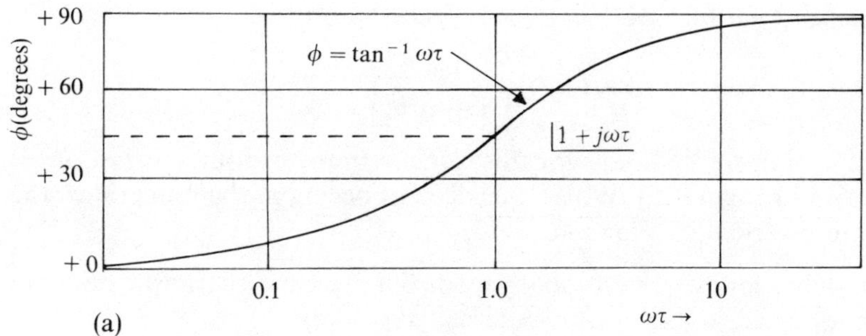

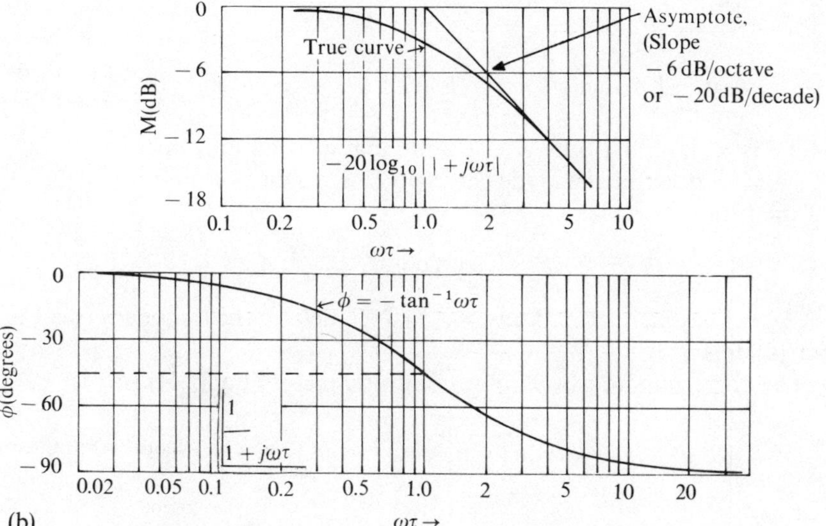

Fig. 2.2 Frequency-normalized Bode plots for $(1 + j\omega\tau)^{\pm 1}$. (a) Plots for $(1 + j\omega\tau)$. (b) Plots for $(1/(1 + j\omega\tau))$. (Reproduced with permission from Thaler, G.J. and Brown, R.G., *Analysis and Design of Feedback Control Systems*. McGraw-Hill, 1960.)

1. K; $M = 20\log_{10}K$ dB.
   $\phi = 0$ degrees.
2. $(j\omega)^N$; $M = 20\log_{10}|(j\omega)^N|$ dB.
   $\phi = 90N$ degrees.

   Plots for $(j\omega)^N$, $N = 1, 2, 3$ in the numerator and denominator of $G(j\omega)$ are shown in Figs. 2.1(a), (b).
3. $(1 + j\omega\tau)$; $M = 20\log_{10}|1 + j\omega\tau|$ dB.
   $\phi = \tan^{-1}\omega\tau$.

   Frequency-normalized plots for $(1 + j\omega\tau)$ in the numerator and denominator of $G(j\omega)$ are shown in Figs. 2.2(a), (b).
4. $(1 + (j\omega)a + (j\omega)^2 b)$; $M = 20\log|1 + j\omega a + (j\omega)^2 b|$ dB

$$\phi = \tan^{-1}\left[\frac{\omega a}{1 - \omega^2 b}\right].$$

   Frequency-normalized plots for this element in the denominator of $G(j\omega)$ are shown in Figure 2.3. Where the element occurs in the numerator, the plots are inverted (cf. Fig. 2.2).

Plots of this kind are traditionally made out in terms of a transfer function denominator quadratic factor, or pole pair (Sec. 1.5):

$$\frac{1}{\omega_n^2}(s^2 + 2\zeta\omega_n s + \omega_n^2)|_{s=j\omega}$$

$$= \frac{1}{\omega_n^2}\left(s + \zeta\omega_n + j\omega_n\sqrt{(1-\zeta^2)}\right)\left(s + \zeta\omega_n - j\omega_n\sqrt{(1-\zeta^2)}\right)\Big|_{s=j\omega}$$

where ζ is a 'damping ratio', ω_n is an 'undamped natural frequency.

ζ and ω_n describe the mode corresponding to the pole pair (Sec. 1.5). This is of the form:

$$Ae^{-\zeta\omega_n t}\cos[(\omega_n\sqrt{(1-\zeta^2)}t].$$

$\zeta\omega_n$ defines the rate of decay; $\omega_n\sqrt{1-\zeta^2}$ defines the frequency (see Figs. 1.5(c), (d), (e)).

The corresponding Bode plot quadratic factor (Eq. (2.1)) is:

$$[1 + (j\omega)a + (j\omega)^2 b] = b\left[(j\omega)^2 + (j\omega)\frac{a}{b} + \frac{1}{b}\right]$$

$$= \frac{1}{\omega_n^2}(s^2 + 2\zeta\omega_n s + \omega_n^2)|_{s=j\omega}.$$

Consequently, in Fig. 2.3,

$$\omega_n^2 = \frac{1}{b}; \qquad 2\zeta\omega_n = \frac{a}{b}.$$

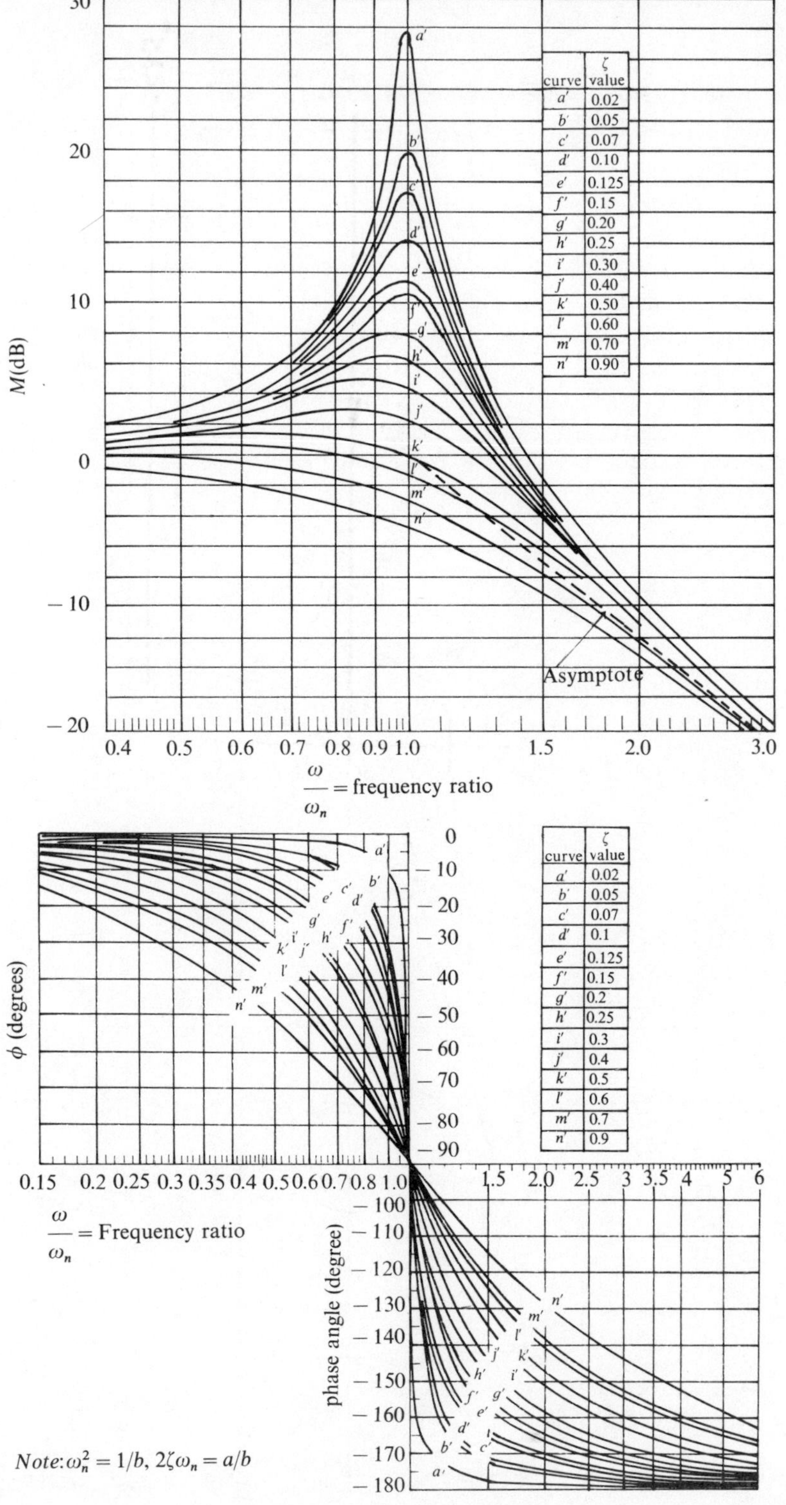

Fig. 2.3 Frequency-normalized Bode plots for $1/[1 + (j\omega)a + (j\omega)^2 b]$. (Reproduced with permission from Thaler, G.J. and Brown, B.S., *Analysis and Design of Feedback Control Systems*. McGraw-Hill, 1960.)

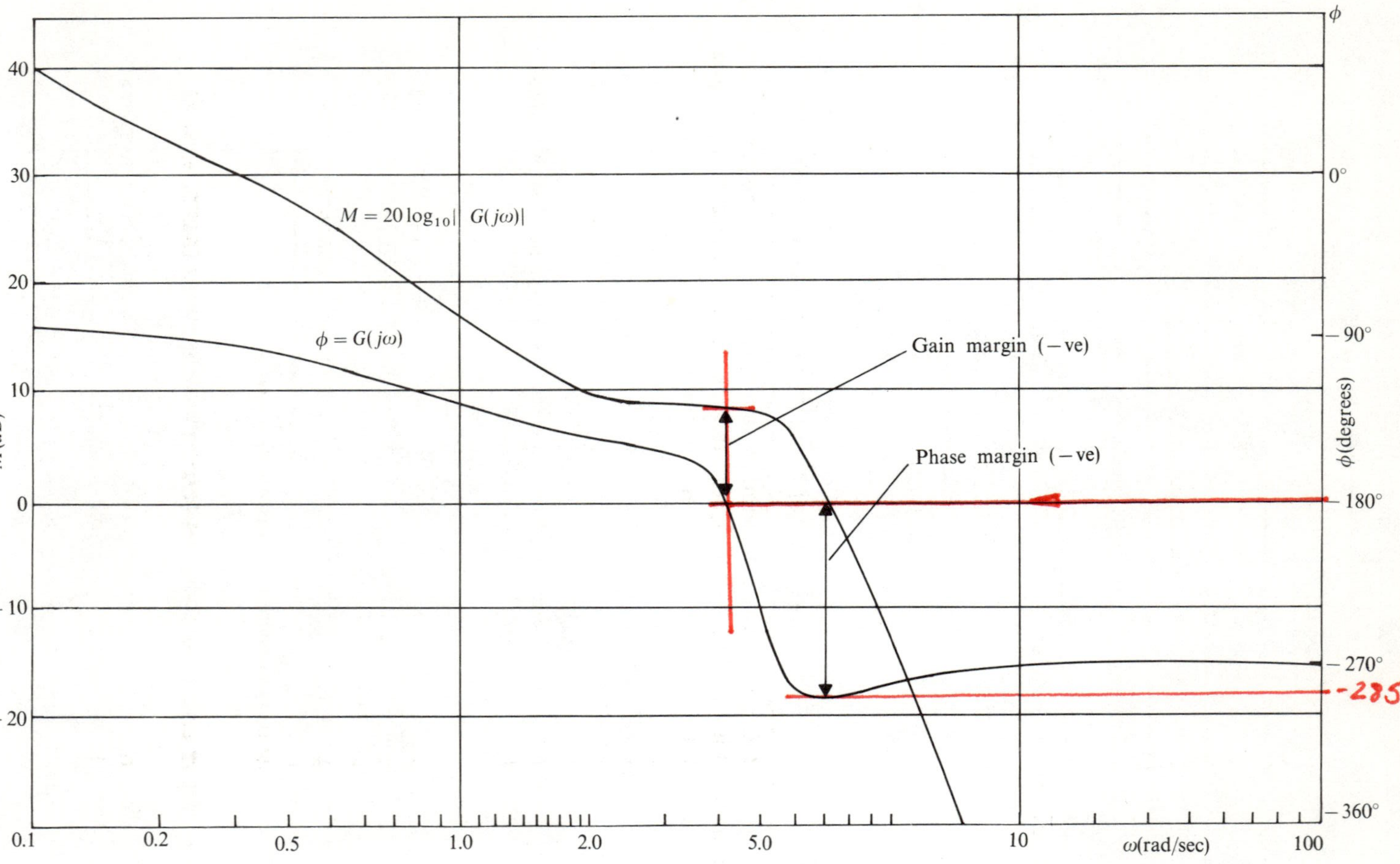

Fig. 2.4 Bode plot (example).

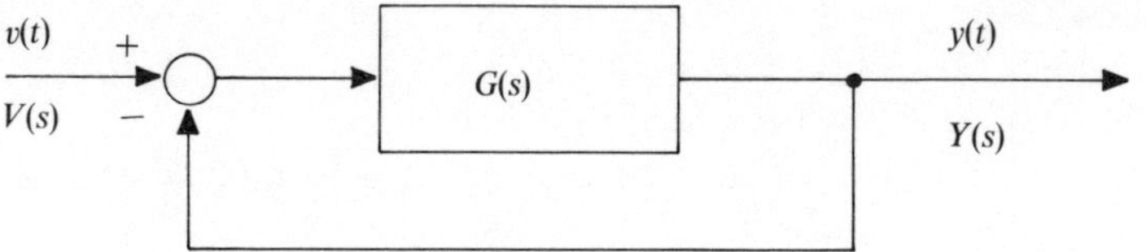

Fig. 2.5 A closed-loop system.

EXAMPLE

Consider:

$$G(s) = \frac{200(s+2.5)}{s^4 + 6.5s^3 + 28s^2 + 12.5s}$$

$$= \frac{40(1+0.4s)}{s(1+2s)(1+0.24s+0.04s^2)}$$

$$G(j\omega) = \frac{40(1+j\omega 0.4)}{j\omega(1+j\omega 2)(1+j\omega 0.24 + (j\omega)^2 0.04)}.$$

The Bode plot of $G(j\omega)$ is found by summing the asymptote plots for the elements (Figs. 2.1, 2.2, 2.3), and adjusting the result to agree with the 'true' curves of Fig. 2.2. This is most simply done by offsetting the curves from the asymptotes by about ± 3 dB at the 'breakpoints.' $M = 20\log_{10}|G(j\omega)|$ and $\phi = \underline{|G(j\omega)}$ are plotted in Fig. 2.4.

The behavior of a closed-loop system incorporating $G(s)$, as shown in Fig. 2.5, may be assessed from the Bode plot of $G(j\omega)$.

The *Bode criterion* for closed-loop stability, assuming no unstable poles of $G(s)$ (Sec. 1.5), is that:

$$M = 20\log_{10}|G(j\omega)| < 0\,\text{dB}, \qquad \text{where } \phi = -180^\circ.$$

The *gain margin* is defined as the value of $(-M)$ where $\phi = -180^\circ$, and the *phase margin* is the value of $(\phi + 180^\circ)$ where $M = 0$ dB. For closed-loop stability, both must the positive.

In Fig. 2.4, the gain margin is -8 dB and the phase margin is -105°. An uncompensated closed-loop incorporating this system would be unstable since the gain and phase margins are negative.

2.3 ROOT LOCUS DIAGRAMS (CONTINUOUS TIME) (REFS. 1, 2, 7)

The numerator and denominator of the transfer function $G(s)$ of a system can be factorized to give (possibly complex) singular points (Sec. 1.5):

$$G(s) = \frac{\bar{K}(s + a_{n1} + jb_{n1})(s + a_{n1} - jb_{n1})\cdots(s + c_{n1})\cdots}{s^N(s + a_{d1} + jb_{d1})(s + a_{d1} - jb_{d1})\cdots(s + c_{d1})\cdots}. \tag{2.2}$$

The points $s = (-a_{n1} \pm jb_{n1}), (-a_{n2} \pm jb_{n2}), -c_{n1}, \ldots$ are *zeros* of $G(s)$ (values of s for which $G(s) = 0$), while $s = (-a_{d1} \pm jb_{d1}), (-a_{d2} \pm jb_{d2}), -c_{d1}, \ldots$ are *poles* of $G(s)$ (values of s for which $G(s) = \infty$).

These poles and zeros may be plotted on a (complex) s plane Argand diagram, or 'pole–zero diagram.'

EXAMPLE

Consider:

$$G(s) = \frac{200(s+2.5)}{s^4 + 6.5s^3 + 28s^2 + 12.5s}$$

$$= \frac{200(s+2.5)}{s(s+0.5)(s+3+j4)(s+3-j4)}.$$

The s plane pole–zero diagram is in Fig. 2.6.

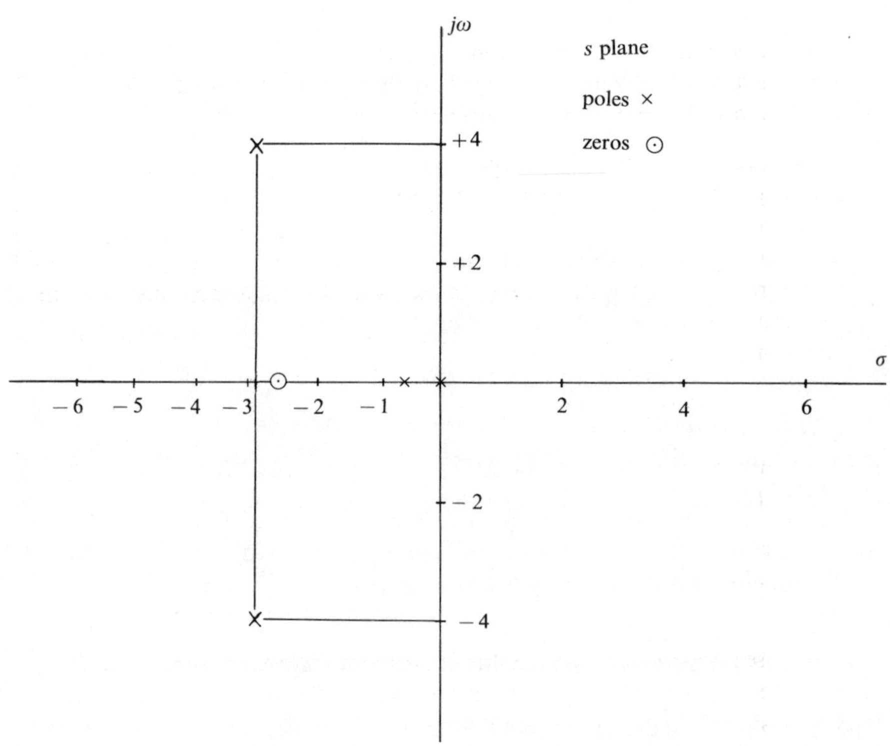

Fig. 2.6 Pole–zero diagram (example).

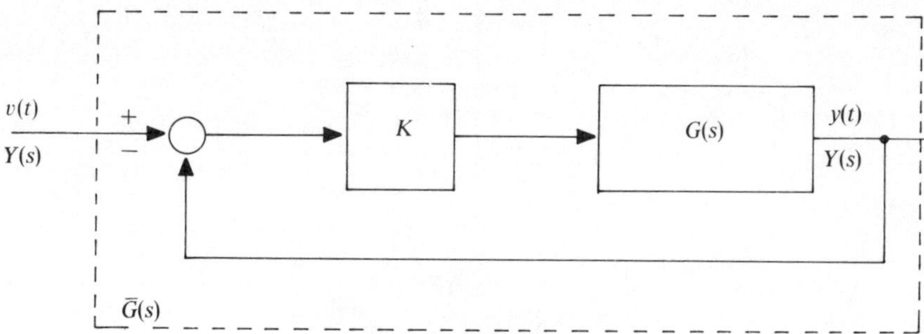

Fig. 2.7 Closed-loop system.

A closed-loop system incorporating $G(s)$ (Fig. 2.7) has the transfer function:

$$\bar{G}(s) = \frac{KG(s)}{1 + KG(s)}. \tag{2.3}$$

The locus of the poles of $\bar{G}(s)$, i.e. the closed-loop poles, $0 \leqslant K \leqslant \infty$, on the s plane, known as the *root locus*, is found by locating all the points on the s plane for which $1 + KG(s) = 0$.

Some common root loci are in Fig. 2.8.

Some rules which help in the construction of a root locus for a system with P open-loop poles and Z open-loop zeros are:

1. There are $(P - Z)$ root locus segments.
2. The segments 'start' at open-loop poles ($K = 0$) and 'finish' at open-loop zeros ($K = \infty$).
3. If vectors are drawn from any point on the root locus to the open-loop poles and zeros, the sum of the angles subtended at the zeros minus the sum of those subtended at the poles is a multiple of 180°.
4. Asymptotes cross the real axis at:

$$\frac{(1 + 2m)180^\circ}{P - Z} \qquad m = 0, \pm 1, \pm 2, \ldots$$

5. Segments of the real axis to the left of an odd number of poles or zeros are on the root locus. This is not true if there are right half plane open-loop zeros.

These rules are illustrated in Fig. 2.8.

The value of K determines the positions of the closed-loop poles on the root locus.

1. The closed-loop system is unstable for values of K which correspond to root locus segments in the closed right half plane.
2. The positions of closed-loop left half plane poles are often considered in terms of damping ratio, ζ, and undamped natural frequency, ω_n (cf. Sec. 2.2, Fig. 2.3).

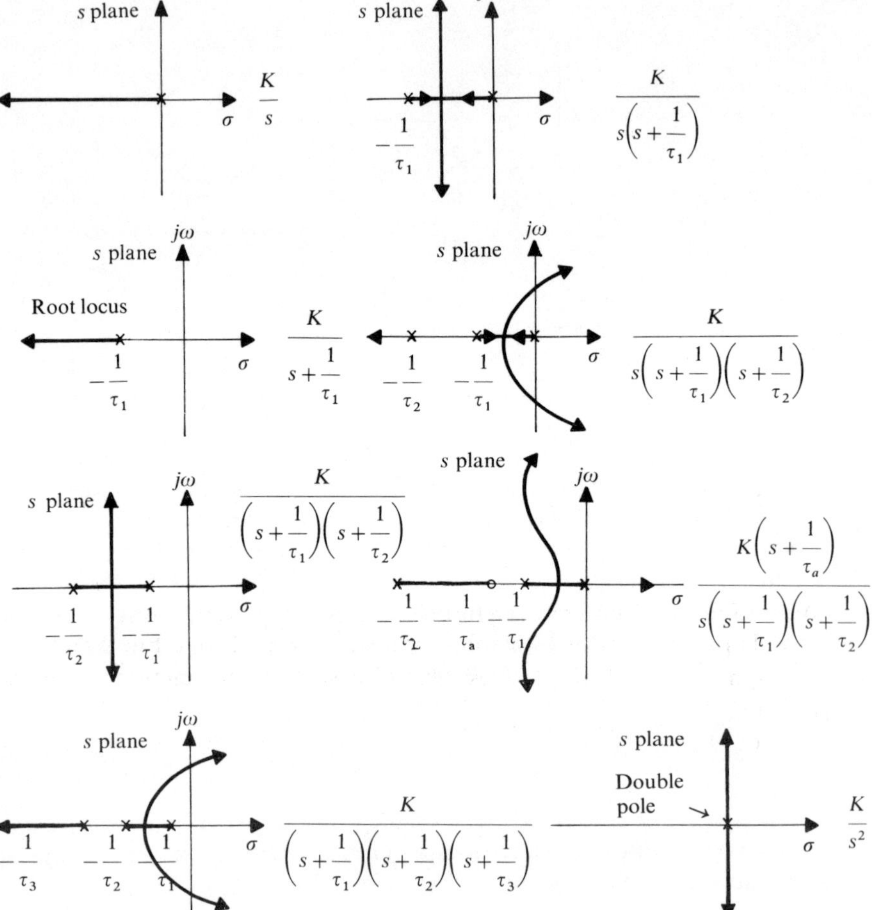

Fig. 2.8 Some common root loci.

Consider a pole pair corresponding to the roots of a second-order transfer function denominator term:

$$\left(\frac{1}{s^2 + 2\zeta\omega_n s + \omega_n^2}\right)$$

i.e.

$$s = \sigma \pm j\omega = -\zeta\omega_n \pm j\omega_n\sqrt{1-\zeta^2}.$$

This pole pair corresponds to a time domain transient response (Sec. 1.5) of the form:

$$Ae^{-\zeta\omega_n t}\cos\left[\omega_n(\sqrt{1-\zeta^2})t\right]$$

ζ and ω_n determine the frequency and decay rate of this transient.

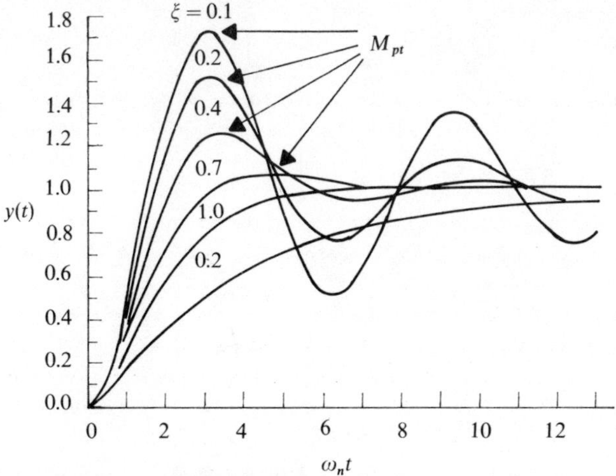

Fig. 2.9 Second-order system step function responses. (Reproduced with permission from D'Azzo, J.J. and Houpis, C.H., *Feedback Control System Analysis and Synthesis.* McGraw-Hill 1966.)

A set of normalized unit step function response curves corresponding to such a pole pair for various damping ratios is in Fig. 2.9 (cf. Fig. 1.6(c)).

A graph of the peak amplitudes, M_{pt}, of these unit step function responses versus damping ratio, ζ, is in Fig. 2.10.

The locus of pole pairs exhibiting a constant damping ratio ζ is given by

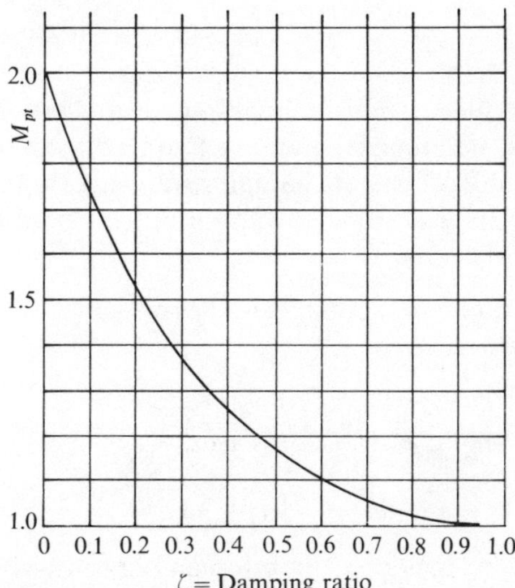

Fig. 2.10 Peak overshoot versus damping ratio. (Reproduced with permission from D'Azzo, J.J. and Houpis. C.H., *Feedback Control System Analysis and Synthesis.* McGraw-Hill 1966.)

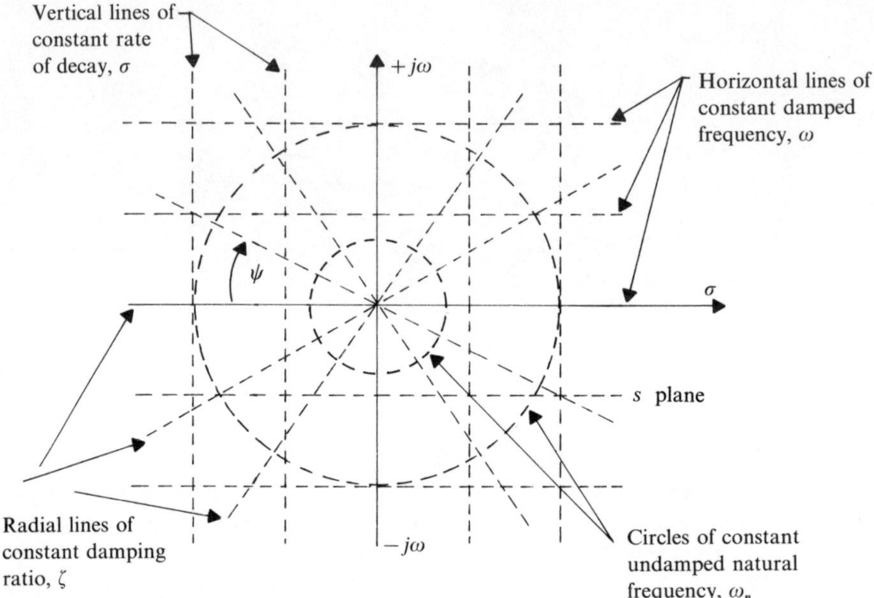

Fig. 2.11 Loci of constant damping ratio and undamped natural frequency. (Reproduced with permission from D'Azzo, J.J. and Houpis, C.H., *Feedback Control System Analysis and Synthesis.* McGraw-Hill, 1966.)

the relationship between σ and ω:

$$\sigma = \frac{\pm\zeta\omega}{\sqrt{1-\zeta^2}}, \qquad \text{or} \qquad \zeta = \cos\psi \qquad \text{(Fig. 2.11)}. \tag{2.4}$$

Several such loci, or 'lines of constant damping ratio,' are shown in Fig. 2.11. Circles of constant undamaged natural frequency are also shown. The intersections of the lines of constant damping ratio and the root locus, together with Figs. 2.9 and 2.10, give some indication of the closed-loop behavior for appropriate values of K.

EXAMPLE

Consider:

$$KG(s) = \frac{K(s+2.5)}{s^4 + 6.5s^3 + 28s^2 + 12.5s}$$

$$= \frac{K(s+2.5)}{s(s+0.5)(s+3+4j)(s+3-4j)}.$$

The root locus and line of constant damping ratio, $\zeta = 0.707$ ($\psi = \pi/4$) are shown in Fig. 2.12.

Selecting $K = 34$, for example, gives closed-loop $\zeta = 0.25$, $\omega_n = 2.3$, and a

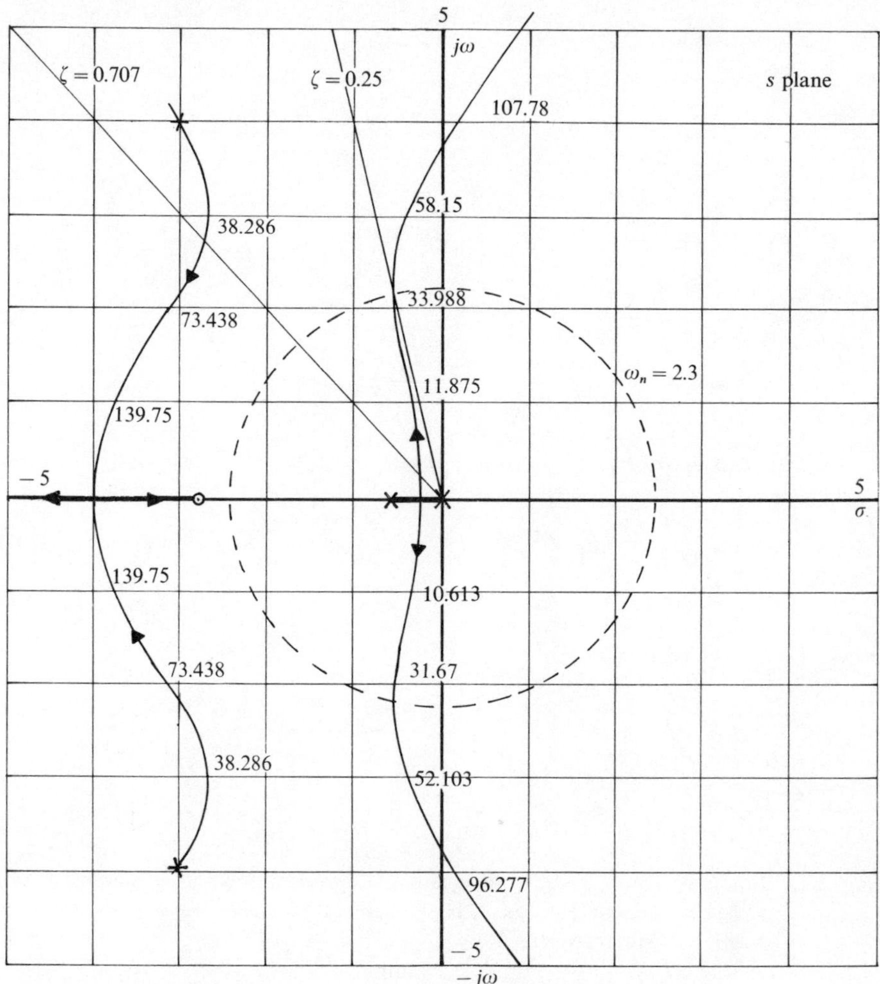

Fig. 2.12 Root locus (example).

unit step response with peak overshoot of approximately 1.4 (Fig. 2.10). This estimate of overshoot must be treated with caution since Fig 2.10 strictly applies to a system with two poles only. Here there are four.

2.4 NYQUIST DIAGRAMS (CONTINUOUS TIME)

The frequency response, $G(j\omega)$, $0 \leqslant \omega \leqslant +\infty$, of a system may be plotted as a 'Nyquist diagram' on a $G(s)$ plane Argand diagram. This can be developed

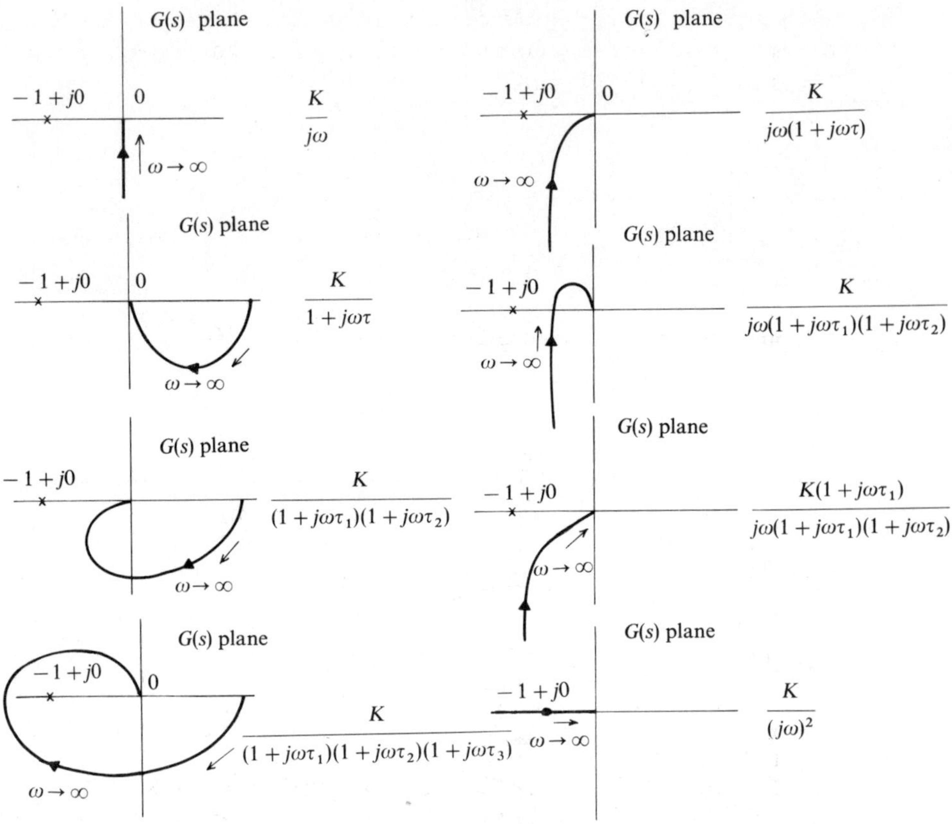

Fig. 2.13 Some common Nyquist diagrams. (Reproduced with permission from D'Azzo, J.J. and Houpis, C.H., *Feedback Control System Analysis and Synthesis*. McGraw-Hill, 1966.)

from experiments on the system, from a Bode plot (Sec. 2.1), by CAD methods (Sec. 2.8), or directly from $G(s)|_{s=j\omega}$ if $G(s)$ is known.

Nyquist diagrams for some common transfer functions are in Fig. 2.13.

The behavior of a closed-loop system incorporating $G(s)$ (Fig. 2.5) may be assessed from the Nyquist diagram of $G(j\omega)$.

For closed-loop stability, the contour $G(j\omega)$, $0 \leqslant \omega \leqslant +\infty$, must 'encircle' the point $(-1 + j0)$ counterclockwise the same number of times that there are open right half plane poles of $G(s)$. Any clockwise encirclement (even partially) indicates closed-loop instability.

Loci of constant closed-loop gains and phase shifts, known as M and N circles, may be drawn on the $G(s)$ plane as in Fig. 2.14(a).

The intersections of the M and N circles with the Nyquist contour $G(j\omega)$ give the closed-loop gains and phase shifts at various frequencies, and so give some indication of closed-loop behavior. The peak closed-loop frequency response is given by the highest M circle which the $G(j\omega)$ contour touches, $M_{p\omega}$; this is related to the step function response peak magnitude M_{pt} (Figs. 2.9, 2.10) as shown in Fig. 2.14(b).

EXAMPLE

Consider

$$G(s) = \frac{200(s+2.5)}{s^4 + 6.5s^3 + 28s^2 + 12.5s}.$$

By developing the Bode plot (Sec. 2.2) and reading from it $|G(j\omega)|$, $\underline{/G(j\omega)}$, or

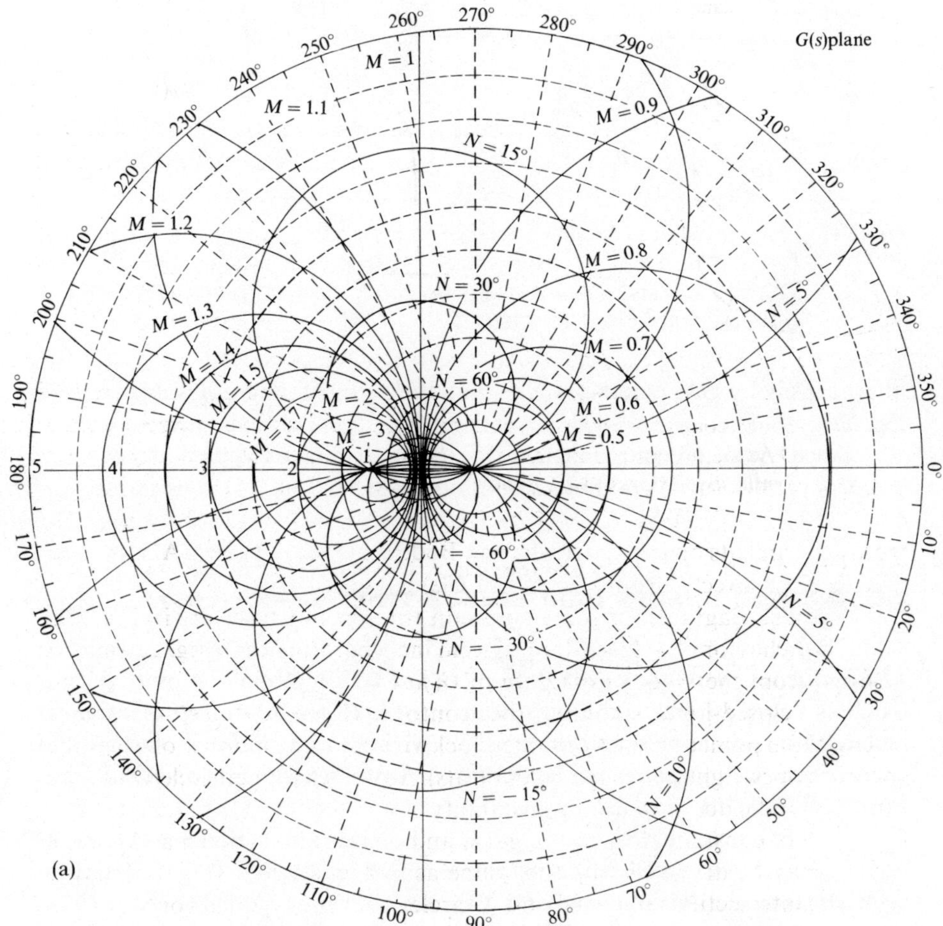

Fig. 2.14 Nyquist diagram design criteria (a) M and N circles.

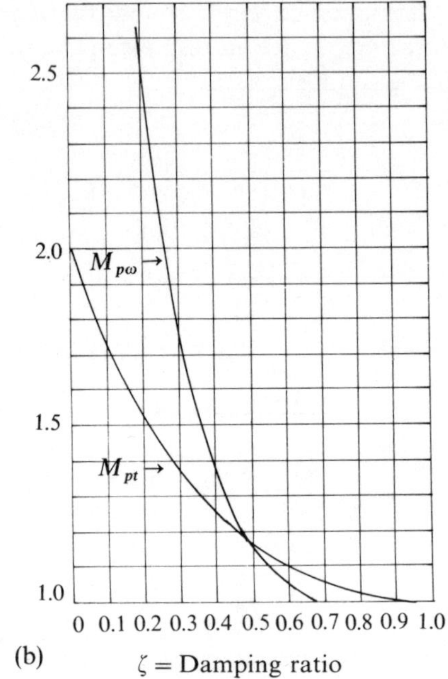

Fig. 2.14 Nyquist diagram design criteria. (a) M and N circles. (b) Frequency and step function response peak amplitudes. (Reproduced with permission from D'Azzo, J.J. and Houpis, C.H. *Feedback Control System Analysis and Synthesis*. McGraw-Hill, 1966.)

alternatively by CAD methods (Sec. 2.8), the Nyquist diagram is drawn as in Fig. 2.15.

Since the $G(j\omega)$ contour encircles $(-1+j0)$ clockwise, partially at least, $0 \leqslant \omega \leqslant \infty$, the corresponding closed-loop system (Fig. 2.5) is unstable.

2.5 NICHOLS CHARTS (CONTINUOUS TIME) (REFS. 1, 2, 7)

The frequency response $G(j\omega)$, $0 \leqslant \omega \leqslant \infty$, of a system may be plotted on a 'Nichols chart' in which the open-loop gain and phase are represented on rectangular coordinates, and the closed-loop gain and phase on appropriate curves. The Nichols chart can be developed from experiments on the system, from a Bode plot (Sec. 2.2), by CAD methods (Sec. 2.8), or directly from $G(s)|_{s=j\omega}$ if $G(s)$ is known.

The behavior of a closed-loop system incorporating $G(s)$ (Fig. 2.5) may be assessed from the Nichols chart for $G(j\omega)$.

For closed-loop stability, the contour $G(j\omega)$, $0 \leqslant \omega \leqslant \infty$, must pass to the right of the open-loop (0 dB, $-180°$) point on the chart.

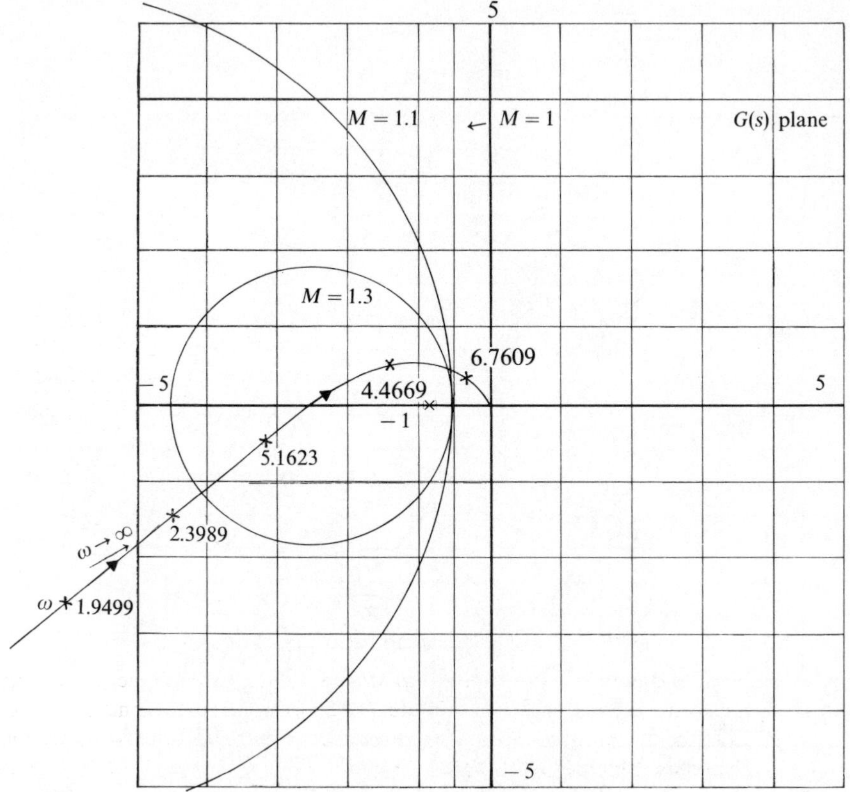

Fig. 2.15 Nyquist diagram (example).

The loci of constant closed-loop gain (M contours) and phase shifts (N contours) are provided on standard charts (Fig. 2.16(a)), and the intersections of these with the $G(j\omega)$ contour give the closed-loop gains and phase shifts at various frequencies. This provides some indication of the closed-loop behavior. As on the Nyquist diagram, the peak closed-loop frequency response is given by the highest M contour which the $G(j\omega)$ contour touches ($M_{p\omega}$); this is related to the step function response peak amplitude M_{pt} (Figs. 2.9, 2.10) as shown in Fig. 2.14(b).

EXAMPLE

Consider

$$G(s) = \frac{200(s + 2.5)}{s^4 + 6.5s^3 + 28s^2 + 12.5s}.$$

By developing the Bode plot (Sec. 2.2) and reading off $|G(j\omega)|$, $\underline{/G(j\omega)}$, or

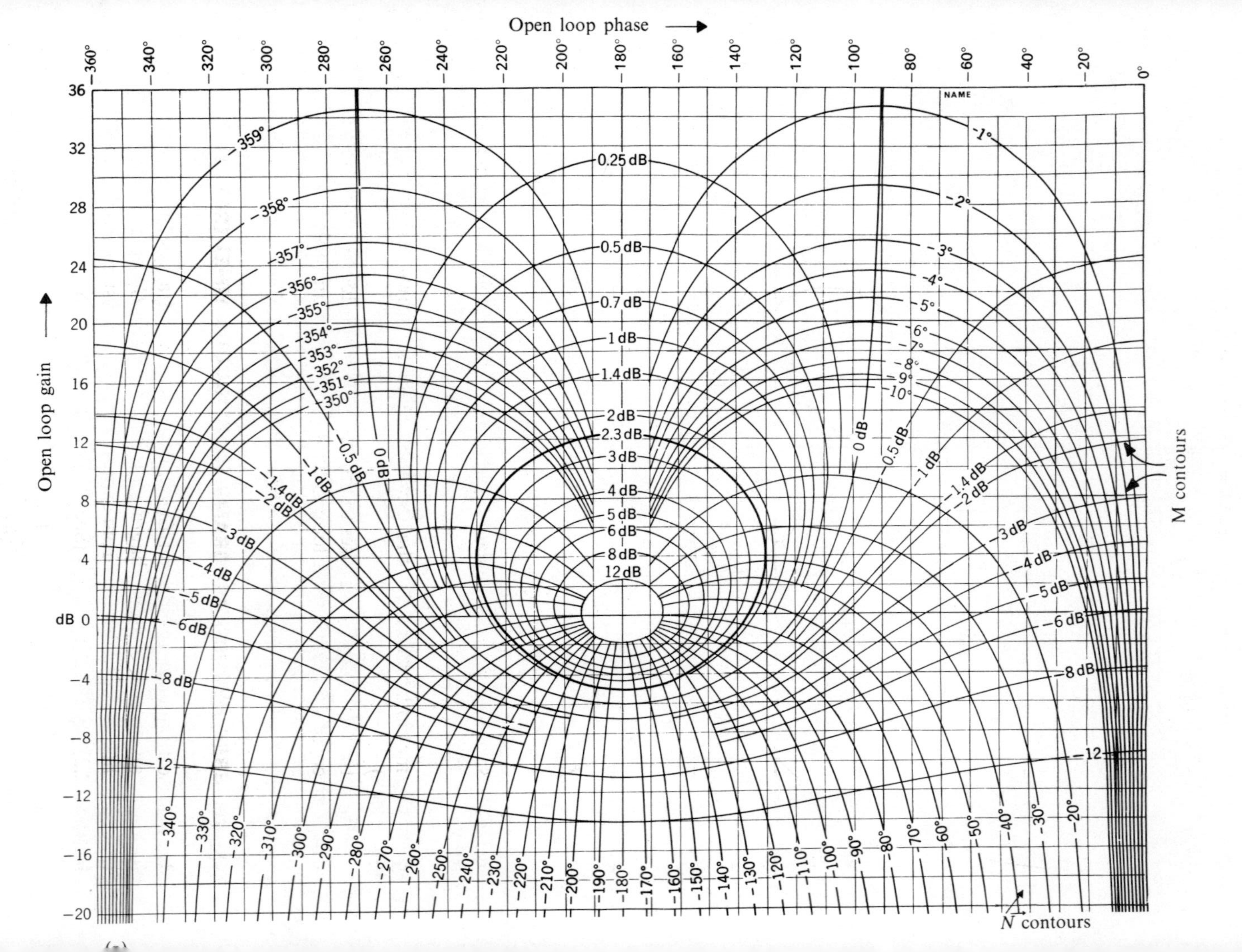
Open loop phase
Open loop gain
M contours
N contours
NAME

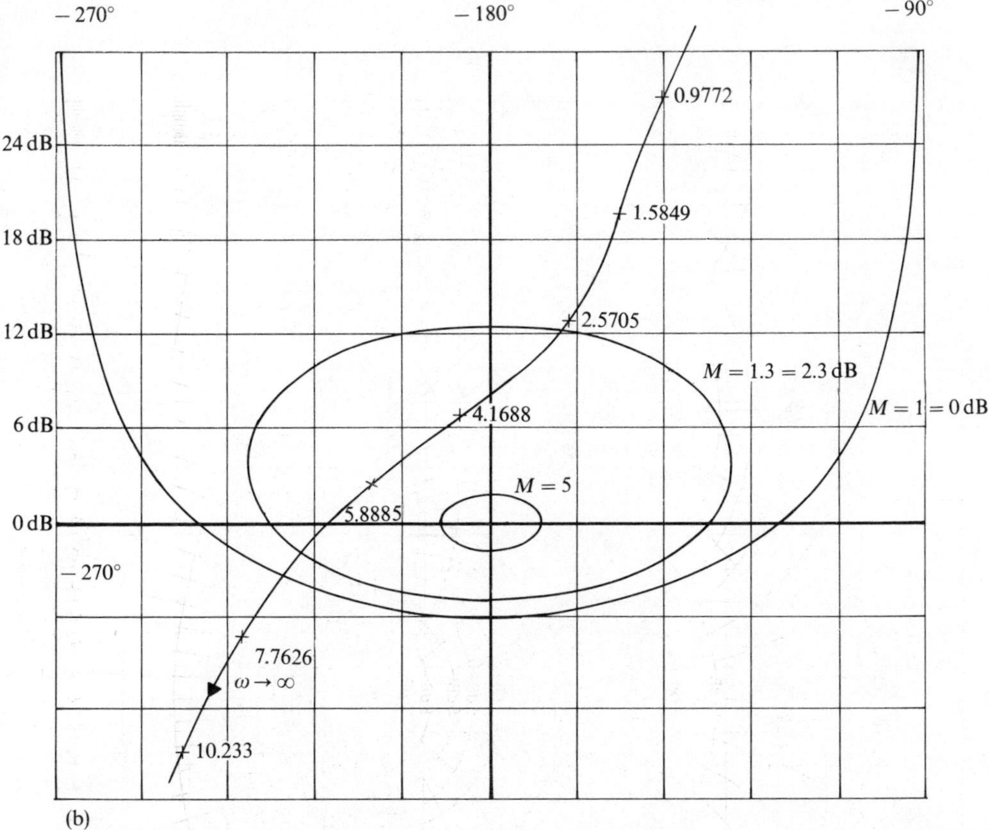

Fig. 2.16 Nichols charts. (a) Standard Nichols chart. (b) Nichols chart (example).

by CAD methods (Sec. 2.8), the Nichols chart is drawn as shown in Fig. 2.16(b).

Since the $G(j\omega)$ contour passes to the left of the open-loop (0 dB, – 180°) point, the corresponding closed-loop system (Fig. 2.5) is unstable.

2.6 INVERSE NYQUIST DIAGRAMS (CONTINUOUS TIME) (REF. 1)

The inverse frequency response $1/G(j\omega)$, $0 \leqslant \omega \leqslant +\infty$ of a system may be plotted as an inverse Nyquist diagram on a (complex) $1/G(s)$ plane Argand diagram. This can be developed from experiments on the system, from a Bode plot (Sec. 2.2), by CAD methods (Sec. 2.8), or directly from

$$\left.\frac{1}{G(s)}\right|_{s=j\omega}$$

if $G(s)$ is known.

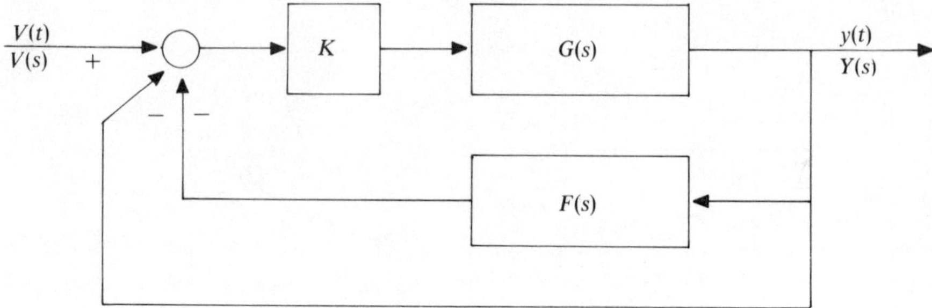

Fig. 2.17 Non-unity feedback closed-loop system.

These diagrams are useful for designing parallel feedback closed-loop systems of the kind shown in Fig. 2.17.

The subsidiary closed-loop system incorporating $K, G(s)$, and $F(s)$ only (Fig. 2.17) has the transfer function:

$$\bar{G}(s) = \frac{KG(s)}{1 + KG(s)F(s)}. \tag{2.5}$$

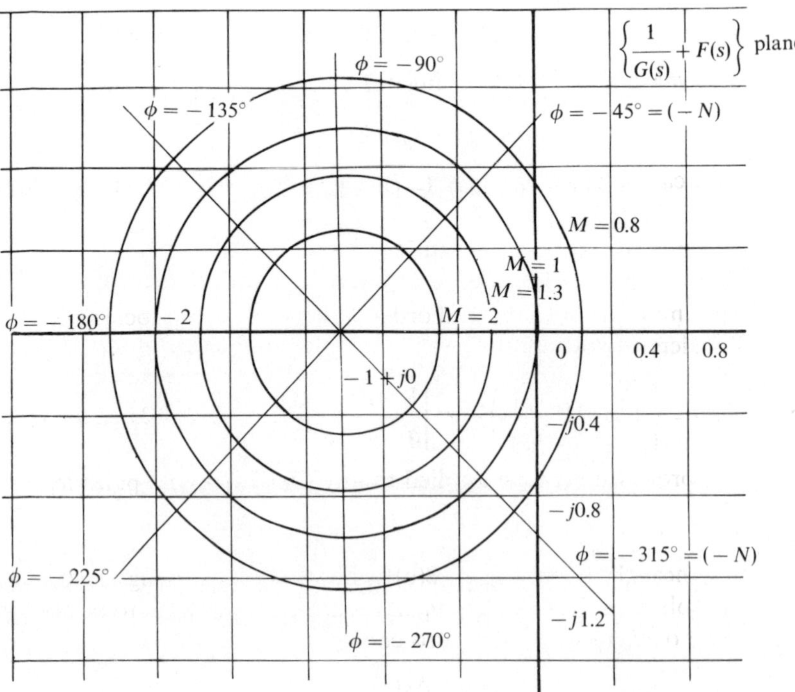

Fig. 2.18 (*M*) and (− *N*) contours on the inverse Nyquist diagram.

Hence

$$\frac{1}{\bar{G}(s)} = \frac{1}{KG(s)} + F(s).$$

The behavior of the whole closed-loop system may be assessed from the contour $\{1/KG(j\omega) + F(j\omega)\}$, $0 \leqslant \omega \leqslant +\infty$, on the $\{1/KG(s) + F(s)\}$ plane Argand diagram.

For closed-loop stability, this contour must encircle $(-1 + j0)$, $0 \leqslant \omega \leqslant \infty$, counterclockwise the same number of times that there are open right half plane poles of $\bar{G}(s)$. Any clockwise encirclement, even partially, indicates closed-loop instability. Loci of constant closed-loop gains and phase shifts, the (M) and $(-N)$ contours, may be drawn as shown in Fig. 2.18, and the intersections of the contour $\{1/KG(j\omega) + F(j\omega)\}$, $0 \leqslant \omega \leqslant \infty$, with these loci give some indication of closed-loop behavior. Closed-loop design procedure involves selecting $F(s)$ and K so that these intersections are satisfactory. An example is considered in Sec. 2.7(v).

2.7 SERVOMECHANISM DESIGN USING BODE PLOTS, ROOT LOCUS DIAGRAMS, NYQUIST DIAGRAMS, AND NICHOLS CHARTS (REFS, 1, 2, 7).

Servomechanisms of the kind shown in Fig. 2.19 may be designed using the diagrams of Secs. 2.2 to 2.6.

An open-loop system transfer function is $G(s)$, and the design requirement is to devise a shaping network, transfer function $F(s)$, which ensures good closed-loop system behavior. More complex closed-loop systems may also be designed using the same basic techniques (refs. 2, 7).

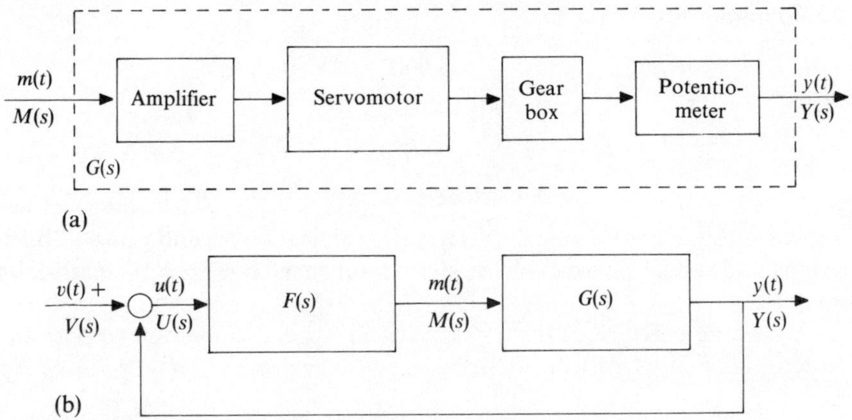

Fig. 2.19 A servomechanism. (a) Open-loop system. (b) Closed-loop system.

The design of a shaping network $F(s)$ for a controlled system comprising an amplifier, servomotor, gearbox, and potentiometer (Fig. 2.19(a)) with transfer function

$$G(s) = \frac{45{,}000}{s(s+2)(s+30)}$$

is considered in the following paragraphs.

If $G(s)$ has a pole at $s = 0$, as in this case, the direct use of a transfer function analyzer (TFA) to develop a Bode plot $G(j\omega)$ and thence to postulate a transfer function $G(s)$ (Sec. 2.2, Example) may be impractical due to experimental drift. Fortunately, it is usually fairly easy to devise experimentally a stable, though sluggish, closed-loop system (Fig. 2.7, K small) of which the Bode plot, $\bar{G}(j\omega)$ and transfer function $\bar{G}(s)$ can be found by means of a TFA. $G(s)$ is then easily found:

$$G(s) = \frac{\bar{G}(s)}{K(1 - \bar{G}(s))}.$$

(I) Bode Plot Design (Continuous Time)

The Bode plot for $G(j\omega)$ is drawn (Sec. 2.2) either from a knowledge of $G(s)$ or by using data found by experiment on the system (see above). The design strategy is to devise a Bode plot for $F(j\omega)$ which when added to that of $G(j\omega)$ results in a 'well-shaped' plot for $G(j\omega)F(j\omega)$ exhibiting good gain margin (> 5 dB), phase margin ($> 45°$), high open-loop gain below the 0 dB *crossover frequency*, and as high a crossover frequency as possible. $F(s)$ is then mechanized in circuit or other form.

EXAMPLE

The Bode plot for

$$G(j\omega) = \frac{45{,}000}{j\omega(j\omega+2)(j\omega+30)}$$

$$= \frac{750}{j\omega(1 + j\omega 0.5)(1 + 0.033 j\omega)}$$

is shown in Fig. 2.20(a). This exhibits negative phase ($-45°$) and gain (-30 dB) margins and would be unstable in an uncompensated ($F(s) = 1$) closed-loop (Fig. 2.19(b)).

A shaping network of the common 'lead' type, containing one pole and one zero, is often worth considering:

$$F(s) = \frac{K(1 + 10as)}{(1 + as)}.$$

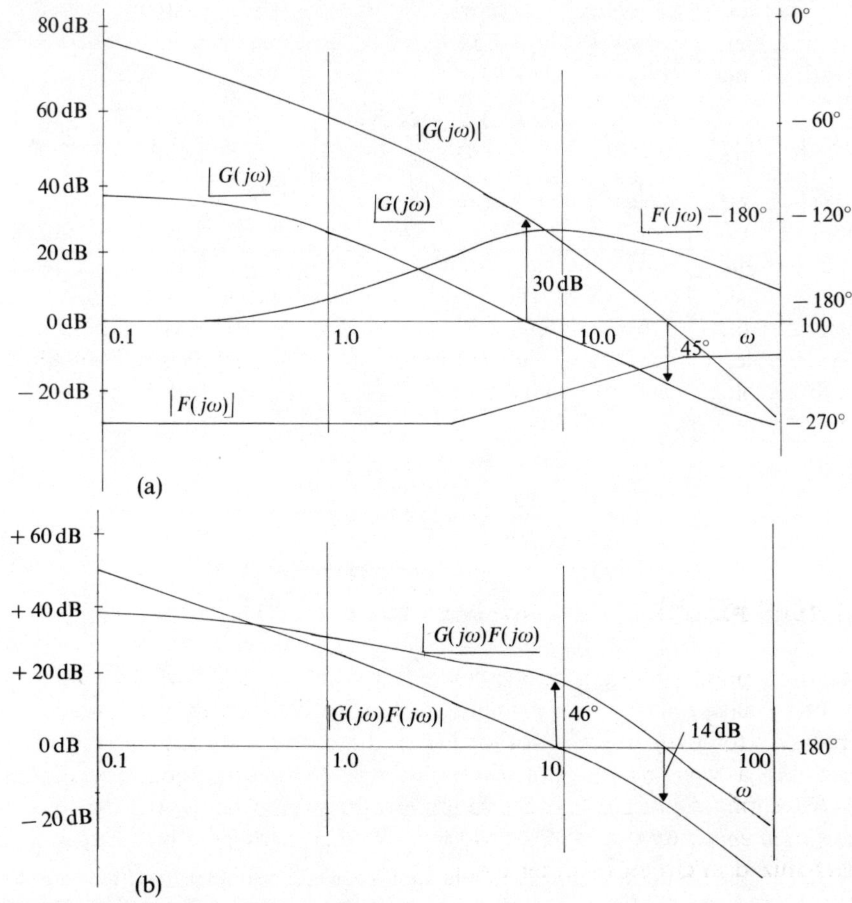

Fig. 2.20 Bode plot (example). (a) Bode plots for $G(j\omega)$ and $F(j\omega)$. (b) Bode plot for $G(j\omega)F(j\omega)$.

A Bode plot of such a network,

$$F(j\omega) = F(s)|_{s=j\omega}$$
$$= 0.03\left(\frac{1 + 0.25j\omega}{1 + 0.025j\omega}\right)$$

is shown in Fig. 2.20(a), and of the compensated system $G(j\omega)F(j\omega)$ in Fig. 2.20(b). The choice of $F(j\omega)$ so that the maximum phase advance (about 50°) is effective where $\angle G(j\omega) = -180°$ is worthy of note.

The compensated system has satisfactory gain margin (14 dB), phase margin (46°), crossover frequency (9.5 rad s^{-1}) and low frequency gain.

The shaping network could be mechanized using operational amplifiers as shown in Fig. 2.21(a) (cf. Fig. 1.7). A closed-loop system unit step function response is in Fig. 2.21(b).

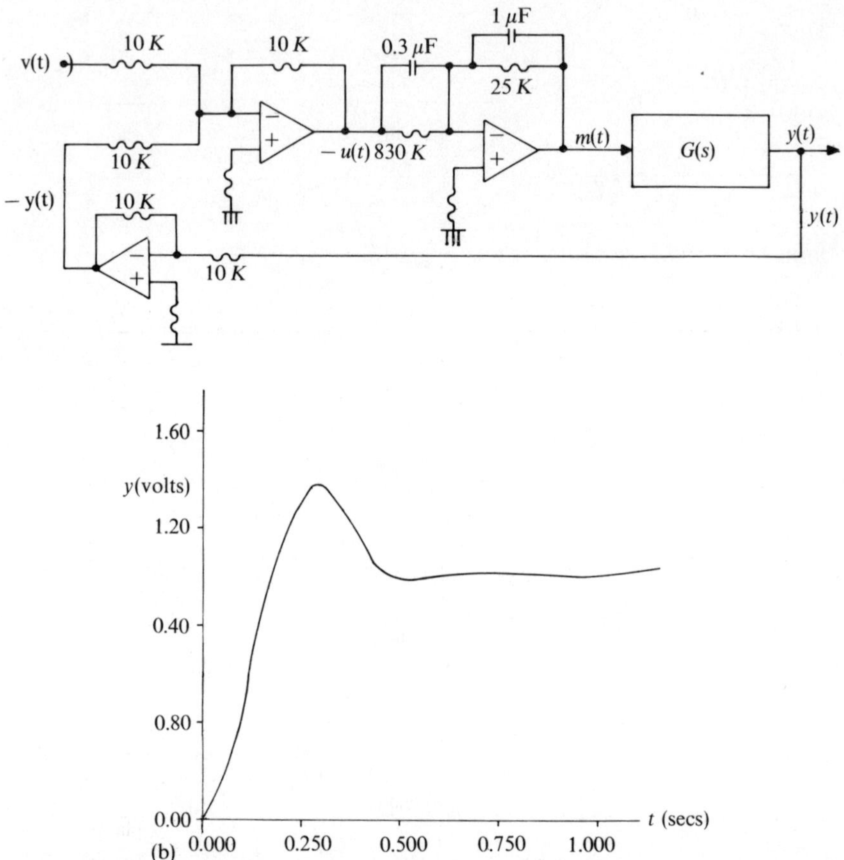

Fig. 2.21 Closed-loop mechanization and step response (example). (a) Circuit mechanization of shaping network. (b) Unit step function response.

The peak unit step function response (Fig. 2.21(b)), $M_{pt} \approx 1.4$ might suggest too low a closed-loop damping ratio on reference to Fig. 2.10. However, it should be recalled that Fig. 2.10 refers to a system with only two poles, whereas the closed-loop system under consideration here has four poles and one zero. The simple criteria of Figs. 2.9, 2.10 are therefore not strictly relevant though of course they provide useful guidance and are quite accurate if two poles are 'dominant' over the others.

(ii) Root Locus Diagram Design (Continuous Time)

The root locus for $KG(s)$ is drawn, $0 \leqslant K \leqslant \infty$ (Sec. 2.3). This requires knowledge of $G(s)$ which might first have to be established by experiment. The

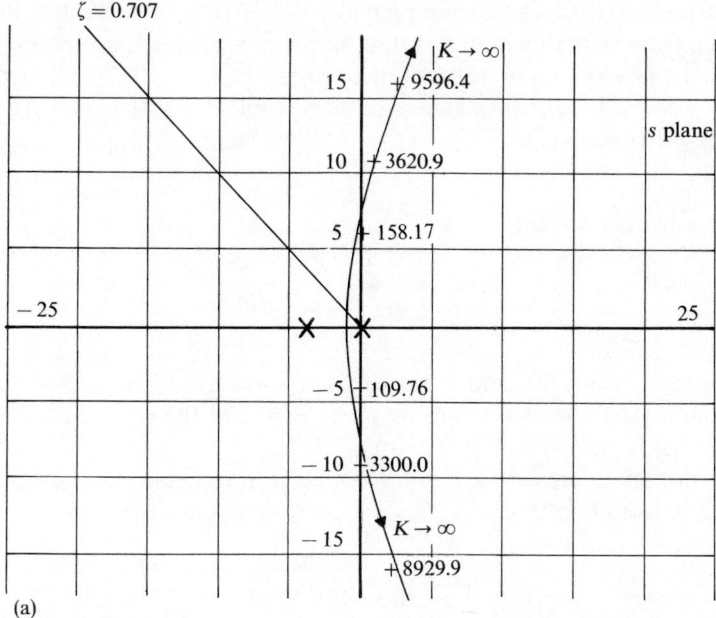

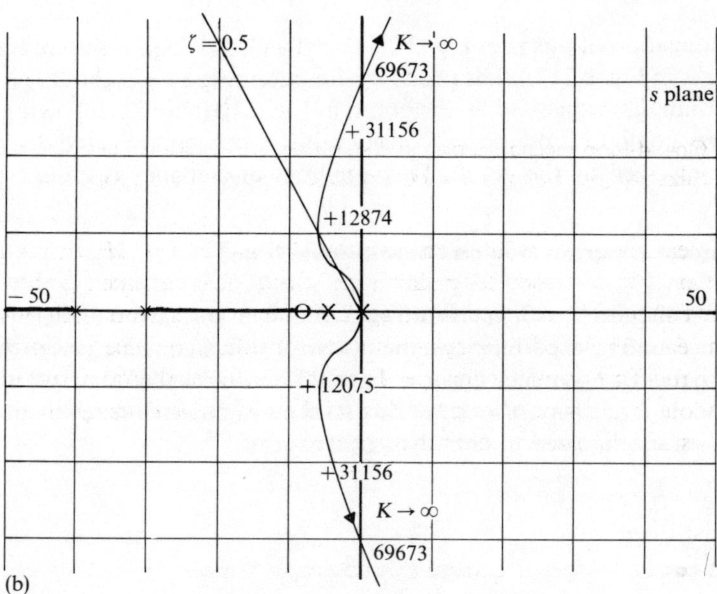

Fig. 2.22 Root locus diagrams (example). (a) Root locus for $KG(s)$. (b) Root locus for $KG(s)F(s)$.

design strategy is to devise a shaping network $F(s)$ (Fig. 2.19(b)) such that the root locus for $KG(s)F(s)$ allows the dominant closed-loop poles to be positioned on good lines of constant damping ratio (say $\zeta = 0.7$) as far from the origin as possible for as high a value of K as possible. $F(s)$ is then mechanized in circuit or other form.

EXAMPLE

A root locus diagram for

$$KG(s) = \frac{K}{s(s+2)(s+30)}$$

is in Fig. 2.22.(a).

A pole at $s = -40$, and a zero at $s = -4$ are added to the diagram, corresponding to a lead shaping network $F(s)$, and the root locus diagram for $KG(s)F(s)$ is shown in Fig. 2.22(b).

The dominant closed-loop poles can be seen to lie on the $\zeta = 0.5$ line at $K = 13{,}500$. Recalling that

$$G(s) = \frac{45{,}000}{s(s+2)(s+30)}$$

this implies a shaping network:

$$F(s) = \frac{0.3(s+4)}{(s+40)} = \frac{0.03(1+0.25s)}{(1+0.025s)}\ .$$

A mechanization of this is in Fig. 2.21(a) and a closed-loop unit step function response is in Fig. 2.21(b). The peak step function response is about 1.4, which is more than that suggested by the graph in Fig. 2.10 (about 1.16) owing to the pole zero configuration being considerably more complicated than that strictly described by the graph. This apparent discrepancy is quite typical.

(iii) Nyquist Diagram Design (Continuous Time)

A Nyquist diagram for $G(j\omega)$ is drawn (Sec. 2.4), usually from a Bode plot, or from data found by experiment on the system. The design strategy is to devise a shaping network $F(s)$ which ensures that the contour $G(j\omega)F(j\omega)$ is tangential to a good M circle (say $M = 1.3 = 2.3$ dB) at as high a frequency as possible. $F(s)$ is then mechanized in circuit or other form.

EXAMPLE

A Nyquist diagram for

$$G(j\omega) = \frac{45{,}000}{j\omega(j\omega+2)(j\omega+30)}$$

is shown in Fig. 2.23(a).

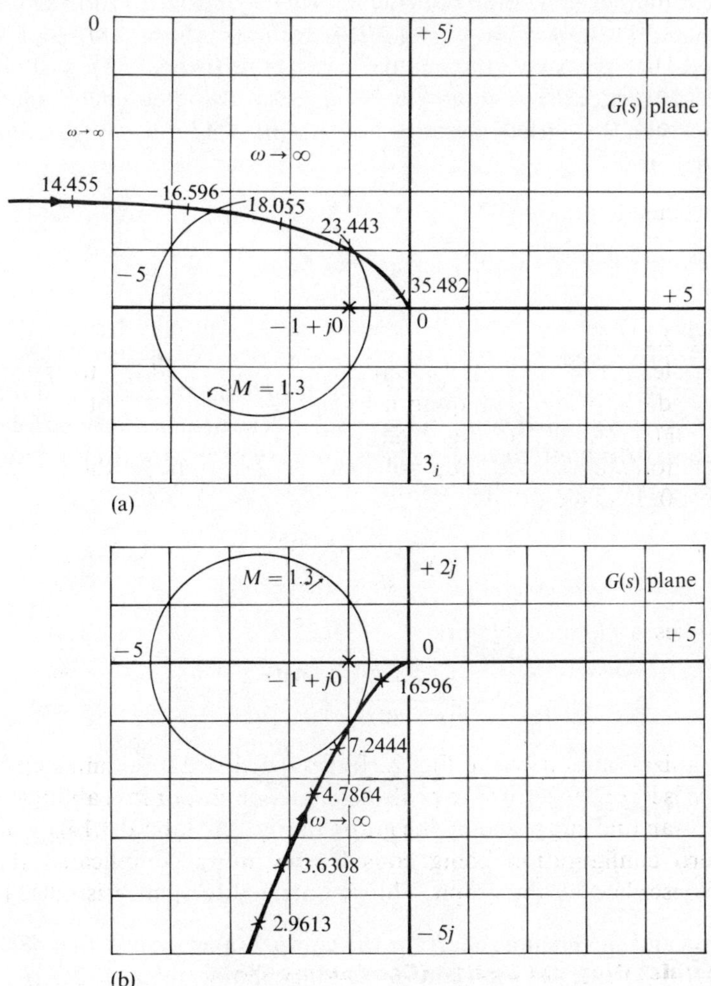

Fig. 2.23 Nyquist diagrams (example). (a) Nyquist diagram for $G(j\omega)$. (b) Nyquist diagram for $G(j\omega)F(j\omega)$.

The $G(j\omega)$ contour encircles $(-1 + j0)$ on the right, $0 \leqslant \omega \leqslant \infty$, indicating uncompensated ($F(s) = 1$) closed-loop instability.

The shaping network

$$F(s) = 0.03\left(\frac{1 + 0.25s}{1 + 0.025s}\right)$$

is devised and a Nyquist diagram for the compensated system $G(j\omega)F(j\omega)$ is shown in Fig. 2.23(b). The $G(j\omega)F(j\omega)$ contour is tangential to the $M = 1.3$ circle at $\omega \approx 10$, indicating satisfactory closed-loop performance.

A circuit mechanization of $F(s)$ is shown in Fig. 2.21(a) and a closed-loop step function response in Fig. 2.21(b). In Fig. 2.14(b), ($M_{p\omega} = 1.3$), it is suggested that the step function peak response (overshoot) is about 1.25, whereas in practice this is about 1.4. This is because the system has more poles and zeros than that strictly described by 2.14(b). This apparent discrepancy is quite typical.

(iv) Nichols Chart Design (Continuous Time)

The Nichols chart for $G(j\omega)$ is drawn (Sec. 2.5), usually from a Bode plot (Sec. 2.2) or from data found by experiment on the system. The design strategy is to devise a shaping network $F(s)$ which ensures that the $G(j\omega)F(j\omega)$ contour is tangential to a good M contour ($M = 1.3 = 2.3$ dB) and otherwise follows the closed-loop 0 dB contour as closely as possible over as wide a bandwidth as possible. $F(s)$ is then mechanized in circuit or other form.

EXAMPLE

A Nichols chart for

$$G(j\omega) = \frac{45{,}000}{j\omega(j\omega + 2)(j\omega + 30)}$$

is shown in Fig. 2.24(a).

The $G(j\omega)$ contour passes to the left of the (0 dB, $-180°$) point, indicating uncompensated ($F(s) = 1$) closed-loop instability.

The shaping network

$$F(s) = 0.03\left(\frac{1 + 0.25s}{1 + 0.025s}\right)$$

is devised and the Nichols chart for the compensated system $G(j\omega)F(j\omega)$ is in Fig. 2.24(b). The $G(j\omega)F(j\omega)$ contour is tangential to the $M = 2.3$ dB contour at $\omega \approx 10$ and otherwise follows the closed-loop 0 dB contour reasonably closely, $0 \leqslant \omega \leqslant 15$.

A circuit mechanization of $F(s)$ is shown in Fig. 2.21(a) and a closed-loop unit step function response in Fig. 2.21(b). The quite typical discrepancy between this and the performance suggested by Fig. 2.14(b) is explained in (iii) above.

(v) Inverse Nyquist Diagram Design (Continuous Time) (ref. 1)

The inverse Nyquist diagram for $1/G(j\omega)$ is drawn (Sec. 2.6), usually from a Bode plot (Sec. 2.1) or from data found by experiment on the system. The design strategy is to devise a gain, K, and shaping network, $F(s)$ (Fig. 2.17),

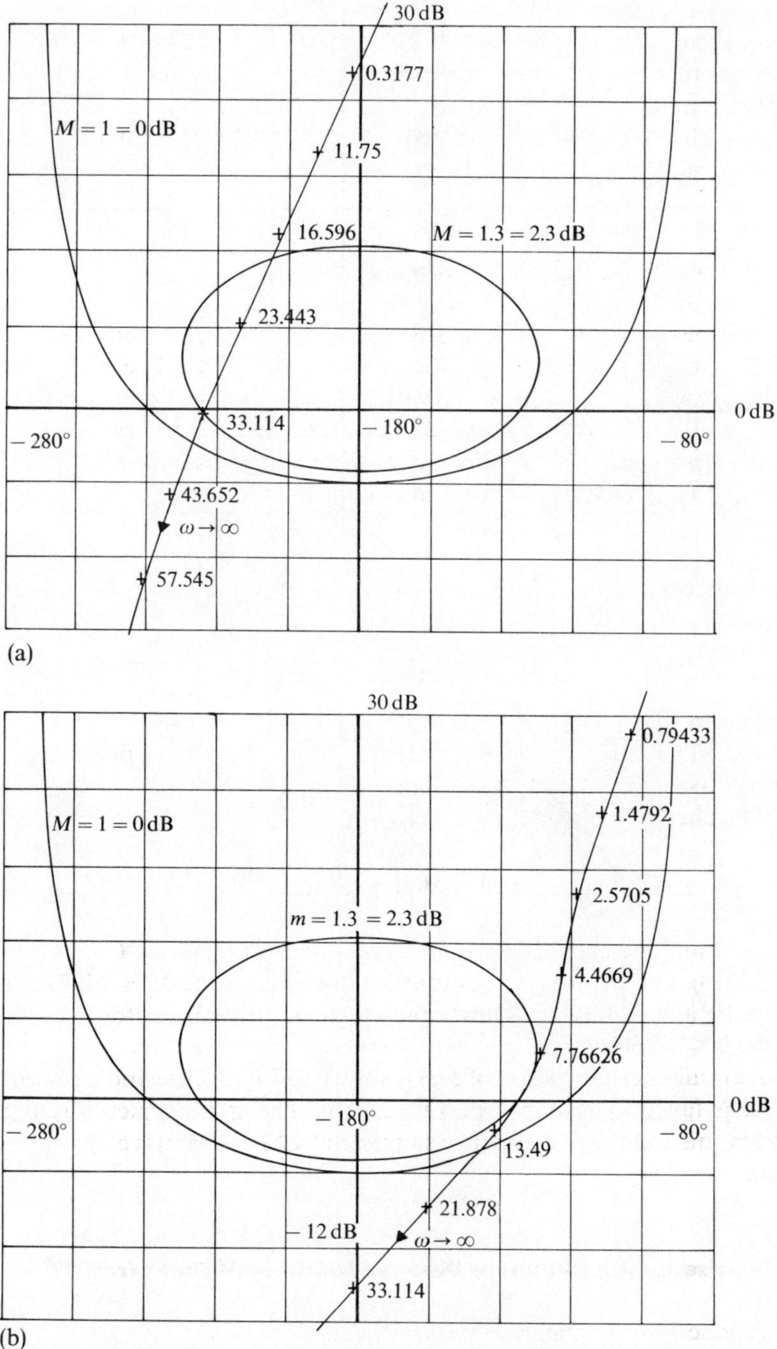

Fig. 2.24 Nichols charts (example). (a) Nichols chart for $G(j\omega)$. (b) Nichols chart for $G(j\omega)F(j\omega)$.

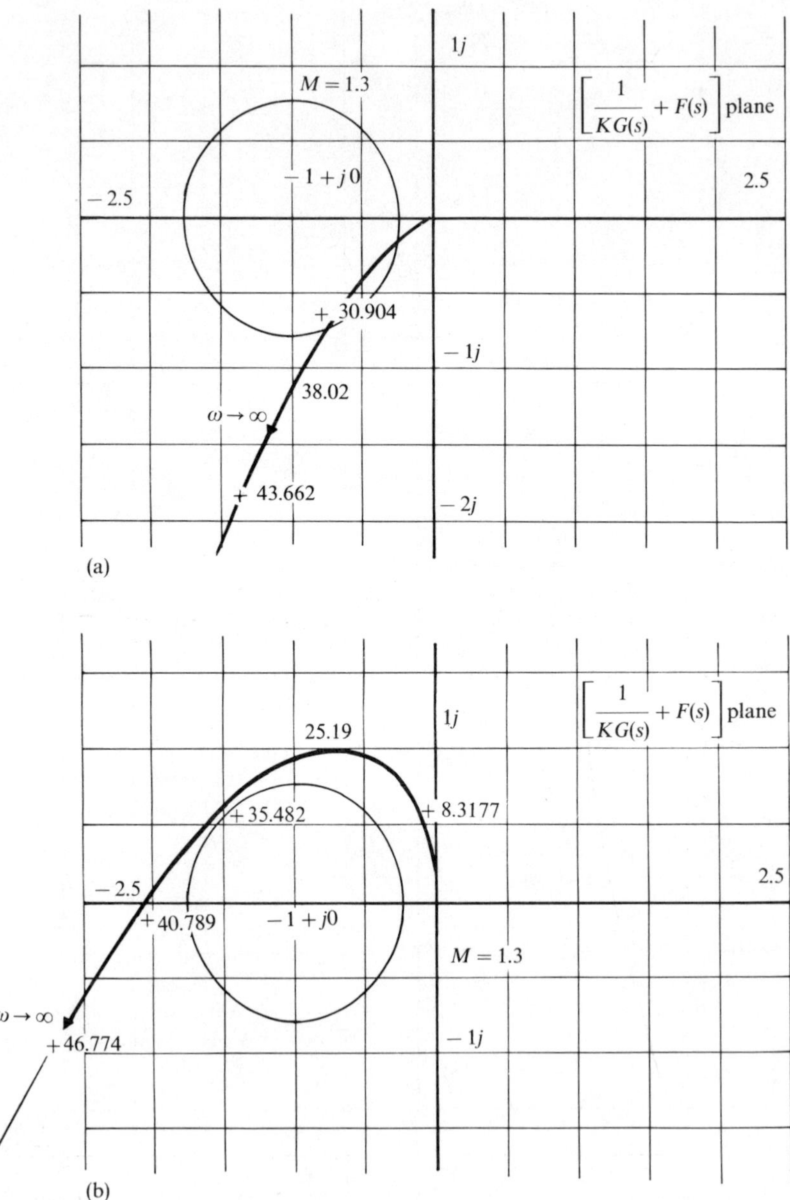

Fig. 2.25 Inverse Nyquist diagram (example). (a) Inverse Nyquist diagram for open-loop system. (b) Inverse Nyquist diagram for compensated system.

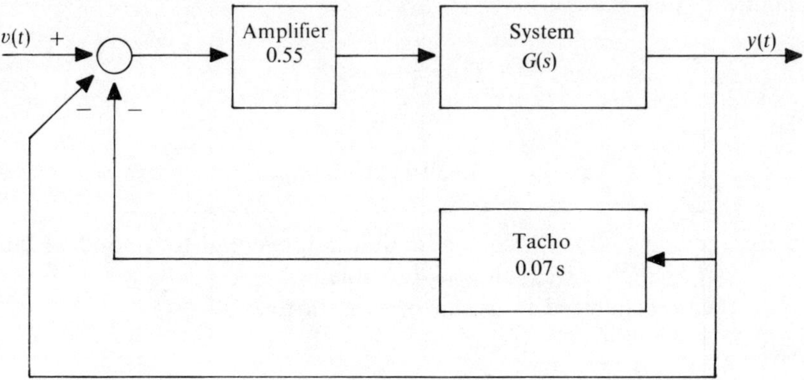

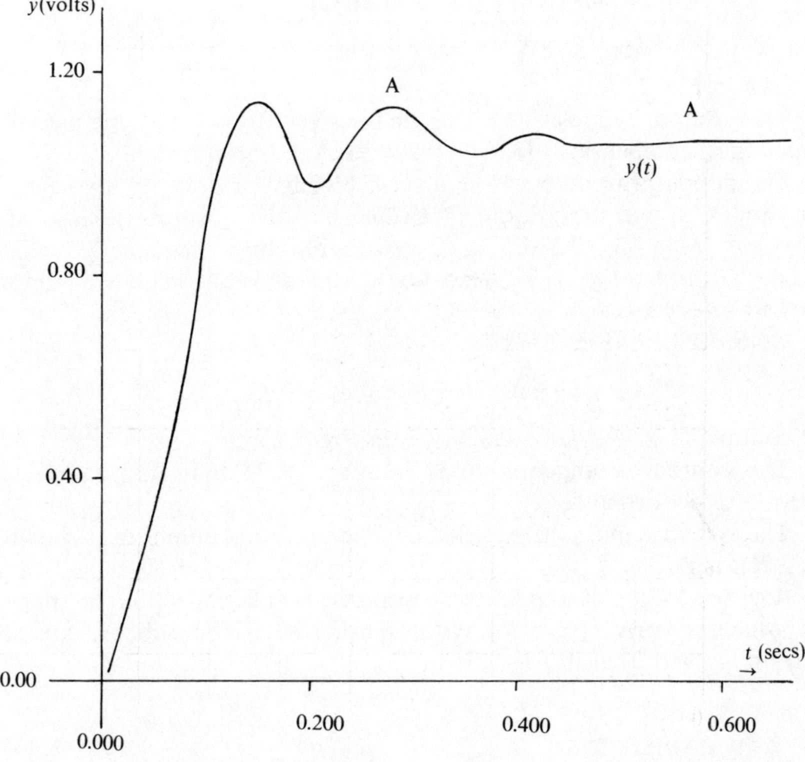

Fig. 2.26 Closed-loop system (example). (a) System mechanization. (b) Step function response.

which ensure that the contour

$$\left\{\frac{1}{KG(j\omega)}+F(j\omega)\right\}, \qquad 0\leqslant\omega\leqslant+\infty,$$

on the

$$\left\{\frac{1}{KG(s)}+F(s)\right\} \text{ plane}$$

does not encircle $(-1+j0)$ clockwise and is tangential to a good M circle (Fig. 2.18) at as high a frequency as possible.

$F(s)$ then mechanized in circuit or other form.

EXAMPLE

An inverse Nyquist diagram for the amplifier–servomotor system (Fig. 2.19(a)),

$$\frac{1}{G(j\omega)}=\frac{j\omega(1+j\omega 0.5)(1+j\omega 0.033)}{750}, \qquad 0\leqslant\omega\leqslant\infty.$$

is shown in Fig. 2.25(a).

This contour encircles $(-1+j0)$ on the right, $0\leqslant\omega\leqslant\infty$ (note direction) indicating uncompensated ($K=1$, $F(s)=0$) closed-loop instability.

The design procedure on the inverse Nyquist diagram entails selecting $F(j\omega)$ and K so that the contour $\{1/KG(j\omega)+F(j\omega)\}$ just touches a good M circle (say $M=1.3=2.3\,\text{dB}$) at a reasonably high frequency. 'Velocity feedback', in which $F(j\omega)=\alpha j\omega$ (i.e. $F(s)=\alpha s$), α constant, is quite commonly found to be effective.

Consider here (Fig. 2.17):

$$F(j\omega)=0.07j\omega; \qquad K=0.55.$$

The contour $\{1/0.55G(j\omega)+0.07j\omega\}$ is in Fig. 2.25(b).

This contour is tangential to $M=1.3$ at $\omega=35$, indicating satisfactory closed-loop performance.

The system is mechanized using a tachometer and amplifier as shown in Fig. 2.26(a).

A closed-loop unit step function response is in Fig. 2.26(b). The shape of this, which is fairly typical of systems using velocity feedback, contrasts interestingly with that in Fig. 2.21(b).

2.8 CAD FACILITY*

A good CAD facility is almost essential if the techniques outlined in Secs. 2.2 to 2.7 are to be exploited fully. As far as the design of this facility is concerned, two categories of problem present themselves.

*The algorithm described here is due to P. Atkinson, and V.S. Dalvi, 'An Improved Algorithm for the Automatic Determination of Roots Loci by Digital Computer,' *Radio and Electronic Engineer*, **41**, (1971), 365.

1. The computation and display of the open-loop transfer function of a system and shaping network (Fig. 2.19) for $s = j\omega$ over a specified range of frequencies, $\omega_L \leqslant \omega \leqslant \omega_U$. This applies to the Bode plot, Nyquist diagrams and Nichols chart.
2. The computation and display of the poles of the closed-loop system on an s plane for a specified range of open-loop gains $0 \leqslant K \leqslant K_U$. This applies to the root locus diagram.

First of all, it is convenient to express the system open-loop transfer function in the form:

$$G(s) = G(\sigma + j\omega) = \frac{\bar{K}\prod_{i=1}^{M}[(\sigma - \sigma_{zi}) + j(\omega - \omega_{zi})]}{\prod_{k=1}^{N}[(\sigma - \sigma_{pk}) + j(\omega - \omega_{pk})]}$$

where there are M zeros at $\sigma_{zi} + j\omega_{zi}$, $\quad i = 1, 2, \ldots, M$,

N poles at $\sigma_{pk} + j\omega_{pk}$, $\quad k = 1, 2, \ldots, N$,

and $\bar{K}$ is a gain constant.

Where the zeros and poles are complex, they occur in pairs (Sec. 1.5). Furthermore, some poles and zeros may be multiple.

Rewriting this:

$$G(\sigma + j\omega) = \frac{\bar{K}\prod_{i=1}^{M}[(\sigma - \sigma_{zi}) + j(\omega - \omega_{zi})]\prod_{k=1}^{N}[(\sigma - \sigma_{pk}) - j(\omega - \omega_{pk})]}{\prod_{k=1}^{N}[(\sigma - \sigma_{pk})^2 + (\omega - \omega_{pk})^2]}. \tag{2.6}$$

All the techniques require the evaluation of $G(\sigma + j\omega)$ for various values of σ and ω in the form of a complex number:

$$G(\sigma + j\omega) = \bar{K}[X + jY].$$

This is fairly easily done. Starting with the 'first' pole,

$$X_1 + jY_1 = \frac{(\sigma - \sigma_{p1}) - j(\omega - \omega_{p1})}{(\sigma - \sigma_{p1})^2 + (\omega - \omega_{p1})^2}.$$

Equation (2.6) is evaluated using the recursive formulas:

$$X_k = \frac{(\sigma - \sigma_{pk})X_{k-1} + (\omega - \omega_{pk})Y_{k-1}}{(\sigma - \sigma_{pk})^2 + (\omega - \omega_{pk})^2} \tag{2.7}$$

$$Y_k = \frac{(\sigma - \sigma_{pk})Y_{k-1} - (\omega - \omega_{pk})X_{k-1}}{(\sigma - \sigma_{pk})^2 + (\omega - \omega_{pk})^2}, \tag{2.8}$$

$k = 2, 3, \ldots, N.$

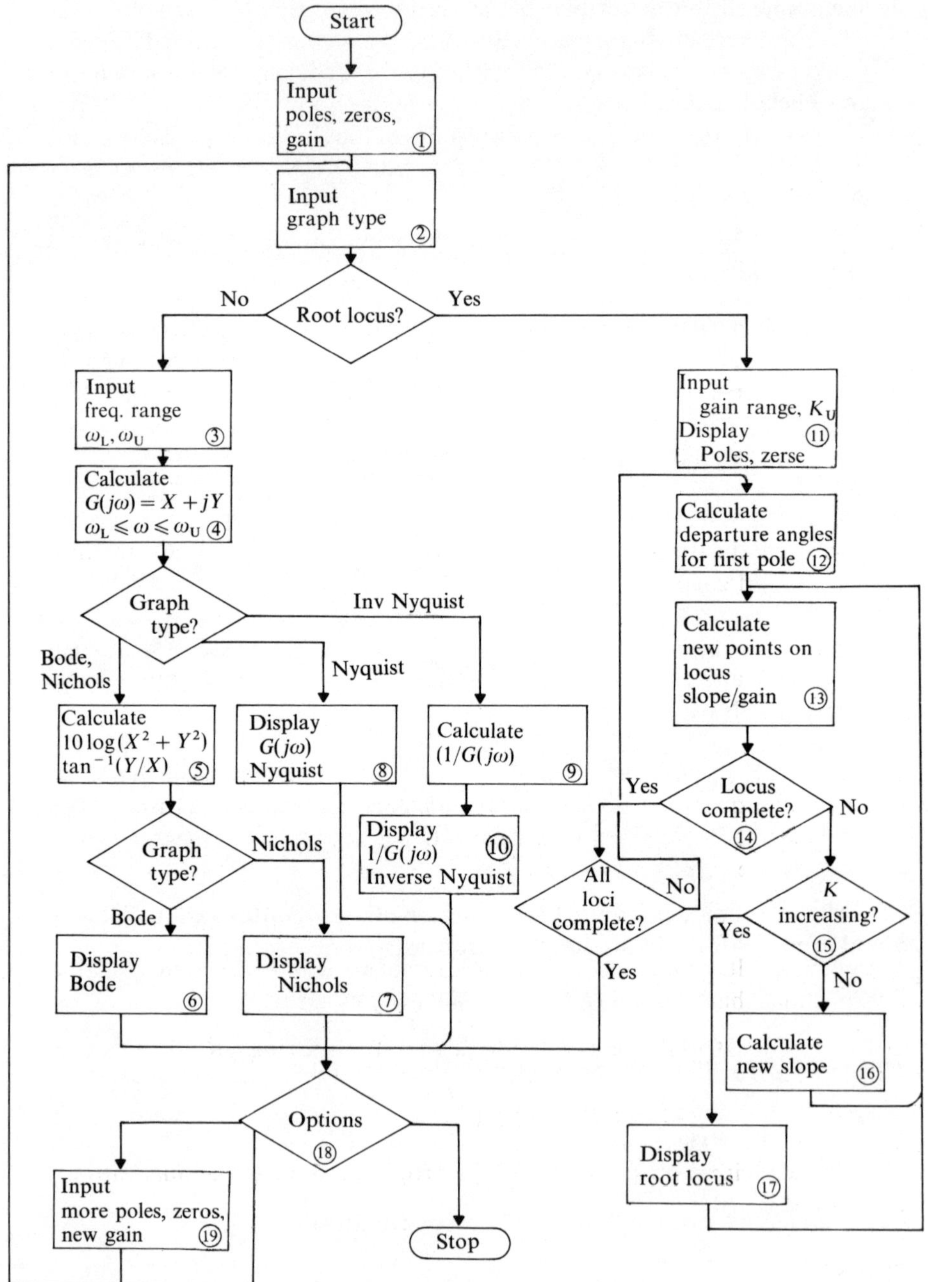

Fig. 2.27 Classical design CAD algorithm.

This is followed by the recursive formulas:

$$X_{i+N} = (\sigma - \sigma_{zi})X_{i+N-1} - (\omega - \omega_{zi})Y_{i+N-1} \tag{2.9}$$

$$Y_{i+N} = (\sigma - \sigma_{zi})Y_{i+N-1} + (\omega - \omega_{zi})X_{i+N-1}, \tag{2.10}$$

$$i = 1, 2, \ldots, M.$$

Then $G(s) = \bar{K}[X_{M+N} + jY_{M+N}]$

Equations (2.7), (2.8) fail where σ, ω are near a pole, and the algorithm must be designed to avoid evaluating $G(s)$ at such points.

A flow diagram of the algorithm is shown in Fig. 2.27. In the interests of clarity, some details have been omitted. In particular, the detection of the proximity of σ, ω to a pole (when X, Y become very large), the scaling of the display diagrams, and the allocation of tasks to subroutines, have all been left to the reader.

The operation of the algorithm is as follows:

BLOCK 1

The details of the transfer function poles, zeros, and gain (Eqn. (2.6)) are input, displayed, and corrected if necessary.

BLOCK 2

The type of graph required (Bode plot, root locus, Nyquist diagram, Nichols chart, inverse Nyquist diagram) is input.

BLOCKS 3, 4

A required frequency range is input (ω_L, ω_U) and $G(j\omega) = X + jY$ is evaluated using Eqns. (2.7), (2.8), (2.9), (2.10) '$\sigma = 0$' for about 100 values of $\omega, \omega_L \leqslant \omega \leqslant \omega_U$. These results are stored.

BLOCK 5

Using the results calculated in Block 4, a table of results is calculated for Bode and Nichols chart display:

$$20 \log|G(j\omega)| = 10 \log_{10}(X^2 + Y^2) + 20 \log_{10}\bar{K}$$

$$\arg(G(j\omega)) = \tan^{-1}\left(\frac{Y}{X}\right).$$

The algorithm must of course take proper account of the quadrature of $\tan^{-1}(Y/X)$.

BLOCKS 6, 7

The results are displayed in Bode or Nichols chart format with straight lines connecting the calculated points. The results are also made available in listed form, either on the screen or in hard copy. A hard copy of the graphical results is very desirable if a graph plotter or suitable printer is available.

BLOCKS 8, 9, 10

For the Nyquist diagram, the results calculated in Block 4 are already in suitable form.

For the inverse Nyquist diagram, the formulas required are:

$$\frac{1}{G(j\omega)} = \frac{1}{X + jY} = \bar{X} + j\bar{Y}$$

where

$$\bar{X} = \frac{X}{X^2 + Y^2}$$

$$\bar{Y} = \frac{-Y}{X^2 + Y^2}.$$

$\bar{X}$ and $\bar{Y}$ are calculated and stored in table form (Block 9).

The results are displayed (Blocks 8, 10) in Nyquist or inverse Nyquist format with straight lines joining the calculated points. The results in listed or hard copy graphical form are also, of course, a desirable option.

BLOCK 11

A maximum value of gain multiplier, $K = K_u$ (as opposed to $\bar{K}$, the open-loop gain), is input, and the poles and zeros are displayed directly (using X and 0 symbols typically) in an s-plane graphical format.

BLOCK 12

To find the root locus, values of $s = \sigma \pm j\omega$ are required which satisfy the equation:

$$1 + KG(s) = 0$$

i.e.

$$1 + K\bar{K}(X + jY) = 0$$

or:

$$X = -\frac{1}{K\bar{K}} \tag{2.11}$$

$$Y = 0. \tag{2.12}$$

The strategy used here is to find values of $s = \sigma \pm j\omega$ which satisfy Eq. (2.12) and use these in Eq. (2.11) to calculate the corresponding value of K.

Since the root loci start at the open-loop poles and terminate either at zeros or asymptotically at ∞ (Sec. 2.3), the search for a locus begins at a selected pole. The 'angle of departure' of the locus from the pole can be found fairly easily. The method is best illustrated by example.

Consider:
$$G(s) = \frac{(s - s_{z1})(s - s_{z2})}{(s - s_{p1})(s - s_{p2})(s - s_{p3})}.$$

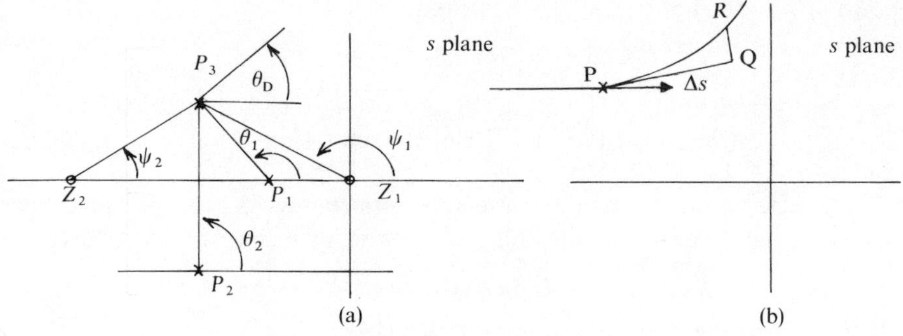

Fig. 2.28 Root locus construction.

The angle of departure, θ_D, of the locus from the pole s_{p3} is such that:

$$(\psi_1 + \psi_2) - (\theta_1 + \theta_2) - \theta_D = (2l + 1)\pi$$

where the angles $\psi_1, \psi_2, \theta_1, \theta_2$ are as shown in Fig. 2.28(a), and

$$l = \pm 1, \pm 2, \pm 3, \ldots \text{ chosen so that:}$$

$$-\pi \leqslant \theta_D \leqslant +\pi.$$

i.e.

$$\theta_D = (\psi_1 + \psi_2) - (\theta_1 + \theta_2) - (2l + 1)\pi.$$

Equivalently, if $G_3(s)$ is defined:

$$G_3(s) = (s - s_{p3})G(s) = X_{p3} + jY_{p3}.$$

θ_D can be calculated using Eqs. (2.7) to (2.10) to find X_{p3}, Y_{p3}, and then:

$$\theta_D = \tan^{-1}\frac{Y_{p3}}{X_{p3}} - (2l + 1)\pi. \tag{2.13}$$

For a multiple pole at s_{p3} of order q, there are q root loci leaving s_{p3} with angles of departure:

$$\theta_{Di} = \left[\tan^{-1}\left(\frac{Y_{p3}}{X_{p3}}\right) + \pi\right]\frac{1}{q} + \frac{2\pi(i-1)}{q}, \qquad i = 1, 2, \ldots, q.$$

This procedure can be applied to any $G(s)$, taking its poles in sequence.

BLOCK 13

A point s_Q near the locus (Fig. 2.28(b)) is now found by taking a step, length Δs, from the pole s_P in the direction θ_D. Δs is arbitrarily chosen to be about 5% of the largest open-loop pole or zero coordinate:

$$s_Q = s_P + \Delta s(\cos\theta_D + j\sin\theta_D). \tag{2.14}$$

A search at right angles to the line ($s_Q - s_P$) is now undertaken to find a point s_R on the root locus (Fig 2.28(b)).

$$s_R = s_Q + L_r(-\sin\theta_D + j\cos\theta_D).$$

This search entails evaluating $G(s) = \bar{K}[X + jY]$ at various points along $(s_R - s_Q)$ – i.e. using Eqs. (2.7) to (2.10) and choosing a value of L_r which minimizes Y (Eq. (2.12)).

A variety of search methods are appropriate (Secs. 5.2 to 5.5); one simple one is the 'iterative secant' formula (ref. 6), which starts with two values, say:

$$L_1 = \Delta s 10^{-4}$$
$$L_2 = 2\Delta s 10^{-4}.$$

Evaluating Y (Eqs. (2.7) to (2.10)) at these points as $Y(L_1)$, $Y(L_2)$, this formula gives a third step length:

$$L_3 = L_1 - \frac{L_1 - L_2}{Y(L_1) - Y(L_2)} Y(L_1). \tag{2.15}$$

This is repeated until $Y(L_r)$ is almost zero (say $< 10^{-6}$) and s_R is then considered to be found.

These calculations give the point s_R. Equation (2.11) can now be used to find the corresponding gain, K. Moreover, if the slope of the locus at s_R can be determined, the sequence can be repeated, in principle at least, until the whole locus is found.

The slope at s_R can be found from:

$$m = \frac{d\omega}{d\sigma} = \frac{\sum_{k=1}^{N} \frac{\omega - \omega_{pk}}{r_{pk}^2} - \sum_{i=1}^{M} \frac{\omega - \omega_{zi}}{r_{zi}^2}}{\sum_{k=1}^{N} \frac{\sigma - \sigma_{pk}}{r_{pk}^2} - \sum_{i=1}^{M} \frac{\sigma - \sigma_{zi}}{r_{zi}^2}} \tag{2.16}$$

where

$$r_{pk}^2 = (\sigma - \sigma_{pk})^2 + (\omega - \omega_{pk})^2$$

and

$$r_{zi}^2 = (\sigma - \sigma_{zi})^2 + (\omega - \omega_{zi})^2.$$

Using the two points s_P, s_R, a new point $s_T = \sigma_T + j\omega_T$ can now be calculated under the (usually almost correct) assumption that the curvature of the locus is unchanged over a small interval:

$$\sigma_T = \sigma_S + \frac{1 - m^2}{1 + m^2}(\sigma_R - \sigma_P) + \frac{2m}{1 + m^2}(\omega_R - \omega_P) \tag{2.17}$$

$$\omega_T = \omega_S + \frac{2m}{1 + m^2}(\sigma_R - \sigma_P) + \frac{1 - m^2}{1 + m^2}(\omega_R - \omega_P). \tag{2.18}$$

These rules can be relaxed if $m < 0.01$ or $m > 100$, when the locus is assumed to be horizontal or vertical respectively.

BLOCK 14
A locus terminates either at a zero, which is easily detected, or at infinity on an asymptote, which is assumed when the calculated value of K exceeds the specified K_U (Block 11).

BLOCKS 15, 16
Breakpoints in the locus are treated in this algorithm as the intersection of two loci, and if the wrong path is followed after such an intersection, the calculated value of K (Eq. (2.11)) decreases (instead of increasing). If a search step produces such a decrease, it is regarded as unsuccessful and a new step is taken at right angles to the unsuccessful one. Since breakpoints almost always occur at right angles, this strategy is usually successful.

BLOCK 17
The locus is displayed, the calculated points being joined with straight lines. Values of K are displayed at appropriate points on the locus.

BLOCKS 18, 19
When the displays are complete, a variety of options allows poles, zeros, and gain to be altered and the program to be re-entered to generate a new diagram not necessarily of the same type.

Comment
A variety of fairly simple additional features can add significantly to the usefulness of this program. These include:

1. a joystick or otherwise operated cursor which can be used to display frequencies and gains on the graphs;
2. the display, at operator request, of lines of constant closed-loop gain and phase for the Nyquist and Nichols charts (Secs. 2.4, 2.5, 2.6);
3. the display, at operator request, of lines of constant damping ratio and undamped natural frequency for the root locus diagram (Sec. 2.3).

An algorithm with some of these features has been used for the design examples and plots appearing throughout this chapter.

An obvious extension to the CAD facility is to provide a means of checking the behavior of the closed-loop system in the time domain. This is perhaps best done by casting the equations for the system in state equation form (Sec. 3.5) and using a Runge–Kutta algorithm (Sec. 1.2) to model its response to step function and other inputs.

2.9 BODE PLOTS (DISCRETE SYSTEMS) (REFS. 3, 4)

The frequency response of a discrete system with transfer function $\bar{G}(z)$ is $\bar{G}(e^{j\beta})$, $0 \leqslant \beta \leqslant \pi$ (Sec. 1.10) where the frequency, β, is in radians/sample. For a sampled data system (Sec. 1.11), $\beta = \omega T$, where ω is in radians/unit time and T is the sampling period.

Factorized into first- and, when necessary, second-order terms:

$$\bar{G}(e^{j\beta}) = \bar{G}(z)|_{z=e^{j\beta}} = \left.\frac{K(z+q_{n1})(z+q_{n2})\cdots(z^2+k_{n1}z+l_{n1})(z^2+k_{n2}z+\cdots}{(z+q_{d1})(z+q_{d2})\cdots(z^2+k_{d1}z+l_{d1})(z^2+k_{d2}z+\cdots}\right|_{z=e^{j\beta}}$$

$$= \frac{K(e^{j\beta}+q_{n1})(e^{j\beta}+q_{n2})\cdots(e^{2j\beta}+k_{n1}e^{j\beta}+l_{n1})\cdots}{(e^{j\beta}+q_{d1})(e^{j\beta}+q_{d2})\cdots(e^{2j\beta}+k_{d1}e^{j\beta}+l_{d1})\cdots}.$$

This does not lend itself to the easy construction of a Bode plot since the characteristics of the factors lack the simplicity which makes the Bode plot attractive (Sec. 2.2). This is immaterial if the Bode plot is plotted using a CAD facility (Sec. 2.14), but it makes manual evaluation impractical. It is common practice therefore to map $\bar{G}(z)$ conformally into $\bar{\bar{G}}(w)$ using the bilinear transformation:

$$z = \frac{1+w}{1-w}, \qquad w = u + jv. \tag{2.19}$$

The 'frequency' response $\bar{\bar{G}}(jv)$ is then:

$$\bar{\bar{G}}(jv) = \bar{\bar{G}}(w)|_{w=jv} = \frac{\bar{\bar{K}}(1+jv\tau_{n1})(1+jv\tau_{n2})\cdots(1+jva_{n1}+(jv)^2b_{n1}\cdots}{(jv)^N(1+jv\tau_{d1})(1+jv\tau_{d2})\cdots(1+jva_{d1}+(jv')^2b_{d1})\cdots}$$

Since this is of the same algebraic form as Eqn. 2.1, the construction of a Bode plot follows the same procedure as the continuous time case (Sec. 2.2) with the following extensions for the fairly common occurrence in $\bar{\bar{G}}(jv)$ of 'non-minimum phase' elements:

1. $(1-jv\tau), \tau > 0; M = 20\log|1-jv\tau|\,\text{dB}$

   $\phi = \tan^{-1}(-v\tau)$ degrees.

Frequency-normalized plots for such elements are as shown in Fig. (2.2(a), (b), but with the phase angle diagrams interchanged, i.e. a numerator term $(1-jv\tau_n)$ is associated with a phase lag $(0 \geqslant \phi \geqslant -90°)$, while a denominator term $(1-jv\tau_d)$ is associated with a phase lead $(0 \leqslant \phi \leqslant 90°)$.

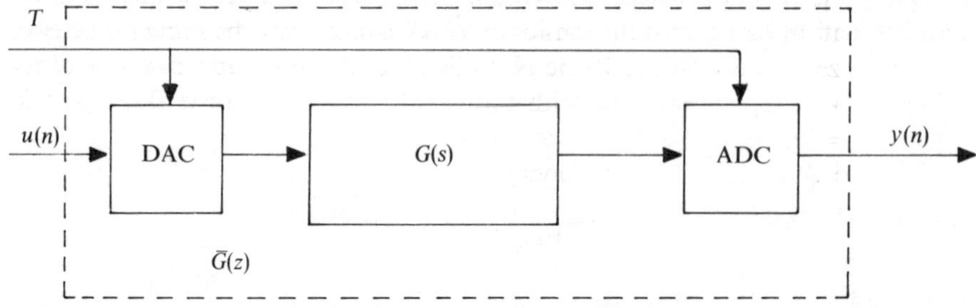

Fig. 2.29 Sampled data system.

2. $(1 - jva + (jv)^2 b)$, $a > 0, b > 0$; $M = 20\log_{\omega}|1 - jva + (jv)^2 b|$ dB

$$\phi = \tan^{-1}(-va/(1 - v^2 b)).$$

Frequency-normalized plots for this element are as shown in Fig. 2.3, but with the phase reversed, i.e. a numerator term of this kind is associated with a phase lag ($0 \geqslant \phi \geqslant -90°$) and a denominator term with a phase lead ($0 \leqslant \phi \leqslant 90°$).

EXAMPLE

Consider the sampled data system in Fig. 2.29.

$$G(s) = \frac{1500}{s(s+2)}, \qquad T = 0.1$$

$$\bar{G}(z) = Z\left\{\frac{1 - e^{-sT}}{s}\frac{1500}{s(s+2)}\right\} \qquad \text{(Sec. 1.11)}$$

$$= (1 - z^{-1})Z\left\{\frac{1500}{s^2(s+2)}\right\} \qquad \text{(Eq. (1.34))}$$

$$= \frac{7.125(z + 0.905)}{(z-1)(z-0.819)} \qquad \text{(Fig. 1.13)}$$

$$\bar{\bar{G}}(w) = \bar{G}(z)|_{z=(1+w)/(1-w)} \qquad \text{(Eq. (2.19))}$$

$$= \frac{75(1 + 0.05w)(1 - w)}{w(1 + 10.050w)}$$

$$\bar{\bar{G}}(jv) = \frac{75(1 + 0.05\,jv)(1 - jv)}{jv(1 + 10.050\,jv)}$$

The Bode plot of this (Figs. 2.1, 2.2) is in Fig. 2.30.

The behavior of a closed-system incorporating $\bar{G}(z)$ as shown in Fig. 2.31 may be assessed using the Bode plot of $\bar{\bar{G}}(jv)$.

The Bode criteria for closed-loop stability, assuming no unstable poles of $\bar{G}(z)$ (Sec. 1.10), are:

1. $M = 20\log_{10}|\bar{\bar{G}}(jv)| < 0$ dB, where $\phi = -180°$.
2. $M < 0$ dB, $v \to \infty$.

The gain margin is defined as $(-M)$ where $\phi = -180°$, while the phase margin is $(\phi + 180°)$ where $M = 0$ dB. As a consequence of the bilinear transformation being zero dimensional, $\bar{\bar{G}}(w)$ usually has the same number of poles and zeros, and on the Bode plot $|\bar{\bar{G}}(jv)|$ and $\underline{/\bar{\bar{G}}(jv)}$ approach constant values, $v \to \infty$. This contrasts with the usual continuous time Bode plot in which $M = 20\log_{10}|G(j\omega)| \to -\infty$, $\omega \to \infty$.

In the Bode plot, the 'frequency,'

$$v = \tan\frac{\beta}{2} \qquad (\beta\text{, frequency, rad/sample})$$

$$= \tan\frac{\omega T}{2} \qquad (\omega\text{, frequency, rad/unit time}).$$

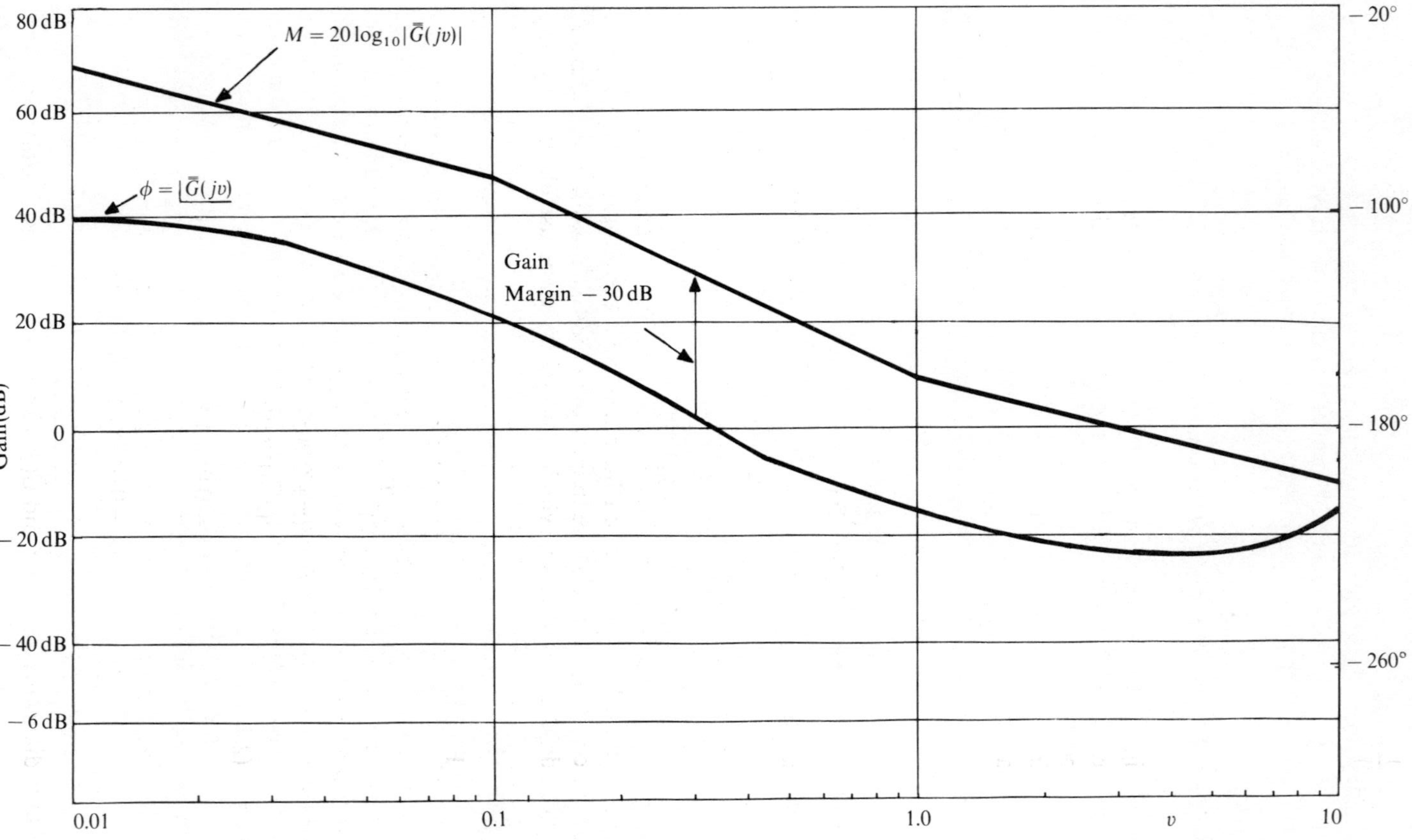

Fig. 2.30 Bode plot (example).

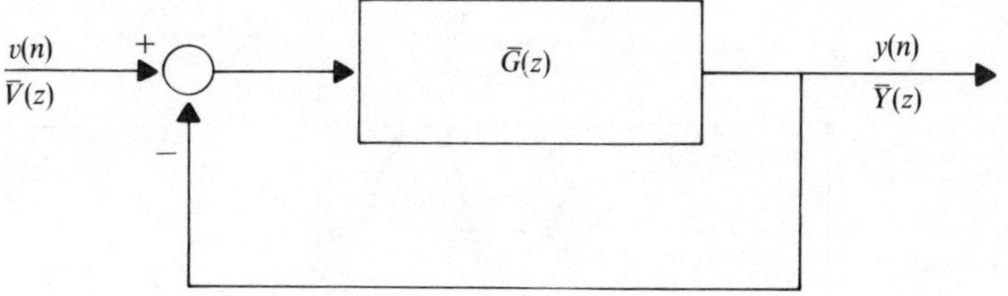

Fig. 2.31 A closed-loop system.

In Fig. 2.30, the gain margin is -30 dB, and $M > 0$, $v \to \infty$. An uncompensated closed-loop (Fig. 2.31) incorporating this system would therefore be unstable.

An interesting modification to this method (ref. 4) uses the transformation:

$$w' = \frac{2}{T} w, \qquad T \text{ the sampling period}, \quad w' = u' + jv'$$

or

$$z = \frac{1 + (T/2)w'}{1 - (T/2)w'} \tag{2.20}$$

In this case, the 'frequency,'

$$v' = \frac{2}{T} \tan\frac{\beta}{2} = \frac{2}{T} \tan\frac{\omega T}{2}$$

$$\approx \omega, \qquad \omega \text{ small.}$$

The Bode plot 'frequencies' then have the same meaning, at least for low values, as continuous time Bode frequencies, and this gives an easy intuitive understanding of the diagram which can sometimes be useful.

EXAMPLE

Consider (Fig. 2.29):

$$G(s) = \frac{1500}{s(s+2)}, \; T = 0.1$$

$$\bar{G}(z) = \frac{7.125(z + 0.905)}{(z-1)(z-0.819)}.$$

From Eq. (2.20)

$$\bar{\bar{G}}(w') = \frac{750(1 - 0.05w')(1 + 2.5 \times 10^{-3}w')}{w'(1 + 0.5w')}$$

$$\bar{\bar{G}}(jv') = \frac{750(1 - 0.05jv')(1 + 2.5 \times 10^{-3}jv')}{jv'(1 + 0.5jv')}$$

The Bode diagrams for $G(j\omega)$ and $\bar{\bar{G}}(jv')$ are shown in Figs. 2.32(a), (b). They

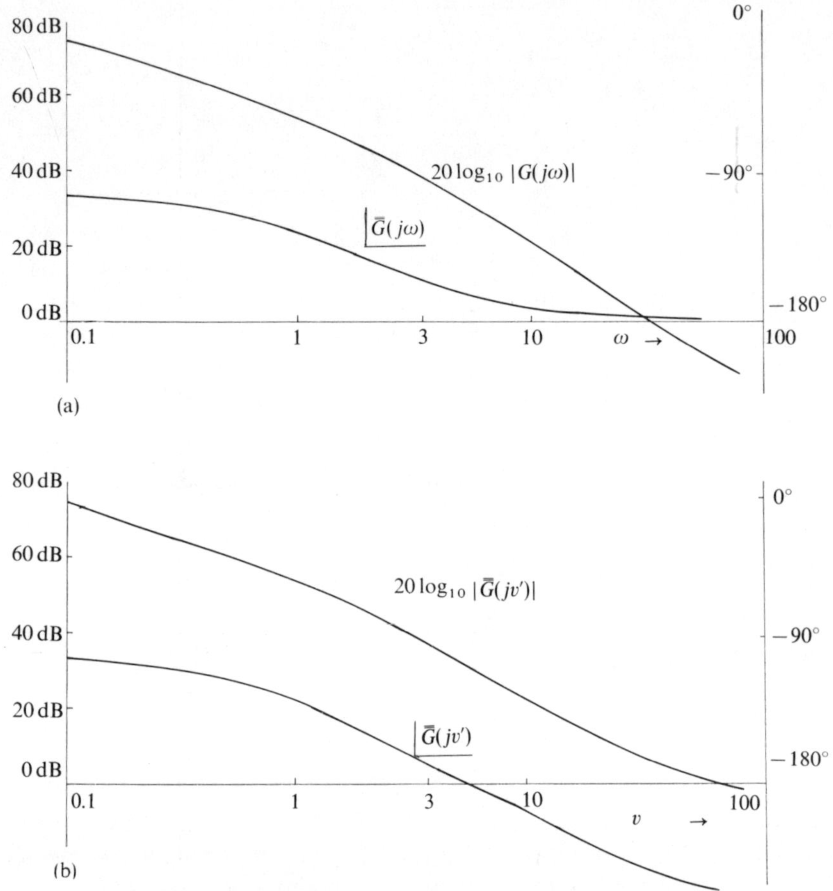

Fig. 2.32 Bode plot (examples). (a) Continuous system – s plane. (b) Sampled data system – w' plane.

are very similar at low frequencies (say $\omega < 3$), but the differences above this are quite substantial. In certain limited cases this similarity allows design of discrete systems directly from continuous time Bode plots, using the transformation of Eq. (2.20) and its inverse. With CAD facilities available, of course, the value of this is small.

Figures 2.30 and 2.32(b) have negative gain margins and indicate that the uncompensated closed–loop incorporating this system (Fig. 2.31) would be unstable. It is noteworthy that the 'corresponding' continuous time closed-loop system incorporating $G(s)$ (Fig. 2.32(a)) would not be unstable. The discrete system has the disadvantage of sampling, and, of course, this disadvantage increases with sampling period.

2.10 ROOT LOCUS DIAGRAMS (DISCRETE SYSTEMS) (REFS. 3, 4)

The numerator and denominator of the transfer function $\bar{G}(z)$ of a system may be factorized to give (possibly complex) singular points.

$$\bar{G}(z) = \frac{\bar{K}(z + a_{n1} + jb_{n1})(z + a_{n1} - jb_{n1})\cdots(z + c_{n1})(z + c_{n2})\cdots}{z^N(z + a_{d1} + jb_{d1})(z + a_{d1} - jb_{d2})\cdots(z + c_{d1})(z + c_{d2})\cdots}.$$

The points $z = -a_{n1} \pm jb_{n1}, -a_{n2} \pm jb_{n2}, \ldots, -c_{n1}, \ldots$ are zeros of $\bar{G}(z)$ (values of z for which $\bar{G}(z) = 0$), while $z = -a_{d1} \pm jb_{d1}, \; -a_{d2} \pm jb_{d2}, \ldots, \; -c_{d1}, \ldots,$ $z = 0$, are poles (values of z for which $\bar{G}(z) = \infty$).

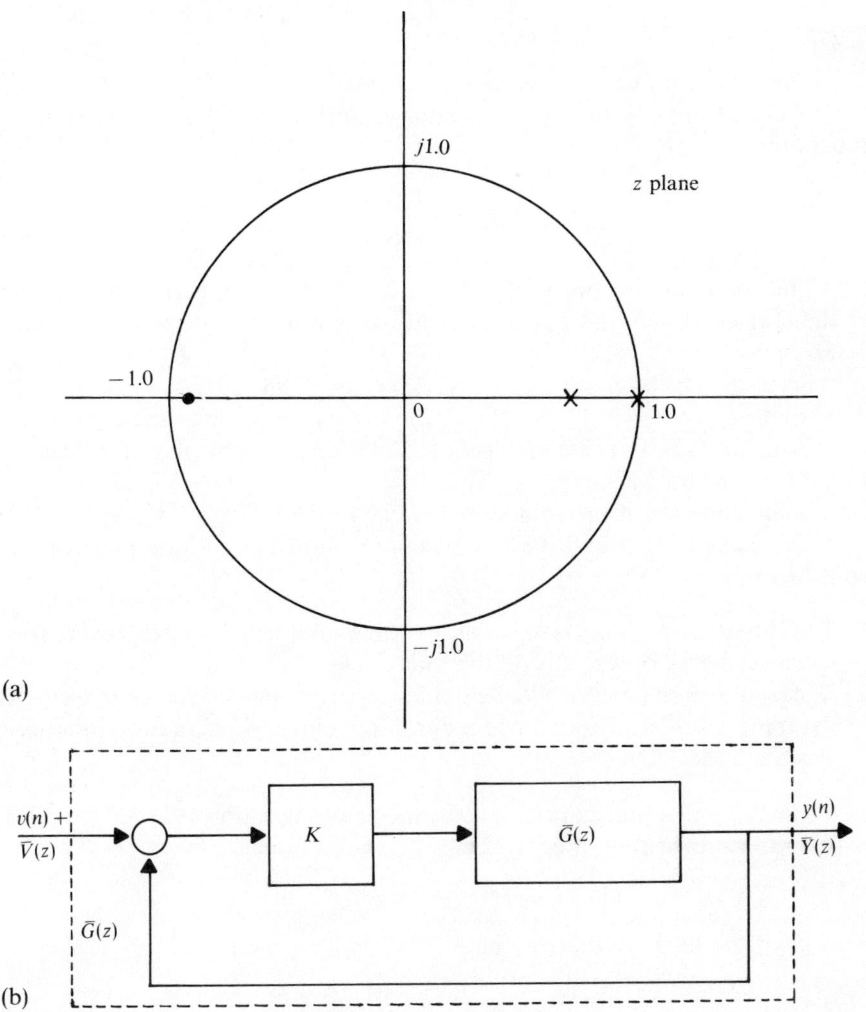

Fig. 2.33 Pole–zero diagram and closed-loop system. (a) Pole–zero diagram (b) Closed-loop system.

The poles and zeros may be plotted on a (complex) z plane Argand diagram.

EXAMPLE

Consider:

$$G(s) = \frac{1500}{s(s+2)}, \qquad T = 0.1$$

$$\bar{G}(z) = Z\left(\frac{1-e^{-sT}}{s}\frac{1500}{s(s+2)}\right) \qquad \text{(Sec. 1.11)}$$

$$= \frac{7.125(z+0.905)}{(z-1)(z-0.819)}.$$

The pole–zero diagram is in Fig. 2.33(a).

A closed-loop system incorporating $\bar{G}(z)$ (Fig. 2.33(b)) has the transfer function:

$$\bar{G}_c(z) = \frac{K\bar{G}(z)}{1+K\bar{G}(z)}. \tag{2.21}$$

The locus of the poles of $\bar{G}_c(z)$, $0 \leqslant K < \infty$, on the z plane, known as the *root locus*, is found by locating all the points on the z plane for which $1 + K\bar{G}(z) = 0$.

Since the algebraic form of Eq. (2.21) is identical to that of Eq. (2.3), the construction of the root locus follows the same procedure as the continuous time case (Sec. 2.3). The shapes of root loci in the z plane are of course different from those in the s plane.

Some common root loci are in Fig. 2.34.

The value of K determines the positions of the closed-loop poles on the root locus.

1. The closed-loop system is unstable for values of K which correspond to root locus segments on or outside the unit circle.
2. The positions of poles inside the unit circle are often considered in terms of damping ratio ζ and undamped natural frequency, ω_n, as in the continuous time case (Sec. 2.3).

The locus of a pole pair with constant damping ratio ζ is found from Eq. (2.4) by the substitution (Sec. 1.12):

$$z = re^{j\beta} = e^{sT} = e^{(\alpha+j\omega)T}$$

This gives the locus in the z plane:

$$r = \exp\left[\frac{\pm\beta\zeta}{\sqrt{1-\zeta^2}}\right]. \tag{2.22}$$

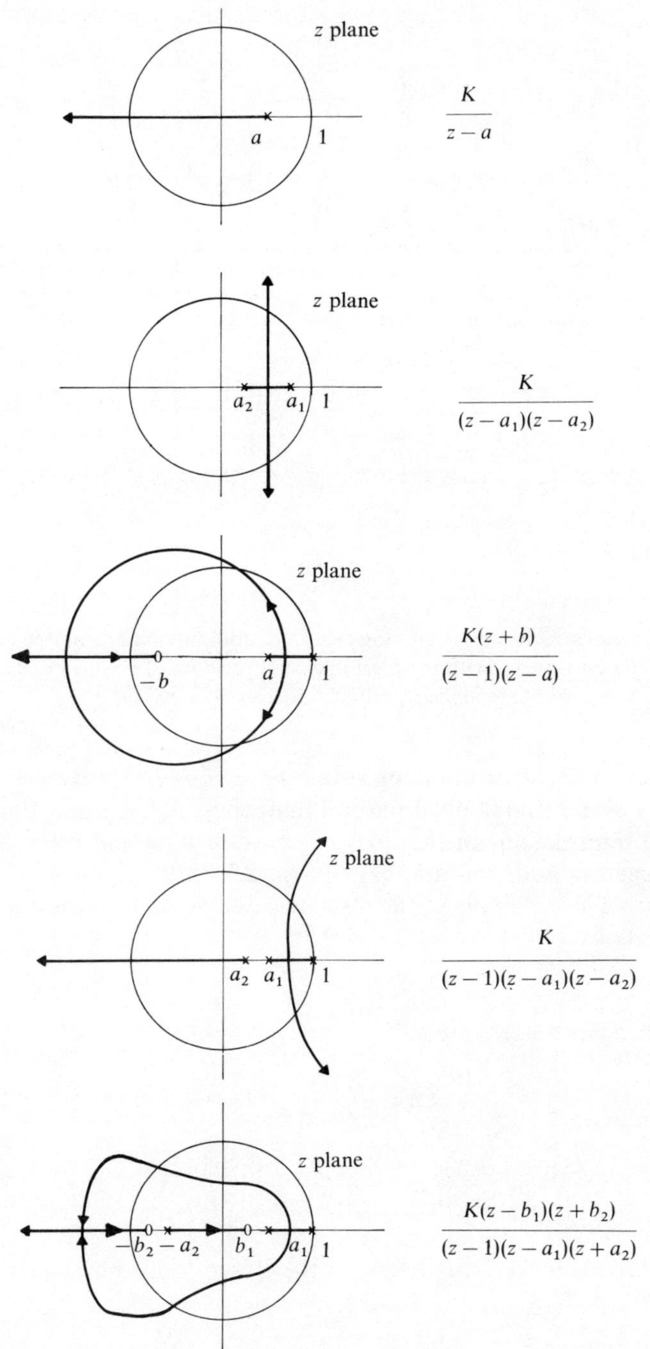

Fig. 2.34 Some common root loci (z plane).

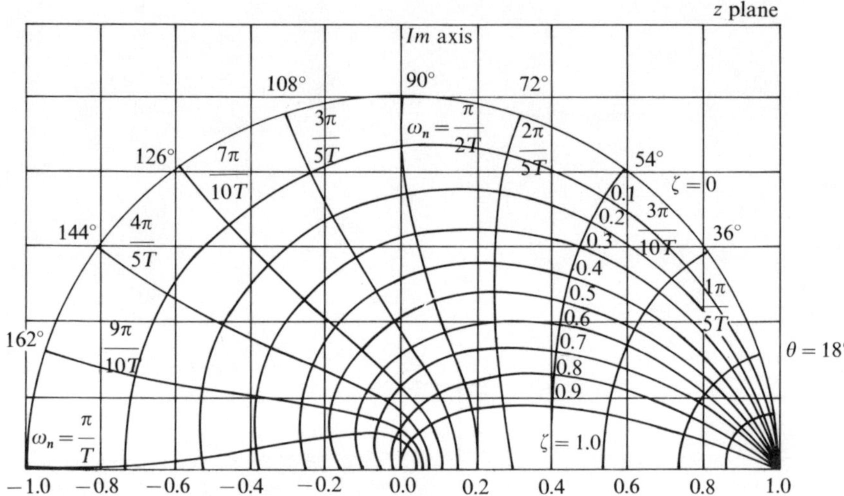

$z =$ plane loci of roots of constant ζ and ω_n

$s = -\zeta\omega_n \pm j\omega_n\sqrt{1-\zeta^2}$

$z = e^{Ts}$

$T =$ sampling period

Fig. 2.35 Lines of constant damping ratio and undamped natural frequency (z plane). (Reproduced with permission from Franklin, G.F. and Powell, J.D., *Digital Control of Dynamic Systems*. Addison-Wesley, 1980.)

Lines of constant damping ratio are in Fig. 2.35, together with loci of poles of constant undamped natural frequency, ω_n (rad/unit time).

The intersections of the lines of constant damping ratio and the root locus, together with the graphs in Figs. 2.9, 2.10, which in sampled form ($t = nT, n = 0, 1, \ldots$) apply to the discrete case, give some indication of closed-loop behavior.

EXAMPLE

Consider again the example:

$$G(z) = \frac{1500}{s(s+2)}, \quad T = 0.1$$

$$\bar{G}(z) = \frac{7.125(z+0.905)}{(z-1)(z-0.819)}$$

Figure 2.36 shows the root locus for the closed-loop characteristic equation:

$$1 + \frac{K7.125(z+0.905)}{(z-1)(z-0.819)} = 0$$

The closed-loop system poles are unstable for $K\bar{K} > 0.17$ and the

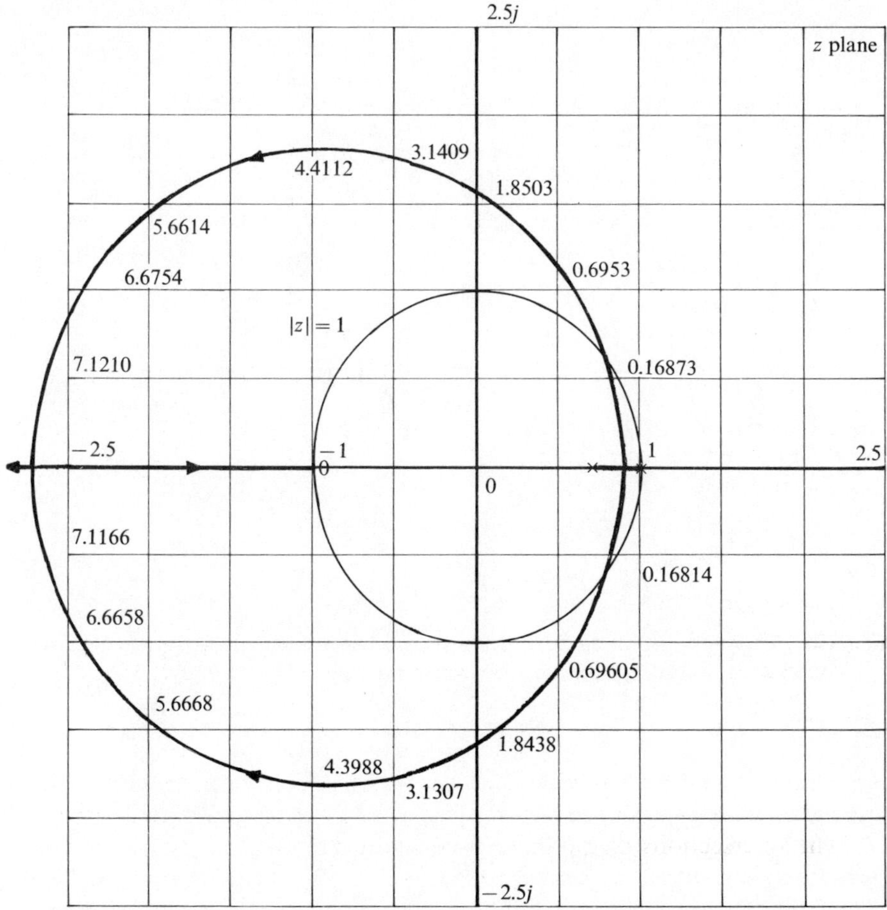

Fig. 2.36 Root locus diagram (example).

uncompensated closed-loop system (Fig. 2.33(b)), $K\bar{\bar{K}} = 7.125$, is therefore unstable.

2.11 NYQUIST DIAGRAMS (DISCRETE SYSTEMS) (REFS. 3, 4)

The frequency response, $\bar{G}(e^{j\beta})$, $0 \leqslant \beta \leqslant \pi$, of a discrete system may be plotted on a $\bar{G}(z)$ plane Argand diagram.

It is feasible to repeat the Bode plot strategy (Sec. 2.9) and map $\bar{G}(z)$ conformally into $\bar{\bar{G}}(w)$ using the bilinear transformation of Eq. 2.19 before developing a Nyquist diagram of $\bar{\bar{G}}(jv)$ in the $\bar{\bar{G}}(w)$ plane. Since the algebras of the s and w planes are identical, the CAD algorithm of Sec. 2.8 may be used

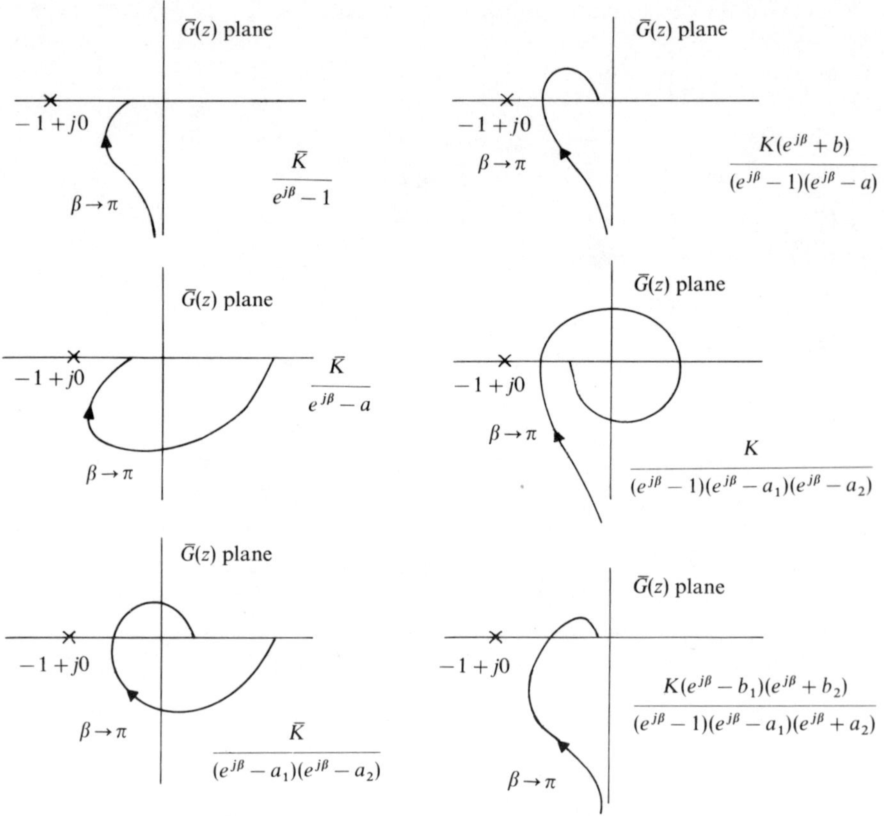

Fig. 2.37 Some common Nyquist diagram (discrete systems).

directly to display this $\bar{\bar{G}}(jv)$ diagram (including the M and N circles). Otherwise, however, there is no significant benefit to be derived from this additional complication, and plots of $\bar{G}(e^{j\beta})$ on the $\bar{G}(z)$ plane, for which a slightly altered CAD algorithm is needed, are used in this section.

The Nyquist diagrams of some common transfer functions are in Fig. 2.37.

The behavior of a closed-loop system incorporating $\bar{G}(z)$ (Fig. 2.31) may be assessed from the Nyquist diagram for $\bar{G}(e^{j\beta})$.

For closed-loop stability, the contour $\bar{G}(e^{j\beta})$, $0 \leqslant \beta \leqslant \pi$, must encircle the point $-1+j0$ counterclockwise the same number of times that there are unstable poles of $\bar{G}(z)$ ($|z| > 1$). Any clockwise encirclement (even partially) indicates closed-loop instability. The loci of constant closed-loop gains and phase shifts, M and N circles, may be drawn on the $\bar{G}(z)$ plane, and since their algebraic derivation is identical to that for the continuous time case, these are as shown in Fig. 2.14(a).

The intersections of the M and N circles with the $\bar{G}(e^{j\beta})$ contour give the closed-loop gains and phase shifts at various frequencies (radians/sample), and so provide some indication of closed-loop behavior. The peak closed-loop frequency response is given by the highest M circle which the $\bar{G}(e^{j\beta})$ contour touches ($M_{p\omega}$); this is related to step function response peak magnitude (M_{pt}) by the curves in Fig. 2.14(b) which in sampled form ($t = nT$, $n = 0, 1, 2, \ldots$) apply directly to the discrete case.

EXAMPLE

Consider:

$$G(s) = \frac{1500}{s(s+2)}, \qquad T = 0.1 \qquad \text{(Fig. 2.29)}$$

$$\bar{G}(z) = Z\left(\frac{1 - e^{-sT}}{s} \frac{1500}{s(s+2)}\right) = (1 - z^{-1})Z\left\{\frac{1500}{s^2(s+2)}\right\} \qquad \text{(Eq. (1.34))}$$

$$= \frac{7.125(z + 0.905)}{(z-1)(z-0.819)}$$

$$\bar{G}(e^{j\beta}) = \frac{7.125(e^{j\beta} + 0.905)}{(e^{j\beta} - 1)(e^{j\beta} - 0.819)}.$$

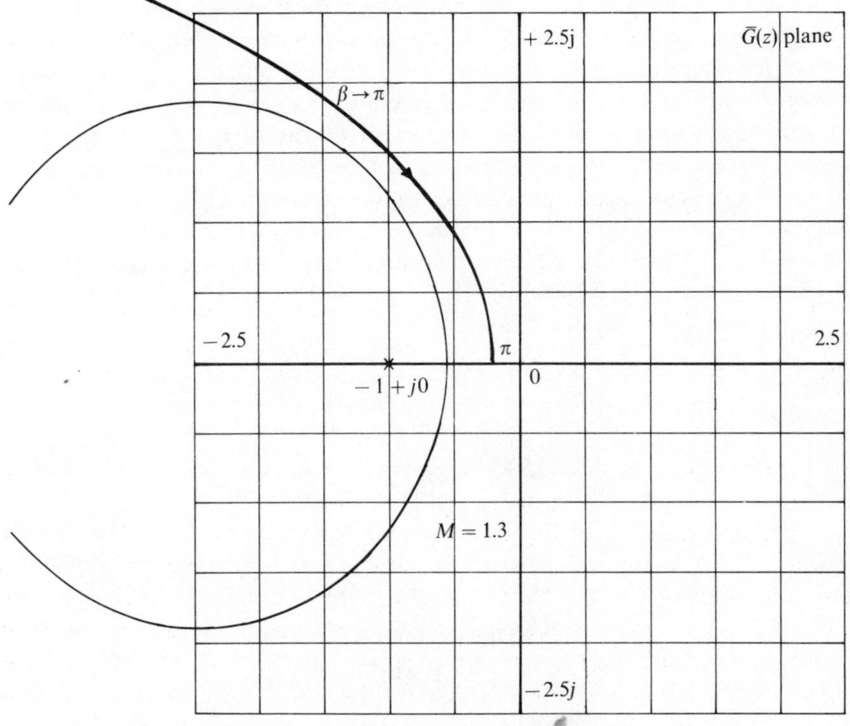

Fig. 2.38 Nyquist diagram (example).

By evaluating $\bar{G}(e^{j\beta})$ for a number of values of β, $0 \leqslant \beta \leqslant \pi$, the Nyquist diagram is drawn as shown in Fig. 2.38.

Since $\bar{G}(e^{j\beta})$ encircles the point $(-1 + j0)$ clockwise, the uncompensated closed-loop system (Fig. 2.31) would be unstable.

2.12 NICHOLS CHARTS (DISCRETE SYSTEMS) (REFS. 3, 4)

The frequency response $\bar{G}(e^{j\beta})$, $0 \leqslant \beta \leqslant \pi$, of a discrete system may be plotted on a Nichols chart, in which the open-loop gain and phase shift are represented on rectangular coordinates (cf. Sec. 2.5). As with the Bode and Nyquist diagrams, a bilinear transformation may be used to map $\bar{G}(z)$ into $\bar{\bar{G}}(w)$, $w = u + jv$, and the Nichols chart developed in terms of $\bar{\bar{G}}(jv)$. The CAD algorithm of Sec. 2.8 is then directly applicable, since the algebras of the w and s planes are identical. There is no other advantage, however, to be gained by introducing the further complication of the w domain, and in this section the $\bar{G}(e^{j\beta})$ Nichols chart, and an appropriately altered CAD algorithm, are used.

The behavior of a closed-loop system incorporating $\bar{G}(z)$ (Fig. 2.31) may be assessed from the Nichols chart for $\bar{G}(e^{j\beta})$.

For closed-loop stability, the contour $\bar{G}(e^{j\beta})$, $0 \leqslant \beta \leqslant \pi$, must pass to the right of the open-loop (0 dB, $-180°$) point on the chart.

As for continuous time systems (Sec. 2.5), the loci of constant closed-loop gains (M contours) and phase shifts (N contours) are provided on standard Nichols charts (Fig. 2.16(a)) and the intersections of these with the $\bar{G}(e^{j\beta})$ contour give the closed-loop gains and phase shifts at various frequencies. This provides some indication of closed-loop behavior. The peak closed-loop frequency response is given by the highest M contour which the $\bar{G}(e^{j\beta})$ contour touches ($M_{p\omega}$); this is related to step function response peak magnitude (M_{pt}) by the graphs in Fig. 2.14(b), which in sampled form ($t = nT$, $n = 0, 1, 2, \ldots$) apply directly to the discrete case.

EXAMPLE

Consider:

$$G(s) = \frac{1500}{s(s+2)}, \qquad T = 0.1 \qquad \text{(Fig. 2.29)}$$

$$\bar{G}(z) = Z\left(\frac{1 - e^{-sT}}{s} \frac{1500}{s(s+2)}\right)$$

$$= \frac{7.125(z + 0.905)}{(z-1)(z-0.819)}$$

$$\bar{G}(e^{j\beta}) = \frac{7.125(e^{j\beta} + 0.905)}{(e^{j\beta} - 1)(e^{j\beta} - 0.819)}.$$

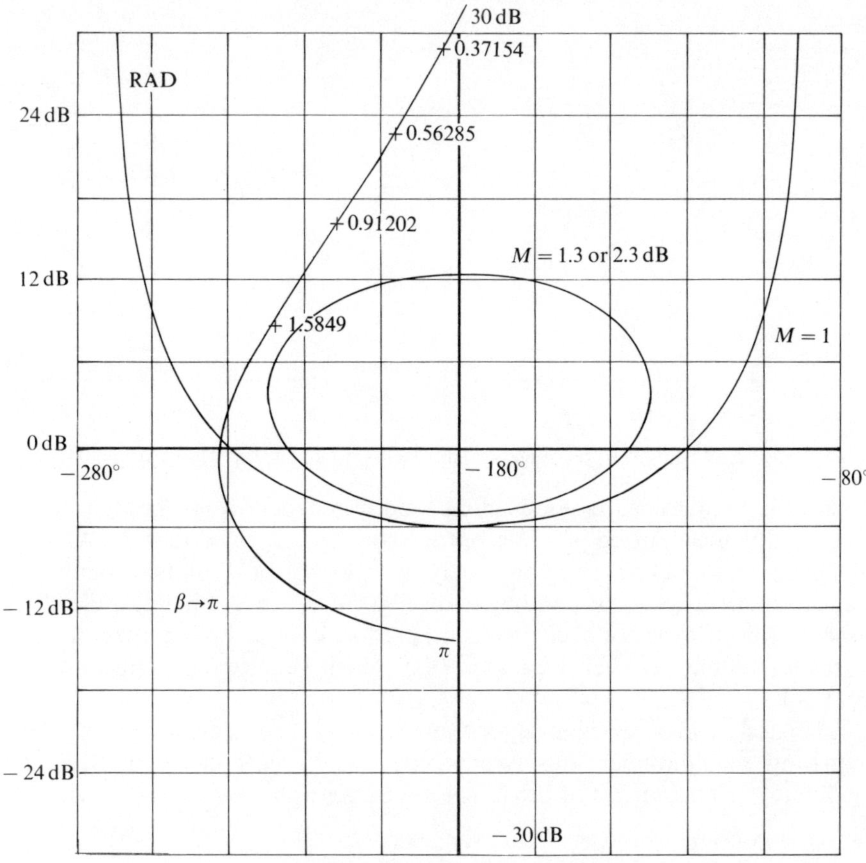

Fig. 2.39 Nichols chart (example).

By evaluating $\bar{G}(e^{j\beta})$ for a number of values of $\beta, 0 \leqslant \beta \leqslant \pi$, the Nichols chart is drawn as shown in Fig. 2.39.

Since the $\bar{G}(e^{j\beta})$ contour passes to the left of the open-loop, 0 dB, $-180°$ point, the uncompensated closed-loop system (Fig. 2.31) would be unstable.

2.13 DIGITAL SERVOMECHANISM DESIGN USING BODE PLOTS, ROOT LOCUS DIAGRAMS, NYQUIST DIAGRAMS, AND NICHOLS CHARTS (REFS. 3, 4)

Servomechanisms of the kind shown in Fig. 2.40 may be designed using the diagrams of Secs. 2.9 to 2.12.

The controlled system (usually continuous time) has a transfer function $G(s)$, and the design requirement is to devise a shaping algorithm, transfer

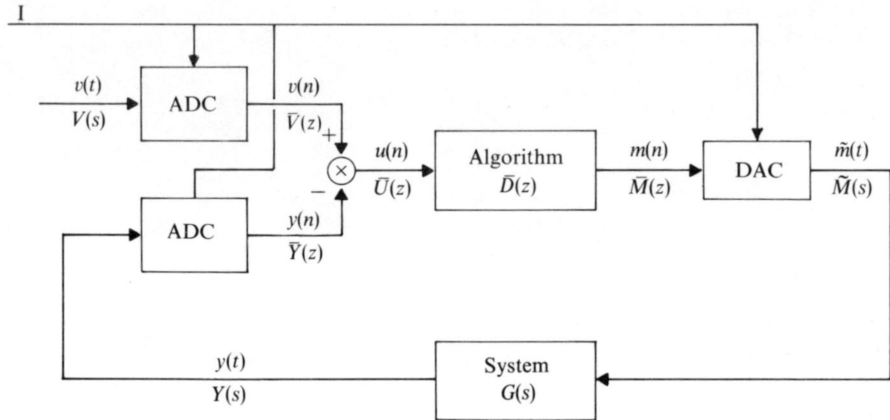

Fig. 2.40 Digital servomechanism with series shaping algorithm.

function $\bar{D}(z)$, which ensures good closed-loop system behavior. The algorithm $\bar{D}(z)$ is readily mechanized in a microprocessor. It is assumed that the ADCs and DACs are 'clocked' simultaneously at intervals of T units, though in practice a small delay corresponding to the algorithm execution time is usually present. If this delay is not negligible compared with the sampling interval, the controlled system transfer function, $G(s)$, must be altered to include it (Sec. 1.14).

The design of a shaping algorithm $\bar{D}(z)$ for a controlled system $G(s)$ comprising the amplifier motor, gearbox, and potentiometer of Sec. 2.7 (Fig. 2.19(a)) is considered in the following paragraphs.

$$G(s) = \frac{45{,}000}{s(s+2)(s+30)}.$$

(i) Design for a Short Sampling Period

A continuous time shaping network, transfer function $F(s)$, may be designed (Sec. 2.7) and an algorithm $\bar{D}(z)$ can then be found whose behavior approximates this, provided the sampling period is short compared with the time constants and natural frequencies of $F(s)$.

Practical experience suggests that a sampling period of one-tenth, and sometimes even one-fifth, of the shortest time constant or undamped natural cycle period (whichever is the shorter) of $F(s)$ can safely be regarded as 'short'.

A variety of methods of finding an algorithm 'equivalent' to $F(s)$ exists (ref. 4); two are described here.

(a) Backward Difference Approximation

A continuous time shaping network is designed (Sec. 2.7, Fig. 2.19):

$$F(s) = \frac{M(s)}{U(s)}.$$

The equivalent differential equation is written down, $m(t)$ replaced with $m(n)$, and any derivatives replaced with backward difference approximations:

$$\frac{\mathrm{d}m}{\mathrm{d}t} \approx \frac{m(n) - m(n-1)}{T} \tag{2.23}$$

$$\frac{\mathrm{d}^2 m}{\mathrm{d}t^2} \approx \frac{\{m(n) - m(n-1)\} - \{m(n-1) - m(n-2)\}}{T^2}$$

$$= \frac{m(n) - 2m(n-1) + m(n-2)}{T^2}. \tag{2.24}$$

$u(t)$ is treated similarly.
This gives a shaping algorithm directly.

EXAMPLE

Recalling the example in Sec. 2.7:

$$F(s) = \frac{M(s)}{U(s)} = 0.03\left[\frac{1 + 0.25s}{1 + 0.025s}\right].$$

Equivalently, in the time domain,

$$33m + 0.83\frac{\mathrm{d}m}{\mathrm{d}t} = u + 0.25\frac{\mathrm{d}u}{\mathrm{d}t}.$$

Since the shortest time constant in $F(s)$ is 0.025 units, $T = 0.003$ may reasonably be regarded as 'short.' The control algorithm is found by substitution from Eq. (2.23):

$$m(n) = \left(33 + \frac{0.83}{T}\right)^{-1}\left[\left(1 + \frac{0.25}{T}\right)u(n) - \frac{0.25}{T}u(n-1) + \frac{0.83}{T}m(n-1)\right]$$

$$= 0.271u(n) - 0.269u(n-1) + 0.895m(n-1).$$

An alternative, and perhaps more elegant method of reaching the same result is to use the formula:

$$\bar{D}(z) = F(s)|_{s=(1-z^{-1})/T}.$$

(b) Approximation by Bilinear Transformation

A continuous time shaping network is designed (Sec. 2.7, Fig. 2.19):

$$F(s) = \frac{M(s)}{U(s)}.$$

A bilinear transformation yields the formula:

$$\bar{D}(z) = F(s)|_{s=(2/T)(1-z^{-1})/(1+z^{-1})}. \tag{2.25}$$

Equation (2.25) is, of course, an empirical formula which depends for its accuracy on T being short and is not to be confused with the exact bilinear transformations of Sec. 2.9.

EXAMPLE

Using the example in the previous subsection:

$$F(s) = 0.03\left[\frac{(1+0.25s)}{(1+0.025s)}\right], \quad T = 0.003.$$

From Eq. (2.25),

$$\bar{D}(z) = \frac{0.283 - 0.281z^{-1}}{1 - 0.886z}.$$

The corresponding algorithm (Eqs. (1.32), (1.33)) is similar to that in subsection (a) above:

$$m(n) = 0.283u(n) - 0.281u(n-1) + 0.886m(n-1).$$

Both these algorithms, mechanized as indicated in Fig. 2.40, give closed-loop step responses which are virtually indistinguishable from that in Fig. 2.21(b).

(ii) Bode Plot Design (Discrete Systems)

The Bode plot for $\bar{\bar{G}}(jv)$ is drawn (Sec. 2.9). The design strategy is to devise an algorithm $\bar{\bar{D}}(jv)$ with a Bode plot which when added to that of $\bar{\bar{G}}(jv)$ gives a composite plot with good gain margin (>5 dB), good 'phase' margin ($>50°$), a high open-loop gain below the 0 dB crossover 'frequency,' and as high a crossover 'frequency' as possible.

$\bar{D}(z)$ is then found using the inverse bilinear transformation: (cf. Eq. (2.19)):

$$w = \frac{z-1}{z+1}. \tag{2.26}$$

Finally, the discrete time domain algorithm is derived from $\bar{D}(z)$ Eqs. (1.32), (1.33), which can easily be manipulated into a variety of convenient forms.

EXAMPLE

Consider the same system as in the previous example, this time with a much longer sampling period, $T = 0.1$. Since T is substantially longer than the shortest time constant in $G(s)$, 0.033 seconds, it is reasonable to ignore the pole ($s = -30$) associated with that time constant. This significantly simplifies the design without noticeable loss of accuracy. It is appropriate to add here that the retention of such 'fast' poles can result in non-minimum phase zeros in $\bar{G}(z)$ (on the negative real z axis, outside the unit circle), which can, often quite unnecessarily, greatly increase the design task.

Adjusting the gain factor appropriately (Eq. (1.10)):

$$G(s) = \frac{45{,}000}{s(s+2)(s+30)} \approx \frac{1500}{s(s+2)}.$$

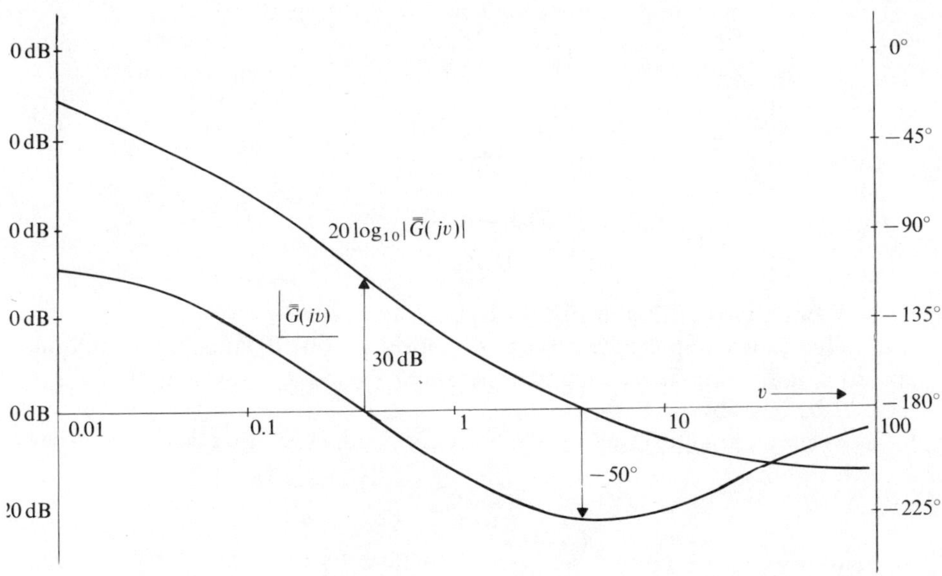

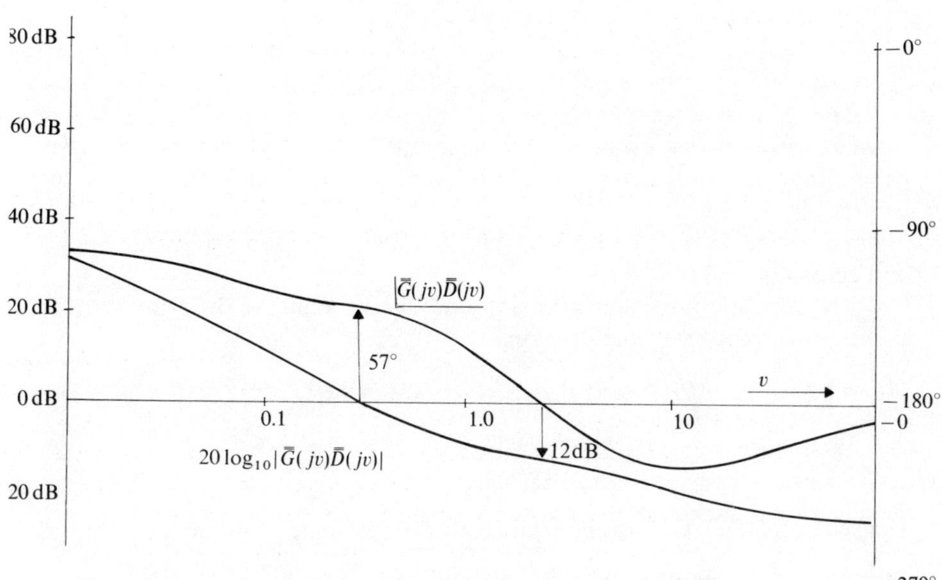

Fig. 2.41 Bode plots (example). (a) Bode plot of $\bar{\bar{G}}(jv)$. (b) Bode plot of $\bar{\bar{G}}(jv)\bar{D}(jv)$.

For a sampling period $T = 0.1$ (Eq. (1.34), Fig. 1.13):

$$\frac{\bar{Y}(z)}{\bar{M}(z)} = \bar{G}(z) = Z\left(\frac{1-e^{-sT}}{s}\frac{1500}{s(s+2)}\right) = \frac{7.125(z+0.905)}{(z-1)(z-0.819)}$$

$$\bar{\bar{G}}(w) = \frac{37.5(1-w)(1+0.05w)}{w(1+10.05w)} \qquad \text{(Eq. (2.19))}$$

$$\bar{\bar{G}}(jv) = \frac{37.5(1-jv)(1+jv0.05)}{jv(1+jv10.05)}.$$

A Bode plot of this in Fig. 2.41(a).

This plot exhibits negative gain margin (− 30 dB) and phase margin (− 50°), indicating that an uncompensated closed-loop system (Fig. 2.31) would be unstable.

A shaping algorithm of the familiar lead type (Sec. 2.7(i)) is often worth trying:

$$\bar{\bar{D}}(jv) = \frac{0.013(1+5.08jv)}{(1+0.25jv)}$$

or

$$\bar{\bar{D}}(w) = \frac{0.013(1+5.08w)}{(1+0.25w)}.$$

A Bode plot of the compensated system $\bar{\bar{G}}(jv)\bar{\bar{D}}(jv)$ is in Fig. 2.41(b).

The gain margin (+ 12 dB) and phase margin (+ 57°) of this are satisfactory, as are the crossover 'frequency' (0.3 units) and the low-frequency gain. (Gain and phase margins should generally be higher than those for continuous systems.)

To ensure that $\bar{\bar{D}}(w)$ is realizable it should have the same number of poles and zeros (Sec. 2.9).

The corresponding algorithm is found by converting to the z domain and thence to the discrete time domain:

$$\bar{D}(z) = \bar{\bar{D}}(w)|_{w=(z-1)/(z+1)} \qquad \text{(Eq. (2.26))}$$

$$= \frac{0.06-0.04z^{-1}}{1+0.60z^{-1}}.$$

Hence (Eqs. (1.32), (1.33)), the equivalent discrete time domain algorithm is:

$$m(n) = 0.06u(n) - 0.04u(n-1) - 0.6m(n-1).$$

A closed-loop unit step function response is shown in Fig. 2.42. The 'staircase' form of $\tilde{m}(t)$ is noteworthy. The algorithm changes the servomechanism output position (in response to an input demand) in some 8 steps.

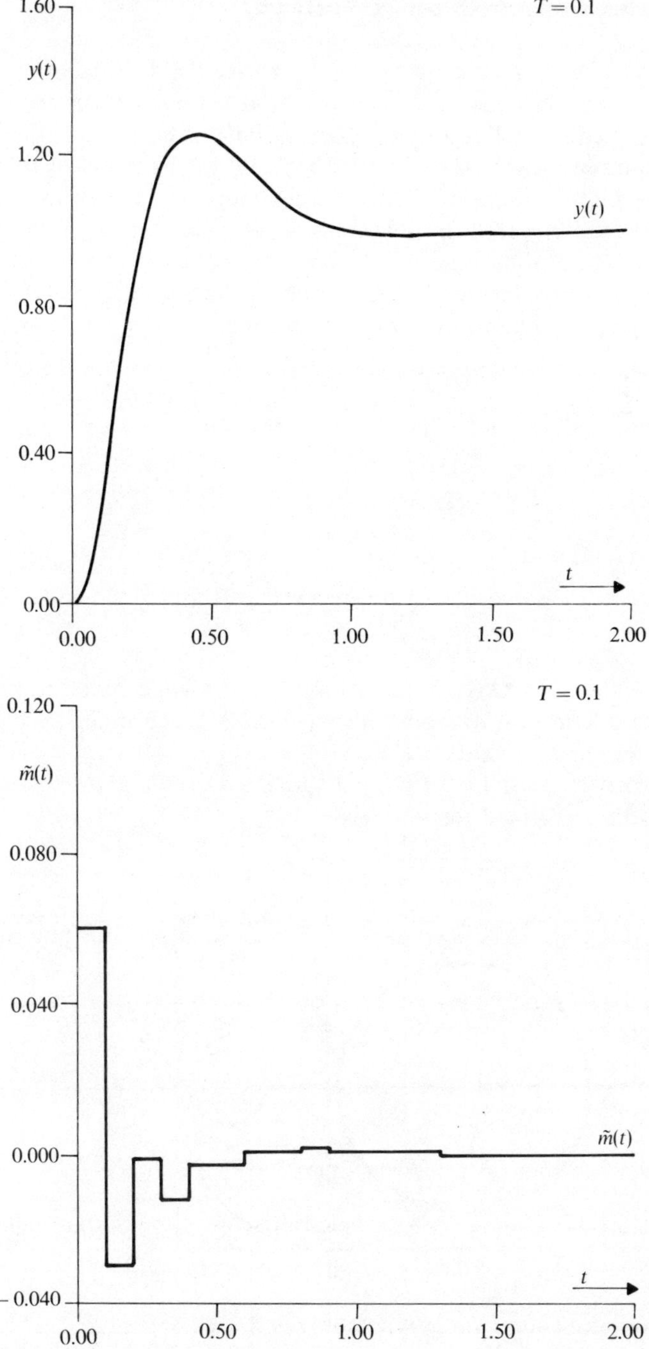

Fig. 2.42 Unit step function response (example). (a) Closed-loop system output. (b) Closed-loop algorithm output.

(iii) Root Locus Design (Discrete Systems)

The root locus for $K\bar{G}(z)$ is drawn $0 \leqslant K \leqslant \infty$ (Sec. 2.10). The design strategy is to devise a shaping algorithm $\bar{D}(z)$ such that the root locus for $K\bar{G}(z)\bar{D}(z)$ allows the dominant closed-loop poles to be positioned on good lines of constant damping ratio (say $\zeta = 0.7$, Fig. 2.35) for as high a value of K as possible. A high undamped natural frequency is also desirable. All this suggests that the closed-loop poles should be placed near the origin where $\beta \to \pi$, but ill-conditioning appears when this policy is carried to extremes. Roughly speaking, poles should the chosen within the sector $\pi/4 \leqslant |\beta| \leqslant 3\pi/4$.

$\bar{D}(z)$ is then mechanized in algorithm form.

EXAMPLE

Consider the example of the previous subsection:

$$G(s) = \frac{1500}{s(s+2)}, \quad T = 0.1 \qquad \text{(Fig. 2.40)}.$$

From Eqs. (1.34) and Fig. 1.13,

$$\frac{\bar{Y}(z)}{\bar{M}(z)} = \bar{G}(z) = \frac{7.125(z+0.905)}{(z-1)(z-0.819)}.$$

The root locus for this (Fig. 2.36) has no worthwhile intersections with the $\zeta = 0.7$ lines of constant damping ratio (Fig. 2.35), and indicates closed-loop instability for the low value $K\bar{K} = 0.169$.

To improve the shape of the root locus, a pole and zero are added at $z = -0.6$ and $z = +0.67$ respectively.

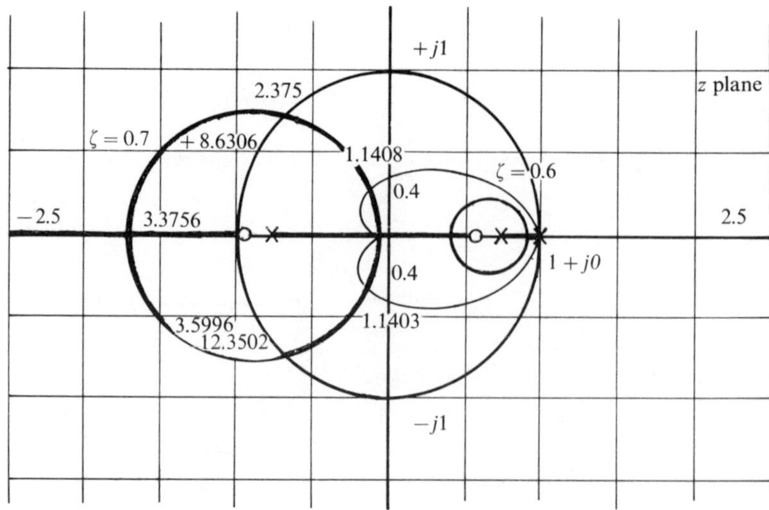

Fig. 2.43 Root locus (example).

The root locus for

$$\frac{K7.125(z+0.905)(z-0.67)}{(z-1)(z-0.819)(z+0.6)}$$

is in Fig. 2.43.

A value $K\bar{K}=0.4$ gives good closed-loop pole positions ($\zeta=0.6$, $\omega_n=4\pi/5T=25.1$ rad sec^{-1}, Fig. 2.35), and, recalling that $\bar{G}(z)$ has a gain constant $\bar{K}=7.125$, this corresponds to a shaping algorithm:

$$\bar{D}(z)=\frac{0.4}{7.125}\frac{(z-0.67)}{(z+0.6)}$$

$$=\frac{0.06-0.04z^{-1}}{1+0.6z^{-1}}.$$

The equivalent discrete time domain algorithm (Eqs. (1.32), (1.33)) is:

$$m(n)=0.06u(n)-0.04u(n-1)-0.6m(n-1).$$

A closed-loop step function response is shown in Fig. 2.42. The concluding remarks of the previous subsection are relevant here also.

(iv) Nyquist Diagram Design (Discrete Systems)

The Nyquist diagram for $\bar{G}(e^{j\beta})$ is drawn (Sec. 2.11). The design strategy is to devise a shaping algorithm $\bar{D}(e^{j\beta})$ which ensures that the contour $\bar{G}(e^{j\beta})\bar{D}(e^{j\beta})$ is tangential to a good M circle (typically $M=1.2$) at as high a frequency as possible.

$\bar{D}(z)$ is then mechanized in algorithm form.

EXAMPLE

Consider the example of the previous subsections:

$$G(s)=\frac{1500}{s(s+2)},\quad T=0.1$$

$$\bar{G}(e^{j\beta})=\frac{7.125(e^{j\beta}+0.905)}{(e^{j\beta}-1)(e^{j\beta}-0.819)}.$$

A Nyquist diagram of this is in Fig. 2.38.

The $\bar{G}(e^{j\beta})$, $0\leqslant\beta\leqslant\pi$, contour 'encircles' the point $-1+j0$ on the right, indicating (uncompensated) closed-loop instability.

Consider the shaping algorithm of the previous subsections:

$$\bar{D}(z)=\frac{0.06(z-0.67)}{(z+0.6)}\quad\text{or}\quad\bar{D}(e^{j\beta})=\frac{0.06(e^{j\beta}-0.67)}{(e^{j\beta}+0.6)}.$$

The Nyquist diagram of the compensated system is shown in Fig. 2.44.

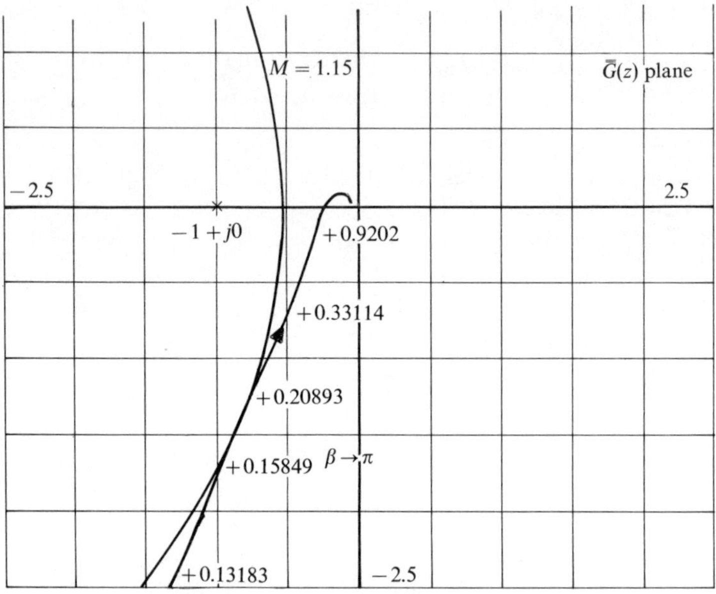

Fig. 2.44 Nyquist diagram (example).

The $G(e^{j\beta})\bar{D}(e^{j\beta})$ contour is almost tangential to the circle $M = 1.15$ at a frequency $\beta = 0.18$ rad/sample. This indicates reasonable closed-loop performance, a 'good' M circle being somewhat lower than for continuous time systems. A step function response is in Fig. 2.42 and the concluding remarks of subsection (ii) also apply here.

(v) Nichols Chart Design (Discrete Systems)

The Nichols chart for $\bar{G}(e^{j\beta})$ is drawn (Sec. 2.12). The design strategy is to devise a shaping algorithm $\bar{D}(e^{j\beta})$ which ensures that the $\bar{G}(e^{j\beta})\bar{D}(e^{j\beta})$ contour is tangential to a good M contour (typically $M = 1.2$), and otherwise follows the closed-loop 0 dB contour as closely as possible over as wide a handwidth as possible.

$\bar{D}(z)$ is then mechanized in algorithm form.

EXAMPLE

Consider the example of the previous subsections:

$$G(s) = \frac{1500}{s(s+2)}, \quad T = 0.1$$

$$\bar{G}(e^{j\beta}) = \frac{7.125(e^{j\beta} + 0.905)}{(e^{j\beta} - 1)(e^{j\beta} - 0.819)}.$$

A Nichols chart of this is in Fig. 2.39.

The $\bar{G}(e^{j\beta})$ contour passes to the left of the open-loop (0 dB, − 180°) point, indicating uncompensated closed-loop instability.

Consider the shaping network of the previous subsections:

$$\bar{D}(z) = \frac{0.06(z - 0.67)}{(z + 0.6)} \qquad \text{or} \qquad \bar{D}(e^{j\beta}) = \frac{0.06(e^{j\beta} - 0.67)}{(e^{j\beta} + 0.6)}.$$

The Nichols chart of the compensated system is shown in Fig. 2.45. The $\bar{G}(e^{j\beta})\bar{D}(e^{j\beta})$ contour is tangential to the $M = 1.15$, 1.21 dB contour at 0.18 rad/sample. This indicates reasonable closed-loop performance. A closed-loop response is in Fig. 2.42. The concluding remarks of subsection (ii) apply to the response and those of (iv) to the choice of M contour.

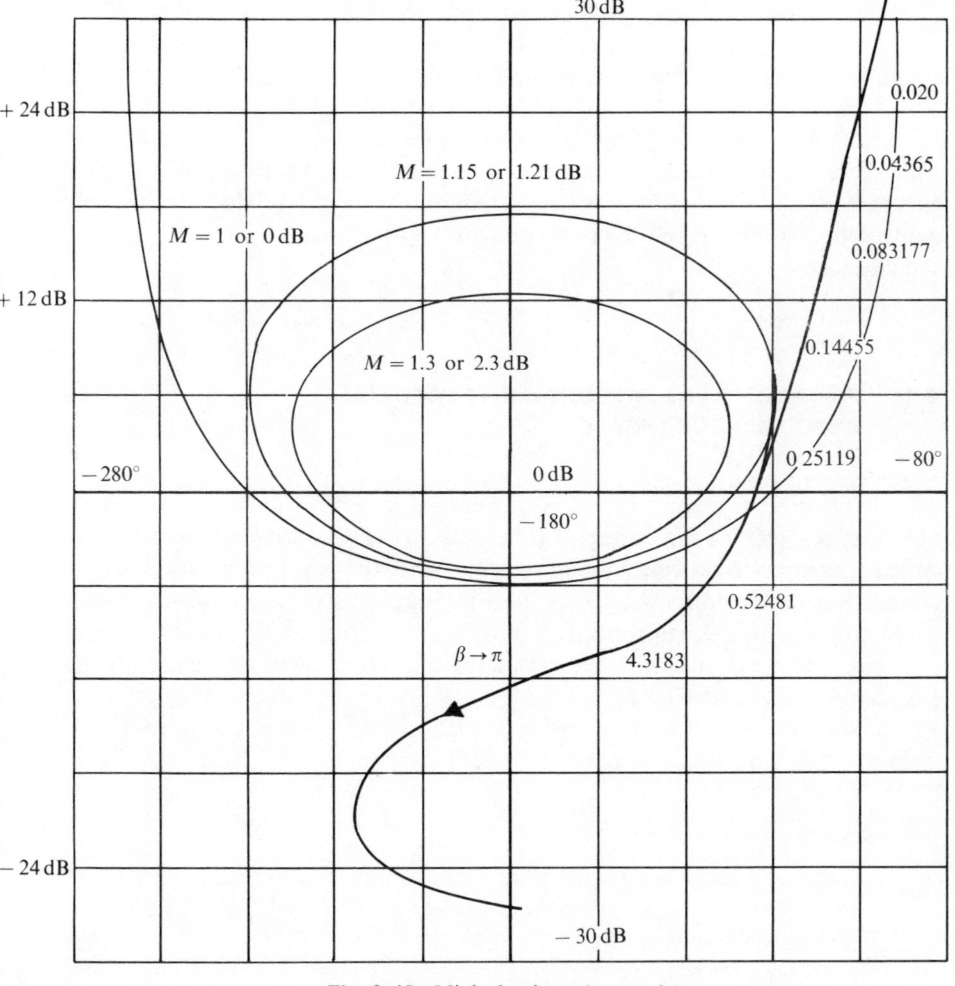

Fig. 2.45 Nichols chart (example).

2.14 CAD FACILITY

In common with the continuous time design methods of Secs. 2.2 to 2.7, a CAD facility is almost essential to the full exploitation of the techniques of Secs. 2.9 to 2.13. The algorithm of Fig. 2.27 is applicable to the discrete system techniques if, instead of the s plane, the w plane is used for the Bode, Nyquist, and Nichols charts. Since the algebra of root locus construction is identical in the s and z (and w) planes, the root locus section of that algorithm is directly applicable to the z domain, though the locus shapes and lines of constant damping ratio and undamped natural frequency are of course different (Sec. 2.10). It is, however, well worthwhile to design a second version of the algorithm to allow design in the z domain throughout (with the possible exception of the Bode plot, which is commonly plotted in the w domain).

A CAD algorithm with some of these features has been used in the example of Sec. 2.13. The plots generated appear in Figs. 2.41 to 2.45.

A useful extension to this CAD facility is to model the system behavior. Perhaps the best way to do this is to cast the continuous time system, $G(s)$, in state equation form (Sec. 3.5) and use a Runge–Kutta algorithm (Sec. 1.2) to generate a solution. A set of difference equations (Sec. 4.2) which model the control algorithm can be used in conjunction with this arrangement to model the closed-loop system.

2.15 PROPORTIONAL INTEGRAL DIFFERENTIAL CONTROLLERS (REF. 5)

Any description of classical control techniques would be incomplete without mention of proportional integral differential (PID) controllers, which find wide application in process control systems. Since the terminology used in this connection differs from the newer terminology of servomechanisms, some explanation of this is appropriate.

A control system using a PID controller is shown schematically in Fig. 2.46 (cf. Fig. 2.19).

The controlled system (transfer function $G(s)$) is usually a process. Its output $y(t)$ is the *controlled variable* (c.v.). This is monitored by a transducer

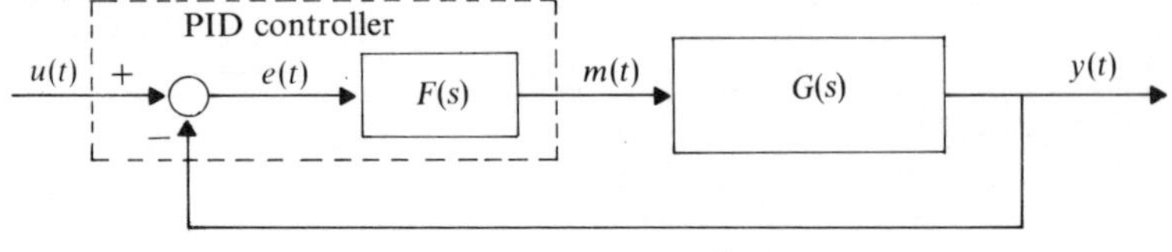

Fig. 2.46 A PID control system.

and compared in the PID controller with the input, $u(t)$, or *set point* (*s.p.*). The difference, or *set point error*, $e(t)$, is shaped by the PID controller (transfer function $F(s)$), and the output of this drives the process input, $m(t)$, or *manipulated variable* (*m.v.*).

The PID controller is a special case of a servomechanism shaping network (Sec. 2.7) and obeys the law:

$$m(t) = K\left[e(t) + \frac{1}{\tau_i}\int_0^t e(t)\mathrm{d}t + \tau_d\frac{\mathrm{d}e(t)}{\mathrm{d}t}\right] \tag{2.27}$$

where K is the *proportional* gain of the controller,
K/τ_i is the *integral* gain of the controller
$K\tau_d$ is the *differential* gain of the controller.

Traditionally, the 'steady state' gain of any element (or set of elements) in a process control system is described by the *proportional band* (PB) of that element. This is the percentage change in the element input which gives rise to 100% change in the output (all steady state).

$$PB = \left[\frac{100}{\text{steady-state gain}}\right]\% .$$

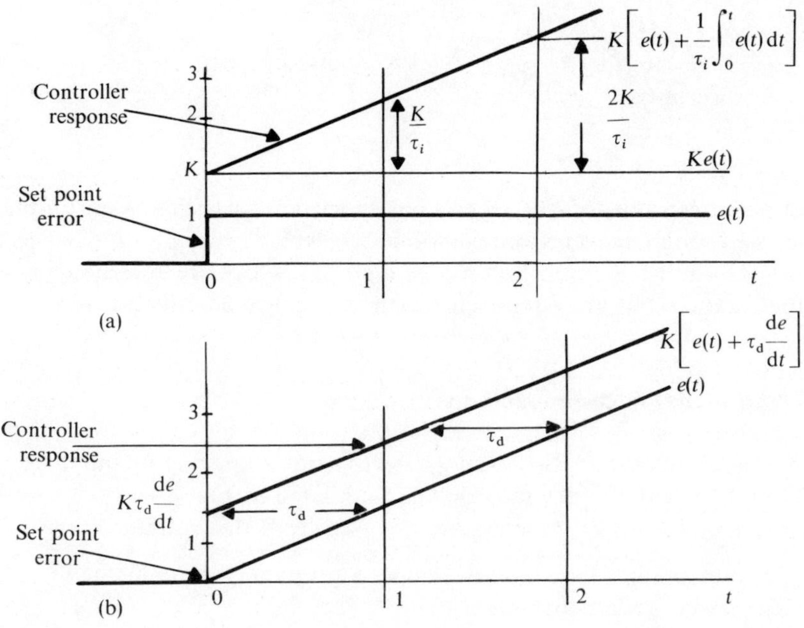

Fig. 2.47 Controller time responses. (a) PI controller. (b) PD controller.

Thus, for the PID controller,

$$PB = \frac{100}{K}$$

The controller *reset rate*, $1/\tau_i$, describes the integral gain (Eq. (2.27)). This is measured in *repeats per unit time*, a reference to the controller response to a step function $e(t)$ ($\tau_d = 0$), as indicated in Fig. 2.47(a).

It can be seen that the integral action 'repeats' the proportional response $1/\tau_i$ times or repeats per unit time.

The controller *rate time*, τ_d, describes the differential gain (Eq. 2.27)). This is measured in units of time (seconds, minutes), a reference to the controller response to a ramp function $e(t)$ ($\tau_i = \infty$) as indicated in Fig. 2.47(b). It can be seen that the differential action 'advances' the proportional response by τ_d units of time.

Basically, a high proportional gain (low *PB*) and high integral gain (high reset rate) are desirable since they lead to a 'tight' control loop and rapid removal of error, $e(t)$. However, increasing both these gains involves decreasing the closed-loop system stability. Increasing the differential gain (the rate time) mitigates this effect, allowing higher *PI* gains than would otherwise be the case.

These properties can be seen by considering the controller transfer function, $F(s)$, found by taking Laplace transforms of Eq. (2.27), and considering the usual case $\tau_i \gg \tau_d$ (typically, $\tau_i \geqslant 4\tau_d$):

$$F(s) = \frac{M(s)}{E(s)} = K\left(1 + \frac{1}{s\tau_i} + s\tau_d\right)$$

$$\approx \frac{K}{s\tau_i}(1 + s\tau_i)(1 + s\tau_d). \tag{2.28}$$

Suitable values of K, τ_i, τ_d can be designed using the methods outlined in Sec. 2.7, though this does entail finding $G(s)$, which is usually an onerous task. Two well-known empirical 'short cut' methods, which are inadequate in some circumstances but are often well worth trying, are described here.

(i) The Reaction Curve Method

The open-loop system, $G(s)$, is subjected to a unit step function input, $m(t)$, and its output response, $y(t)$, recorded as illustrated in Fig. 2.48.

The following parameters are assessed from this *reaction curve*:

1. The maximum reaction curve slope, S (units of gain/unit time).
2. The delay, L (units of time).
3. The steady-state gain $\bar{K}$ (usually volts/volt, p.s.i./p.s.i., etc.) of the system $G(s)$.

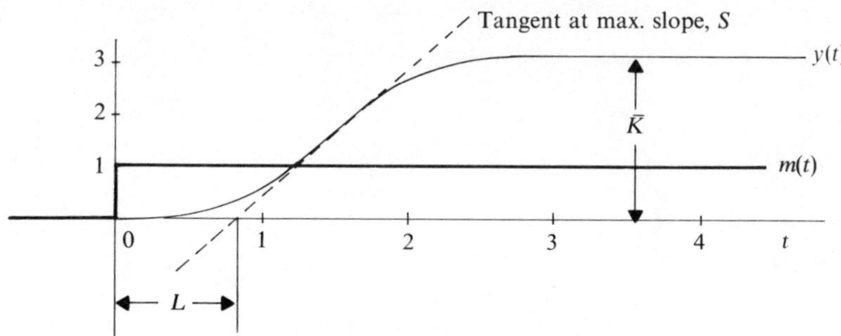

Fig. 2.48 Reaction curve (example).

A recommended set of controller settings is:

$$K = \frac{1.2}{SL\bar{K}}; \quad \tau_i = \frac{L}{0.5}, \quad \tau_d = 0.5L.$$

(ii) The Continuous Cycling Method

In this method, first set $\tau_d = 1/\tau_i = 0$, and the controller PB to its maximum (i.e. K to its minimum) in the closed-loop system: thus the controller is *proportional* only, with low gain. The controller PB is now gradually decreased (K increased) until the closed-loop system becomes just unstable, the period, P, of the resulting oscillation is measured (units of time), and the controller gain, $\hat{K}$, noted.

A recommended set of controller settings is:

$$K = 0.6\hat{K}; \quad \tau_i = \frac{P}{2}; \quad \tau_d = \frac{P}{8}.$$

Before considering an example, a note on the mechanization of PID controllers is appropriate. Pneumatic controllers (ref. 5), offering the advantage of safety, are common in process control situations, while hydraulic controllers are appropriate where suitable power supplies exist. Electronic mechanization is of course much cheaper, and a number of basic circuits are in Fig. 2.49.

'Latchup,' which can occur when an integral controller output saturates on a voltage limit, and the amplifier summing junction ceases to be a virtual earth, is usually prevented by bypass circuitry of the kind shown in Fig. 2.49(d).

EXAMPLE

A useful example is provided by laboratory equipment in which a reservoir tank feeds water via a precisor-driven control valve and delay channel to a

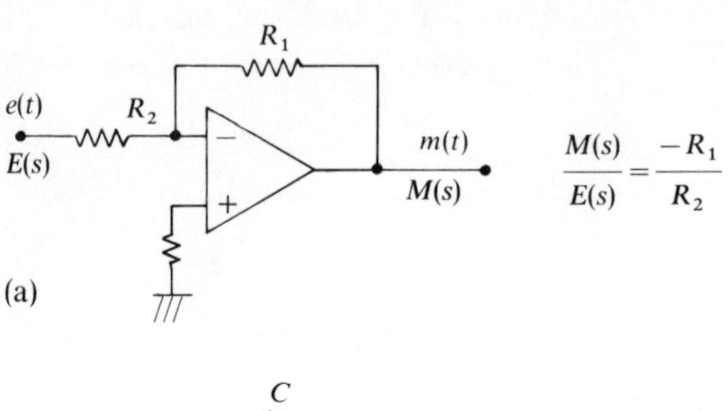

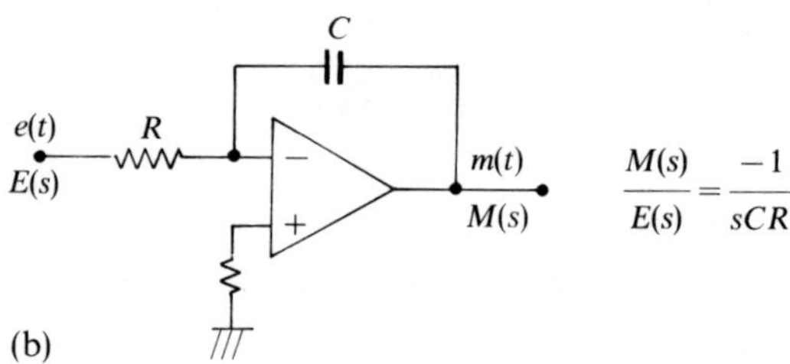

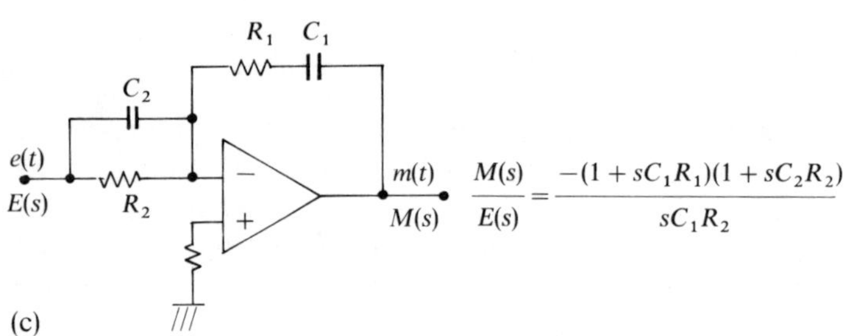

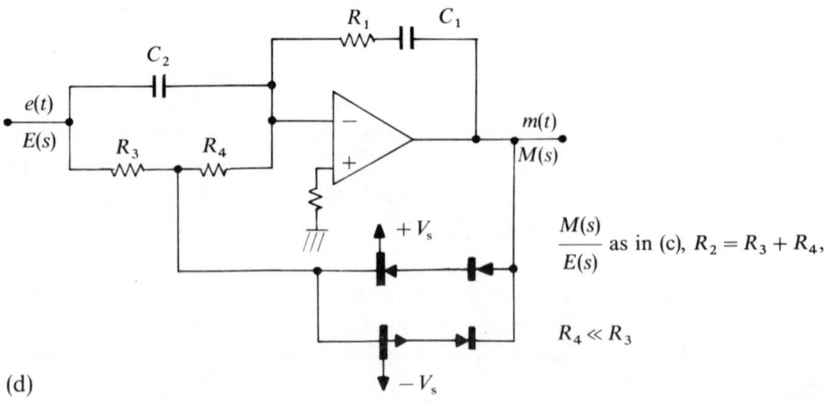

Fig. 2.49 Basic electronic controller circuits. (a) Basic proportional controller. (b) Basic integral controller. (c) Basic proportional integral differential controller. (d) Proportional integral differential controller with latchup prevention circuit.

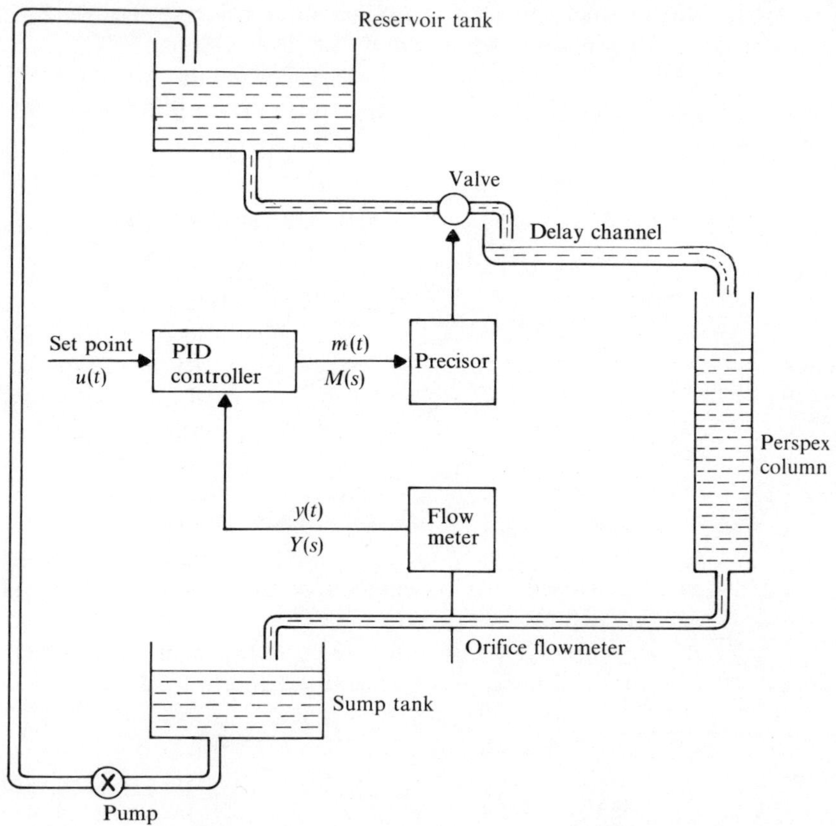

Fig. 2.50 Laboratory process control (example).

perspex column from the bottom of which the water flows through an orifice-type flowmeter to a sump tank. From this it is returned via a pump to the reservoir. A PID controller monitors the column outflow rate (controlled variable), compares this with a set point and drives the valve (manipulated variable) according to the difference.

This arrangement is shown in Fig. 2.50.

The transducers and PID controller could easily be electronic, but in a system of this kind are more typically pneumatic.

A series of tests using a transfer function analyzer and suitable electro-pneumatic transducers gives an open-loop transfer function (time units, minutes):

$$G(s) = \frac{Y(s)}{M(s)} = \frac{1.68e^{-0.25s}}{(1 + 1.1s)(1 + 0.21s)}.$$

The two 'capacitance–resistance' time constants (1.1 and 0.21 min.) are associated with the column and delay channel respectively, while the 'transport lag' (0.25 min.) is associated with the delay channel.

A Bode plot of the system, $G(j\omega)$, (Secs. 2.2, 2.7) and a suitable PID controller (Eq. (2.28)) are shown in Fig. 2.51(a), with the resulting Bode plot of $G(j\omega)F(j\omega)$ in Fig. 2.51(b).

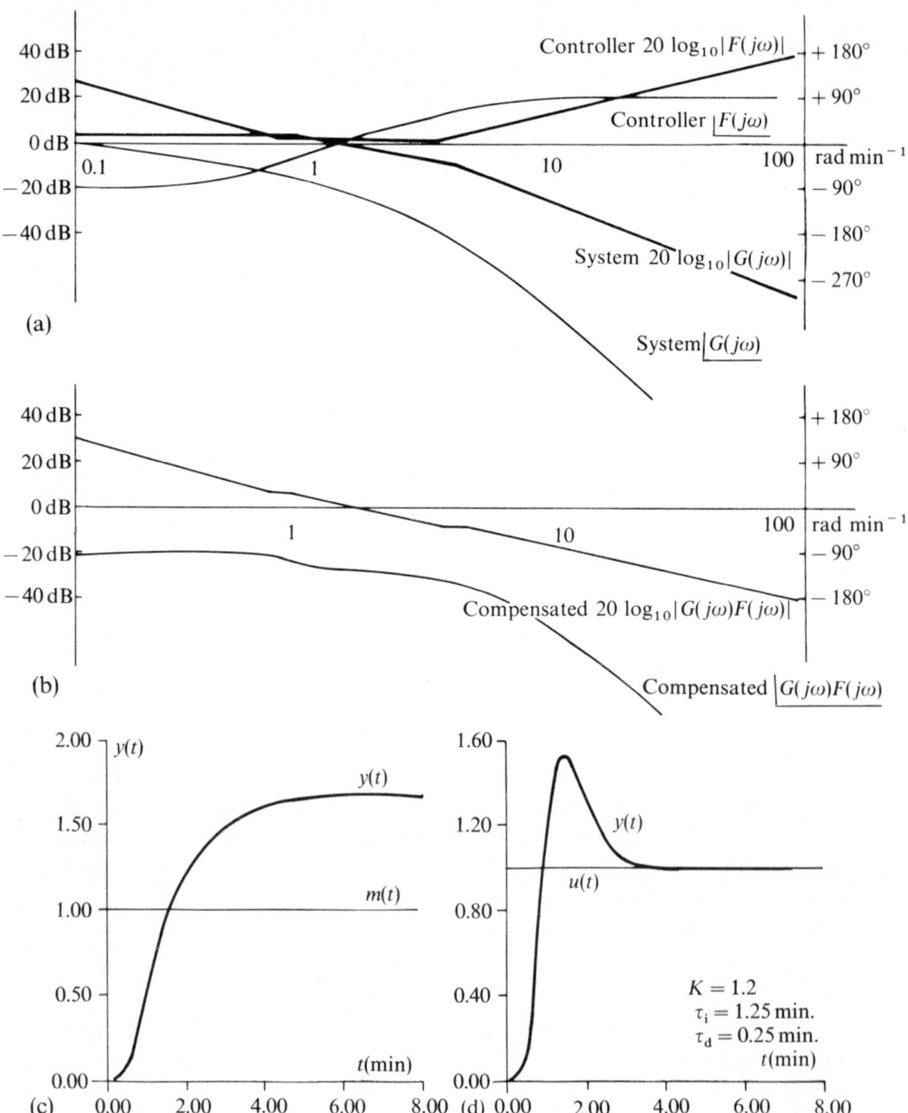

Fig. 2.51 Bode plots and system responses (example). (a) Bode plot – uncompensated system. (b) Bode plot – compensated system. (c) Reaction curve ($m(t) = 1$). (d) Closed-loop response ($u(t) = 1$).

The (quite typical) transport lag element $e^{-0.25s}$ (Eq. (1.8)) is easily represented on frequency response diagrams such as the Bode plot:

$$|e^{-0.25j\omega}| = 20\log_{10} 0\,\mathrm{dB}$$
$$\underline{|e^{-0.25j\omega}} = -0.25\omega.$$

The angle of lag is thus linear w.r.t. ω, and appears on the (logarithmic) Bode plot as a logarithmically increasing angle of lag.

Alternatively, a fairly good second-order approximation to the transport lag element is:

$$e^{-0.25j\omega} = \frac{e^{-0.125j\omega}}{e^{+0.125j\omega}}$$
$$\approx \frac{1 - 0.125j\omega + \frac{1}{2}(0.125j\omega)^2}{1 + 0.125j\omega + \frac{1}{2}(0.125j\omega)^2}.$$

Figure 2.51(a) suggests controller parameters (Fig. 2.51(b)):

$K = 1.2$	or $PB = 83\%$
$\tau_i = 1.25$ min.	or reset rate = 0.8 repeats/min.
$\tau_d = 0.25$ min.	or rate time = 0.25 min.

A closed-loop step function response using this controller is shown in Fig. 2.51(d).

Very similar results are obtained by the empirical methods outlined in subsections (i), (ii) above. A reaction curve used as the basis of the first method, showing the system open-loop response, is in Fig. 2.51(c). This affords interesting comparison with the closed-loop response of Fig. 2.51(d).

Comment

PID controllers, though necessarily limited in their capabilities, can be applied successfully to a surprising variety of systems including nonlinear chemical processes, liquid flow systems, and even gas turbine engines.

REFERENCES

1. D'Azzo, J.J., and Houpis, C.H., *Feedback Control System Analysis and Synthesis*, McGraw-Hill, 1968.
2. Di Stefano, J.J., Stubberud, A.R., and Williams, J.J., *Feedback and Control Systems*, Schaum Outline Series, 1967.
3. Franklin, G.F., and Powell, J.D., *Digital Control of Dynamic Systems.* Addison-Wesley, 1980.
4. Katz, P., *Digital Control using Microprocessors.* Prentice-Hall, 1981.
5. Morris, N.M., *Control Engineering.* McGraw-Hill, 1983.
6. Ralston, A., and Wilf, H.S., *Mathematical Methods for Digital Computers*, Wiley, 1966.
7. Thaler, G.J., and Brown, R.G., *Analysis and Design of Feedback Control Systems*, McGraw-Hill, 1960.

3 CONTINUOUS TIME STATE SPACE DESIGN TECHNIQUES

3.1 INTRODUCTION

The classical design techniques described in Chap. 2 depend on models of system input–output or 'external' behavior. The value of basing design techniques on models of the 'internal' behavior, or states, of systems has become apparent in recent years, and this modern approach is the subject of this chapter, which deals with linear continuous time systems. Discrete systems are dealt with in Chap. 4.

3.2 STATE EQUATIONS

An ordinary differential equation of any order can usually be cast as a set of first-order differential equations (Sec. 1.2). It follows that if the behavior of a dynamical system can be represented by ordinary differential equations, it can also be represented by a set of first-order equations in *canonical*, or standard *state space* form:

$$\dot{\mathbf{x}} = \mathbf{f}(\mathbf{x}, \mathbf{u}, t) \tag{3.1}$$

$$\mathbf{y} = \mathbf{g}(\mathbf{x}, \mathbf{u}, t) \tag{3.2}$$

where $\mathbf{x}$ is the state vector $(x_1 \quad x_2 \quad \dots \quad x_m)^T$, x_i a real variable,
$\mathbf{u}$ is the input vector $(u_1 \quad u_2 \quad \dots \quad u_q)^T$, u_i a real variable,
$\mathbf{y}$ is the output vector $(y_1 \quad y_2 \quad \cdots \quad y_r)^T$, y_i a real variable,
t is the independent variable, time.

This is shown schematically in Fig. 3.1.

Equation (3.1), which is a set of first-order differential equations, is the *state equation*, while Eq. (3.2), which is algebraic, is the *output equation*.

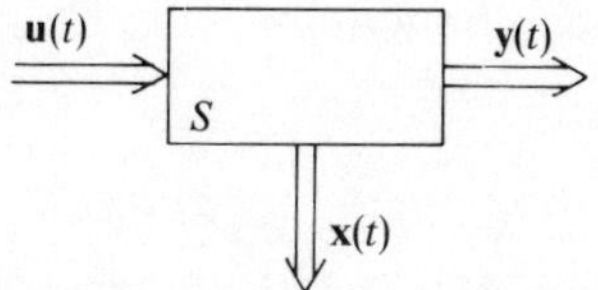

Fig. 3.1 State space representation of dynamical system S.

Linear ordinary differential equations, which represent the behavior of linear systems, can usually be cast in the form:

$$\dot{\mathbf{x}} = \mathbf{A}(t)\mathbf{x} + \mathbf{B}(t)\mathbf{u} \tag{3.3}$$

$$\mathbf{y} = \mathbf{C}(t)\mathbf{x} + \mathbf{D}(t)\mathbf{u} \tag{3.4}$$

where the matrices $\mathbf{A}(t)$, $\mathbf{B}(t)$, $\mathbf{C}(t)$, $\mathbf{D}(t)$ vary with time, t.

Finally, linear constant-coefficient ordinary differential equations, which represent the behavior of linear time-invariant (LTI) systems can usually be cast in the form:

$$\dot{\mathbf{x}} = \mathbf{A}\mathbf{x} + \mathbf{B}\mathbf{u} \tag{3.5}$$

$$\mathbf{y} = \mathbf{C}\mathbf{x} + \mathbf{D}\mathbf{u} \tag{3.6}$$

where $\mathbf{A}$, $\mathbf{B}$, $\mathbf{C}$, $\mathbf{D}$ are constant matrices.

EXAMPLE

Consider the amplifier, servomotor, gearbox, and potentiometer shown in Fig. 3.2 (and Fig. 2.19).

The constant-coefficient linear differential equation governing the behavior of this system is:

$$\frac{d^3y}{dt^3} + 32\frac{d^2y}{dt^2} + 60\frac{dy}{dt} = 45{,}000u$$

where y is the potentiometer output voltage,
u is the amplifier input voltage.

Let

$$x_1 = y$$
$$x_2 = \dot{y}$$
$$x_3 = \ddot{y}.$$

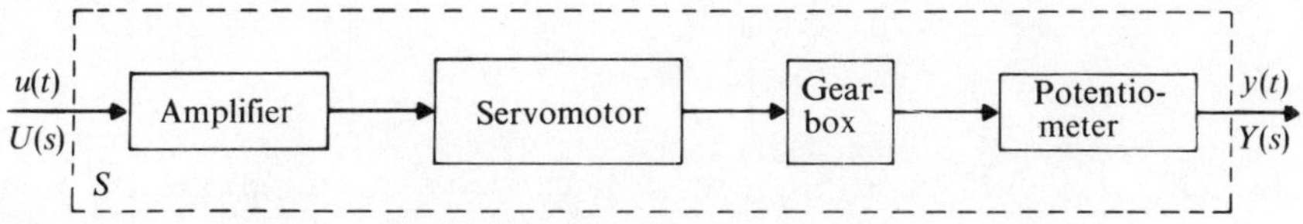

Fig. 3.2 Linear time-invariant system (example).

The differential equation can be written:

$$\dot{x}_1 = x_2$$

$$\dot{x}_2 = x_3$$

$$\dot{x}_3 + 32x_3 + 60x_2 = 45{,}000u$$

or

$$\begin{bmatrix} \dot{x}_1 \\ \dot{x}_2 \\ \dot{x}_3 \end{bmatrix} = \begin{bmatrix} 0 & 1 & 0 \\ 0 & 0 & 1 \\ 0 & -60 & -32 \end{bmatrix} \begin{bmatrix} x_1 \\ x_2 \\ x_3 \end{bmatrix} + \begin{bmatrix} 0 \\ 0 \\ 45{,}000 \end{bmatrix} u$$

$$y = (1 \quad 0 \quad 0) \begin{bmatrix} x_1 \\ x_2 \\ x_3 \end{bmatrix}$$

This is in the form of Eqs. (3.5), (3.6).

The behavior of nonlinear systems for small perturbations around some datum state can often be represented adequately by a set of LTI equations (Eqs. (3.5), (3.6)) obtained by a simple linearization procedure.

In Eqs. (3.1), (3.2), if a 'datum' state $\mathbf{x}_\mathrm{d}$ and corresponding input $\mathbf{u}_\mathrm{d}$ are chosen, the behavior of a small perturbation in the state $\Delta\mathbf{x}$, caused by a small perturbation in the input, $\Delta\mathbf{u}$, is modeled using the first terms of a Taylor expansion:

$$\dot{\mathbf{x}} + \Delta\dot{\mathbf{x}} \approx \mathbf{f}(\mathbf{x}_\mathrm{d}, \mathbf{u}_\mathrm{d}, t) + \left.\frac{\partial \mathbf{f}}{\partial \mathbf{x}}\right|_{\mathbf{x}_d, \mathbf{u}_d} \Delta\mathbf{x} + \left.\frac{\partial \mathbf{f}}{\partial \mathbf{u}}\right|_{\mathbf{x}_d, \mathbf{u}_d} \Delta\mathbf{u}$$

where

$$\frac{\partial \mathbf{f}}{\partial \mathbf{x}} = \begin{bmatrix} \frac{\partial f_1}{\partial x_1} & \frac{\partial f_1}{\partial x_2} & \cdots & \frac{\partial f_1}{\partial x_m} \\ \frac{\partial f_2}{\partial x_1} & & & \vdots \\ \frac{\partial f_m}{\partial x_1} & & & \frac{\partial f_m}{\partial x_m} \end{bmatrix}$$

$$\frac{\partial \mathbf{f}}{\partial \mathbf{u}} = \begin{bmatrix} \frac{\partial f_1}{\partial u_1} & \frac{\partial f_1}{\partial u_2} & \cdots & \frac{\partial f_1}{\partial u_q} \\ \frac{\partial f_2}{\partial u_1} & & & \cdot \\ \vdots & & & \vdots \\ \frac{\partial f_m}{\partial u_1} & & & \frac{\partial f_m}{\partial u_q} \end{bmatrix}$$

Subtracting Eq. (3.1) from this, a set of linearized state equations are:

$$\dot{\mathbf{z}} = \mathbf{A}\mathbf{z} + \mathbf{B}\mathbf{v} \tag{3.7}$$

where

$$\mathbf{z} = \Delta\mathbf{x}; \qquad \mathbf{A} = \left.\frac{\partial \mathbf{f}}{\partial \mathbf{x}}\right|_{\mathbf{x}_d, \mathbf{u}_d}; \qquad \mathbf{B} = \left.\frac{\partial \mathbf{f}}{\partial \mathbf{u}}\right|_{\mathbf{x}_d, \mathbf{u}_d}; \qquad \mathbf{v} = \Delta\mathbf{u}$$

This has the form of the LTI Eq. (3.5). A linearized output equation (Eq. (3.6)) can be constructed by a similar procedure:

$$\mathbf{w} = \mathbf{C}\mathbf{z} + \mathbf{D}\mathbf{v} \tag{3.8}$$

where

$$\mathbf{w} = \Delta\mathbf{y}; \mathbf{C} = \left.\frac{\partial \mathbf{g}}{\partial \mathbf{x}}\right|_{\mathbf{x}_d, \mathbf{u}_d}; \quad \mathbf{D} = \left.\frac{\partial \mathbf{g}}{\partial \mathbf{u}}\right|_{\mathbf{x}_d, \mathbf{u}_d}$$

EXAMPLE

Consider a pendulum consisting of a mass 1 kg attached to a light rod of length 1 m driven by a motor which can supply a torque T N m, friction being negligible. The system is shown in Fig. 3.3.

The behavior of the pendulum is modeled by the equation:

$$1\ddot{\theta} = -9.81 \sin\theta + T$$

where θ is the angular displacement (rad),
T is the applied torque (N m),
and the gravitational constant is taken to be 9.81 ms^{-2}.

Casting this in state equation from (Eqs. (3.1), (3.2)):

$$\dot{\mathbf{x}} = \begin{bmatrix} \dot{x}_1 \\ \dot{x}_2 \end{bmatrix} = \begin{bmatrix} x_2 \\ -9.81 \sin x_1 + u \end{bmatrix}$$

$$y = x_1$$

where

$$x_1 = \theta; \quad x_2 = \dot{\theta}; \quad u = T; \quad y = \theta.$$

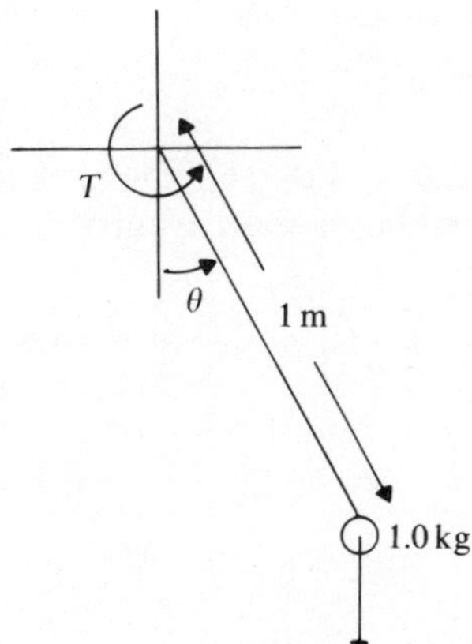

Fig. 3.3 Motor-driven pendulum (example)

The system clearly has an equilibrium point ($\ddot{\theta} = \dot{\theta} = 0$) where $T = 9.81 \sin\theta$,

i.e.
$$\theta_d = \sin^{-1}\left(\frac{T}{9.81}\right).$$

The system's behavior for small perturbations around this point might be of interest:

$$\mathbf{x}_d = \begin{pmatrix} x_{1d} \\ x_{2d} \end{pmatrix} = \begin{pmatrix} \theta_d \\ 0 \end{pmatrix}$$

$$\left.\frac{\partial \mathbf{f}}{\partial \mathbf{x}}\right|_{\mathbf{x}_d, \mathbf{u}_d} = \begin{pmatrix} 0 & 1 \\ -9.81 \cos\theta_d & 0 \end{pmatrix}$$

$$\left.\frac{\partial \mathbf{f}}{\partial u}\right|_{\mathbf{x}_d, \mathbf{u}_d} = \begin{pmatrix} 0 \\ 1 \end{pmatrix}, \quad \left.\frac{\partial g}{\partial \mathbf{x}}\right|_{\mathbf{x}_d, \mathbf{u}_d} = (1 \quad 0), \quad \left.\frac{\partial g}{\partial u}\right|_{\mathbf{x}_d, \mathbf{u}_d} = 0.$$

The linear equations describing the behavior of small perturbations around $\mathbf{x}_d$ are (Eqs. (3.7), (3.8)):

$$\dot{\mathbf{z}} = \begin{pmatrix} 0 & 1 \\ -9.81 \cos\theta_d & 0 \end{pmatrix} \mathbf{z} + \begin{pmatrix} 0 \\ 1 \end{pmatrix} v$$

$$w = (1 \quad 0)\mathbf{z}$$

where $\mathbf{z} = \Delta\mathbf{x}$, $v = \Delta u$, $w = \Delta y$.

3.3 PROPERTIES OF LINEAR SYSTEM STATE EQUATIONS

(i) Solution of the State Equations

The solution of Eq. 3.3, given an 'initial' state $\mathbf{x}(t_0)$, is:

$$\mathbf{x}(t) = \boldsymbol{\phi}(t, t_0)\mathbf{x}(t_0) + \int_{t_0}^{t} \boldsymbol{\phi}(t, \tau)\mathbf{B}(\tau)\mathbf{u}(\tau)\,\mathrm{d}\tau$$

where τ is a dummy variable and $\boldsymbol{\phi}(t, t_0)$, the *state transition matrix*, is defined (ref. 3) as an $(m \times m)$ nonsingular matrix such that:

$$\frac{\mathrm{d}}{\mathrm{d}t}\boldsymbol{\phi}(t, t_0) = \mathbf{A}(t)\boldsymbol{\phi}(t, t_0), \tag{3.9}$$

$$\boldsymbol{\phi}(t, t) = \mathbf{I}.$$

It can readily be shown that:

$$\boldsymbol{\phi}(t, t_0) = \boldsymbol{\phi}^{-1}(t_0, t)$$

and

$$\boldsymbol{\phi}(t_2, t_0) = \boldsymbol{\phi}(t_2, t_1)\boldsymbol{\phi}(t_1, t_0), \qquad t_0 \leqslant t_1 \leqslant t_2.$$

For the LTI case, the solution of Eq. (3.5), $t_0 = 0$, is:

$$\mathbf{x}(t) = e^{\mathbf{A}t}\mathbf{x}(0) + \int_0^t e^{\mathbf{A}\tau}\mathbf{B}\mathbf{u}(\tau)\,\mathrm{d}\tau \tag{3.10}$$

where:

$$e^{\mathbf{A}\tau} = \mathbf{I} + \mathbf{A}\tau + \frac{1}{2!}\mathbf{A}^2\tau^2 + \frac{1}{3!}\mathbf{A}^3\tau^3 + \cdots.$$

In this case, the state transition matrix is:

$$\boldsymbol{\phi}(t, t_0) = e^{\mathbf{A}(t-t_0)}; \qquad \text{for} \quad t_0 = 0, \quad \boldsymbol{\phi}(t, 0) = e^{\mathbf{A}t}.$$

$e^{\mathbf{A}t}\mathbf{x}(0)$ is the *zero input response*, or the state behavior for $\mathbf{u} = \mathbf{0}$, $\mathbf{x}(0) \neq \mathbf{0}$, which describes the natural modes of the system.

$\int_0^t e^{\mathbf{A}\tau}\mathbf{B}\mathbf{u}(\tau)\,\mathrm{d}\tau$ is the *zero initial state response*, which describes the response of the system particular to the input $\mathbf{u}$, for zero initial state, $\mathbf{x}(0) = \mathbf{0}$.

A useful CAD facility for solving Eq. (3.5) for step function inputs using Eq. (3.10) is easily constructed.

CAD Facility

The interactive CAD algorithm shown in Fig. 3.4 calculates the states and output of an LTI system for a specified initial state and a step of specified amplitude at each input.

The operation of the algorithm is as follows.

BLOCK 1

The system parameter matrices **A**, **B**, **C**, **D** are requested, input, displayed and corrected if necessary.

BLOCK 2

A calculation interval T is requested and input. This should be short compared with the system time constants to allow accuracy to be achieved. Since these time constants are generally unknown at this stage, some guesswork may be necessary. The initial states, $\mathbf{x}(0)$, the input step amplitudes, $\mathbf{u}$, and the total number of steps required, N, are input. A printout ratio, R, is also requested and input. This specifies the number of iterations (or calculation intervals) between the values of $\mathbf{x}(nT)$ and $y(nT)$) output to a display or printer, $n = 0, 1, \ldots, N$.

BLOCK 3

The solution of Eq. (3.5) in the nth calculation interval $nT < t \leqslant (n+1)T$, remembering that the inputs are constant ($\mathbf{u}(t) = \mathbf{u}$), is (cf. Eq. (3.10)):

$$\mathbf{x}((n+1)T) = e^{\mathbf{A}T}\mathbf{x}(nT) + \int_0^T e^{\mathbf{A}\tau}\,\mathrm{d}\tau\,\mathbf{B}\mathbf{u}.$$

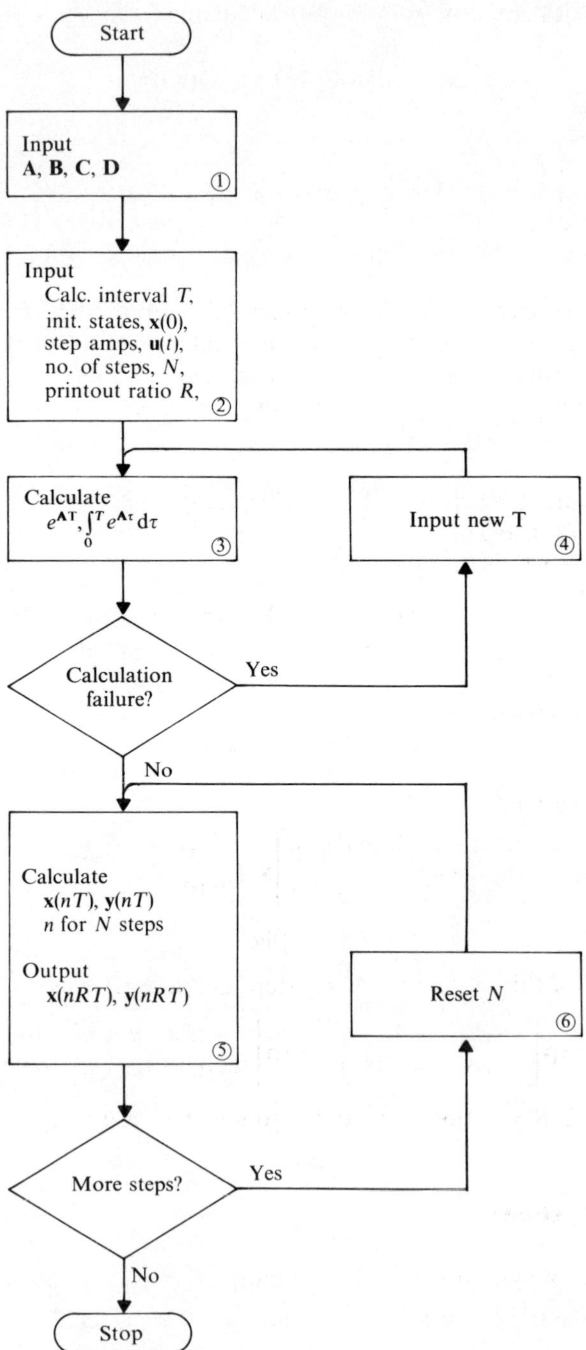

Fig. 3.4 Step function response CAD program.

The matrices $e^{\mathbf{A}T}$ and $\int_0^T e^{\mathbf{A}\tau}\,d\tau$ are calculated from the formulas:

$$e^{\mathbf{A}T} = \mathbf{I} + \mathbf{A}T + \frac{1}{2!}\mathbf{A}^2T^2 + \frac{1}{3!}\mathbf{A}^3T^3 + \cdots$$

and

$$\int_0^T e^{\mathbf{A}\tau}d\tau = \mathbf{I} + \frac{1}{2!}\mathbf{A}T^2 + \frac{1}{3!}\mathbf{A}^2T^3 + \frac{1}{4!}\mathbf{A}^3T^4 + \cdots.$$

BLOCK 4

The numbers of terms taken in these series are limited by taking an upper bound (say 10^{-4}) on any element in the last (matrix) terms taken into account. If, using this criterion, the number of terms required exceeds 20, say, the algorithm requests the user to select a shorter calculation interval T.

BLOCK 5

Using the results from Block 3, the values $\mathbf{x}(nT)$, $\mathbf{y}(nT)$ are calculated and $\mathbf{x}(nRT)$, $\mathbf{y}(nRT)$ displayed.

Setting $t = nT$ in Eq. 3.6:

$$\mathbf{y}(nT) = \mathbf{Cx}(nT) + \mathbf{Du}(nT).$$

BLOCK 6

A facility for increasing N is useful.

EXAMPLE

Consider the system:

$$\dot{\mathbf{x}} = \begin{bmatrix} -2 & 1 \\ -101 & 0 \end{bmatrix}\mathbf{x} + \begin{bmatrix} 0 \\ 101 \end{bmatrix}u$$

$$y = [1 \quad 0]\mathbf{x} + (0)u.$$

The response of this system to a unit step function input is:

$$\mathbf{x}(t) = \exp\begin{bmatrix} -2 & 1 \\ -101 & 0 \end{bmatrix}t + \int_0^t \left[\exp\begin{bmatrix} -2\tau & \tau \\ -101\tau & 0 \end{bmatrix}\right]\cdot\begin{bmatrix} 0 \\ 101 \end{bmatrix} 1\, d\tau.$$

Using $T = 0.02$, $R = 5$, this is evaluated to give the results in Fig. 3.5.

(ii) Similar Systems

The definition of system states is not unique. If $\mathbf{x} = \mathbf{Tz}$ where $\mathbf{T}$ is a square matrix, Eqs. (3.5), (3.6) become:

$$\dot{\mathbf{z}} = \mathbf{T}^{-1}\mathbf{ATz} + \mathbf{T}^{-1}\mathbf{Bu}$$

$$\mathbf{y} = \mathbf{CTz} + \mathbf{Du}.$$

COMPUTING INTERVAL: T = 0.2000E − 01
SYSTEM INPUT STEP MAGNITUDES: U = 1.0000
PRINTOUT RATIO: R = 5 NUMBER OF DATA POINTS: N = 50

TIME (SECS)	Y1	TIME (SECS)	Y1
0.00000E + 00	0.00000E + 00	0.26000E + 01	0.94629E + 00
0.10000E + 00	0.43497E + 00	0.27000E + 01	0.10132E + 01
0.20000E + 00	0.12663E + 01	0.28000E + 01	0.10569E + 01
0.30000E + 00	0.17230E + 01	0.29000E + 01	0.10448E + 01
0.40000E + 00	0.14889E + 01	0.30000E + 01	0.99724E + 00
0.50000E + 00	0.88611E + 00	0.31000E + 01	0.96061E + 00
0.60000E + 00	0.48838E + 00	0.32000E + 01	0.96375E + 00
0.70000E + 00	0.59300E + 00	0.33000E + 01	0.99680E + 00
0.80000E + 00	0.10209E + 01	0.34000E + 01	0.10266E + 01
0.90000E + 00	0.13537E + 01	0.35000E + 01	0.10286E + 01
0.10000E + 01	0.13287E + 01	0.36000E + 01	0.10062E + 01
0.11000E + 01	0.10318E + 01	0.37000E + 01	0.98267E + 00
0.12000E + 01	0.76200E + 00	0.38000E + 01	0.97797E + 00
0.13000E + 01	0.74124E + 00	0.39000E + 01	0.99265E + 00
0.14000E + 01	0.94185E + 00	0.40000E + 01	0.10109E + 01
0.15000E + 01	0.11550E + 01	0.41000E + 01	0.10166E + 01
0.16000E + 01	0.11992E + 01	0.42000E + 01	0.10074E + 01
0.17000E + 01	0.10678E + 01	0.43000E + 01	0.99360E + 00
0.18000E + 01	0.90326E + 00	0.44000E + 01	0.98770E + 00
0.19000E + 01	0.84988E + 00	0.45000E + 01	0.99322E + 00
0.20000E + 01	0.93242E + 00	0.46000E + 01	0.10034E + 01
0.21000E + 01	0.10568E + 01	0.47000E + 01	0.10089E + 01
0.22000E + 01	0.11109E + 01	0.48000E + 01	0.10059E + 01
0.23000E + 01	0.10619E + 01	0.49000E + 01	0.99847E + 00
0.24000E + 01	0.96973E + 00		
0.25000E + 01	0.91972E + 00		

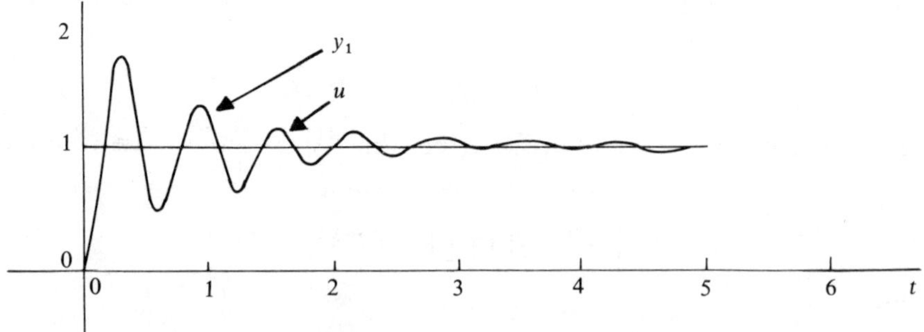

Fig. 3.5 System response (example).

One interesting case is where $\mathbf{T}^{-1}\mathbf{AT}$ is a diagonal matrix containing the eigenvalues of **A**, while **T** is a square matrix consisting of the eigenvectors of **A** (App. A.9.) This is possible if **A** has distinct eigenvectors; otherwise a Jordan form $\mathbf{T}^{-1}\mathbf{AT}$, which is an 'almost' diagonal form (ref. 3) can be obtained.

The solution (cf. Eq. (3.10)) is:

$$\mathbf{z}(t) = e^{\mathrm{T}^{-1}\mathrm{AT}t}\mathbf{z}(0) + \int_0^t e^{\mathrm{T}^{-1}\mathrm{AT}\tau}\mathbf{T}^{-1}\mathbf{Bu}(\tau)\,\mathrm{d}\tau.$$

If **A** has m distinct eigenvectors, (and so, of necessity, m distinct eigenvalues, $\lambda_1, \lambda_2, \ldots, \lambda_m$)

$$\mathbf{T}^{-1}\mathbf{A}\mathbf{T} = \begin{bmatrix} \lambda_1 & & \cdots & 0 \\ 0 & \lambda_2 & \cdots & 0 \\ \vdots & & \ddots & \vdots \\ 0 & & & \lambda_m \end{bmatrix}$$

This yields 'decoupled' zero input responses of the form:

$$z_i(t) = e^{\lambda_i t} z_i(0), \qquad i = 1, 2, \ldots, m. \tag{3.11}$$

If **A** has fewer than m distinct eigenvectors, $\mathbf{T}^{-1}\mathbf{A}\mathbf{T}$ in Jordan form yields zero input responses:

$$z_i(t) = e^{\lambda_i t} z_i(0) + t e^{\lambda_i t} z_{i+1}(0) + t^2 e^{\lambda_i t} z_{i+2}(0) + \cdots. \tag{3.12}$$

Eigenvalues may of course be complex numbers which occur in conjugate pairs, representing sinusoidal responses (Sec. 1.5).

EXAMPLE

Reconsider the system:

$$\dot{\mathbf{x}} = \begin{bmatrix} -2 & 1 \\ -101 & 0 \end{bmatrix} \mathbf{x} + \begin{bmatrix} 0 \\ 101 \end{bmatrix} u$$

$$y = [1 \quad 0]\mathbf{x} + 0u.$$

The eigenvalues of

$$\begin{bmatrix} -2 & 1 \\ -101 & 0 \end{bmatrix}$$

are $-1 + j10$, $-1 - j10$, and the corresponding eigenvectors (App. A.9) are:

$$\begin{bmatrix} 1 \\ 1 + j10 \end{bmatrix}, \begin{bmatrix} 1 \\ 1 - j10 \end{bmatrix}.$$

Thus:

$$\mathbf{T} = \begin{bmatrix} 1 & 1 \\ (1 + j10) & (1 - j10) \end{bmatrix}$$

$$\mathbf{T}^{-1} = \frac{-1}{20j} \begin{bmatrix} 1 - j10 & -1 \\ -1 - j10 & 1 \end{bmatrix} \qquad \text{(App. A.7).}$$

The diagonalized state equations are consequently:

$$\dot{\mathbf{z}} = \begin{bmatrix} -1 + j10 & 0 \\ 0 & -1 - j10 \end{bmatrix} \mathbf{z} + \begin{bmatrix} -j5.05 \\ j5.05 \end{bmatrix} u$$

$$y = [1 \quad 1]\mathbf{z}.$$

The zero input response (Eq. (3.11)), which in this case corresponds to a complex conjugate pair of eigenvalues, is:

$$z_1(t) = e^{(-1+j10)t} z_1(0)$$
$$z_2(t) = e^{(-1-j10)t} z_2(0)$$
$$y(t) = [1 \quad 1]\mathbf{z}$$
$$= Ke^{-t}\cos(10t + L).$$

K, L are real constants involving $x_1(0)$ and $x_2(0)$. The form of this agrees with the results in Fig. 3.5.

(iii) Solution of the State Equations by Laplace Transforms

Taking Laplace transforms (Secs. 1.3, 1.4) of Eqs. (3.5), (3.6) gives:

$$s\mathbf{X}(s) - \mathbf{x}(0) = \mathbf{AX}(s) + \mathbf{BU}(s)$$
$$\mathbf{Y}(s) = \mathbf{CX}(s) + \mathbf{DU}(s).$$

Hence

$$\mathbf{Y}(s) = [\mathbf{C}(s\mathbf{I} - \mathbf{A})^{-1}\mathbf{B} + \mathbf{D}]\mathbf{U}(s) + \mathbf{C}(s\mathbf{I} - \mathbf{A})^{-1}\mathbf{x}(0). \quad (3.13)$$

Setting $\mathbf{x}(0) = \mathbf{0}$, this yields the transfer function matrix (App. A.7):

$$\mathbf{F}(s) = \mathbf{C}(s\mathbf{I} - \mathbf{A})^{-1}\mathbf{B} + \mathbf{D} \quad (3.14)$$
$$= \mathbf{C}\frac{\operatorname{adj}(s\mathbf{I} - \mathbf{A})}{\det(s\mathbf{I} - \mathbf{A})}\mathbf{B} + \mathbf{D}$$

where $\mathbf{Y}(s) = \mathbf{F}(s)\mathbf{U}(s)$.

The behavior of the output $\mathbf{y}$ for any input $\mathbf{u}$ (including $\mathbf{u} = \mathbf{0}$, $\mathbf{x}(0) \neq \mathbf{0}$) is found from the inverse Laplace transform of Eq. (3.13).

The system poles are defined as those values of s for which any element(s) in the transfer function matrix (Eq. (3.14)) is infinite. The poles are therefore roots of the *characteristic equation*:

$$\det(s\mathbf{I} - \mathbf{A}) = 0. \quad (3.15)$$

There may be more roots than there are poles, since cancellation of factors in the 'characteristic polynomial' $\det(s\mathbf{I} - \mathbf{A})$ may occur when the algebra of Eq. (3.14) is worked out. This feature is examined in a discussion of controllability and observability in Secs. 3.4(i), (ii).

All the natural modes of the system (Eq. (3.11)) are described by (all) the roots of the characteristic cquation (Eq. (3.15)), and since this is identical in format to the equation for finding the eigenvalues of $\mathbf{A}$ (App. A.9), it is clear that the system modes are described by these eigenvalues. Equally clearly, the system poles lie at (not necessarily all of) these eigenvalues. If the eigenvalues of $\mathbf{A}$ are λ_i, $i = 1, 2, \ldots, m$, the natural modes of the system are of the form

(Eq. (3.11)):

$$e^{\lambda_i t} z_i(0).$$

Here the eigenvalues are distinct, and the corresponding poles are single. When the eigenvalues are repeated, and the poles multiple, the natural modes are of the form (Eq. (3.12)):

$$e^{\lambda_i t} z_i(0) + t e^{\lambda_i t} z_{i+1}(0) + \cdots.$$

EXAMPLE

Consider the example of the previous two subsections:

$$\dot{\mathbf{x}} = \begin{bmatrix} -2 & 1 \\ -101 & 0 \end{bmatrix} \mathbf{x} + \begin{bmatrix} 0 \\ 101 \end{bmatrix} u$$

$$y = [1 \quad 0]\mathbf{x}.$$

The system transfer function matrix (Eq. (3.14)) is:

$$\mathbf{F}(s) = [1 \quad 0] \begin{bmatrix} s+2 & -1 \\ 101 & s \end{bmatrix}^{-1} \begin{bmatrix} 0 \\ 101 \end{bmatrix} + 0$$

$$= \frac{101}{s^2 + 2s + 101}.$$

The characteristic equation (Eq. (3.15)) is:

$$s^2 + 2s + 101 = 0.$$

This has roots (and the system has poles) at:

$$s = -1 \pm j10,$$

These correspond to system modes of the form (Eq. (3.11)):

$$e^{(-1+j10)t}, \; e^{(-1-j10)t} \tag{3.16}$$

Considered together, these complex modes represent a single real system mode (Sec. 1.5):

$$e^{(-1+j10)t} z_1(0) + e^{(-1-j10)t} z_2(0) = K \cos(10t + L), \qquad K, L \text{ constants.}$$

This agrees with the analysis in subsection (ii).

CAD Facility

The interactive CAD algorithm in Fig. 3.6 generates the transfer function matrix and poles of a system given the state matrices **A**, **B**, **C**, **D**.

The operation of the algorithm is as follows.

BLOCK 1

The system parameters **A, B, C, D** are requested, input, displayed and corrected if necessary.

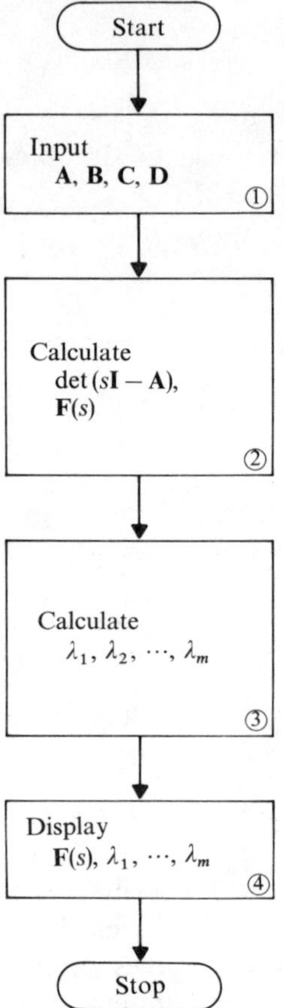

Fig. 3.6 Transfer function and pole determination CAD program.

BLOCK 2

The transfer function matrix (Eq. (3.14)) is:

$$\mathbf{F}(s) = \mathbf{C}(s\mathbf{I} - \mathbf{A})^{-1}\mathbf{B} + \mathbf{D}$$

$$= \frac{1}{\det(s\mathbf{I} - \mathbf{A})}\{\mathbf{C}\,\mathrm{adj}\,(s\mathbf{I} - \mathbf{A})\mathbf{B} + \mathbf{D}\det(s\mathbf{I} - \mathbf{A})\}. \qquad \text{(App. A.7)}$$

One reliable method of calculating this is the 'Faddeev' algorithm:*

Let

$$\mathbf{G}_1 = \mathbf{I}_m; \qquad \mathbf{F}_1 = \frac{-1}{1}\mathrm{tr}\,(\mathbf{A}\mathbf{G}_1)$$

*This algorithm is a modification by D.K. Faddeev of a method by Leverrier, given in D.K. Fadeev and V.N. Faddeeva, *Computational Methods of Linear Algebra* (Freeman and Co., 1963).

$$\mathbf{G}_2 = \mathbf{F}_1\mathbf{I}_m + \mathbf{A}\mathbf{G}_1; \qquad \mathbf{F}_2 = \frac{-1}{2}\operatorname{tr}(\mathbf{A}\mathbf{G}_2)$$

$$\mathbf{G}_3 = \mathbf{F}_2\mathbf{I}_m + \mathbf{A}\mathbf{G}_2; \qquad \mathbf{F}_3 = \frac{-1}{3}\operatorname{tr}(\mathbf{A}\mathbf{G}_3)$$

$$\vdots \qquad\qquad \vdots$$

$$\mathbf{G}_m = \mathbf{F}_{m-1}\mathbf{I}_m + \mathbf{A}\mathbf{G}_{m-1}; \qquad \mathbf{F}_m = \frac{-1}{m}\operatorname{tr}(\mathbf{A}\mathbf{G}_m)$$

where $\mathbf{I}_m$ is the $m \times m$ identity matrix, and $\operatorname{tr}(\mathbf{A}\mathbf{G}_i)$ is the sum of the m diagonal elements of $\mathbf{A}\mathbf{G}_i$.

Then:

$$\det(s\mathbf{I} - \mathbf{A}) = s^m + \mathbf{F}_1 s^{m-1} + \mathbf{F}_2 s^{m-2} + \cdots + \mathbf{F}_m$$

$$\mathbf{F}(s) = \frac{1}{\det(s\mathbf{I} - \mathbf{A})}\{s^m\mathbf{D} + s^{m-1}(\mathbf{C}\mathbf{G}_1\mathbf{B} + \mathbf{F}_1\mathbf{D}) + s^{m-2}(\mathbf{C}\mathbf{G}_2\mathbf{B} + \mathbf{F}_2\mathbf{D}) + \cdots + s^0(\mathbf{C}\mathbf{G}_m\mathbf{B} + \mathbf{F}_m\mathbf{D})\}.$$

BLOCK 3

The roots of the characteristic equation (Eq. (3.15)) are found using the algorithm of Fig. 1.3.

BLOCK 4

The transfer function matrix is displayed in the form of a set of matrix elements and the characteristic polynomial ($s\mathbf{I} - \mathbf{A}$), the roots of which are also displayed.

3.4 PROPERTIES OF LTI SYSTEMS

(i) Controllability

A system S is *controllable* if all the modes can be excited, or controlled, by the input. If the state equations are cast in diagonal form (Sec. 3.3(ii)) and if the ith row of $\mathbf{T}^{-1}\mathbf{B}$ is all zeros, then clearly the corresponding state z_i is not affected by the input $\mathbf{u}$; that state is uncontrollable from the input and the system is said to be *uncontrollable*.

In the calculation of the system transfer function (Eq (3.14)), this property manifests itself in the cancellation of factors of the characteristic polynomial $\det(s\mathbf{I} - \mathbf{A})$ when $(s\mathbf{I} - \mathbf{A})^{-1}$ is postmultiplied by $\mathbf{B}$.

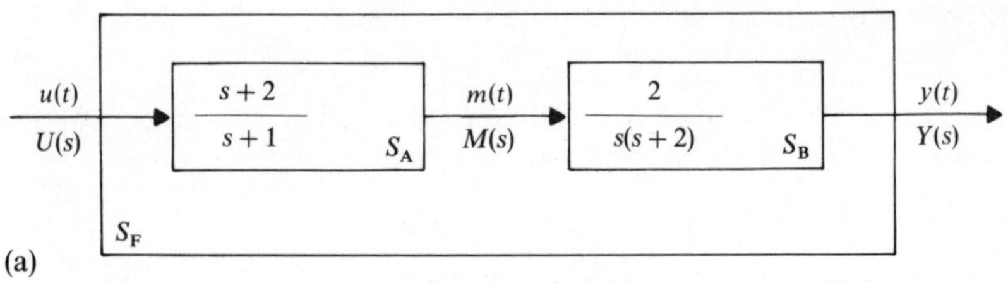

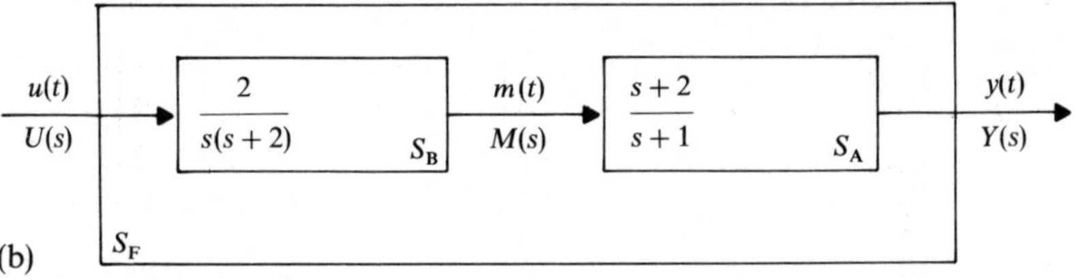

Fig. 3.7 Reducible systems (example). (a) Uncontrollable system. (b) Unobservable system.

In terms of the state equations (Eq. (3.5)), the equivalent condition is that S is controllable if and only if:

$$\mathbf{P} = (\mathbf{B} \quad \mathbf{AB} \quad \mathbf{A}^2\mathbf{B} \quad \cdots \quad \mathbf{A}^{m-1}\mathbf{B}) \tag{3.17}$$

has rank m (App. A.5).

EXAMPLE

Consider the system in Fig. 3.7(a).

The single-input, single-output system S_F comprises two subsystems S_A, S_B, which are governed by the differential equations:

$$\frac{dm}{dt} + m = \frac{du}{dt} + 2u$$

and

$$\frac{d^2y}{dt^2} + 2\frac{dy}{dt} = 2m.$$

Letting $x_1 = m - u$; $x_2 = y$; $x_3 = dy/dt$, S_F may be modeled:

$$\begin{bmatrix} \dot{x}_1 \\ \dot{x}_2 \\ \dot{x}_3 \end{bmatrix} = \begin{bmatrix} -1 & 0 & 0 \\ 0 & 0 & 1 \\ +2 & 0 & -2 \end{bmatrix} \begin{bmatrix} x_1 \\ x_2 \\ x_3 \end{bmatrix} + \begin{bmatrix} 1 \\ 0 \\ 2 \end{bmatrix} u$$

$$y = [0 \quad 1 \quad 0] \begin{bmatrix} x_1 \\ x_2 \\ x_3 \end{bmatrix} + 0u.$$

Casting this in the diagonal form (Sec. 3.3(ii)):

$$\dot{\mathbf{z}} = \begin{bmatrix} 0 & 0 & 0 \\ 0 & -1 & 0 \\ 0 & 0 & -2 \end{bmatrix} \mathbf{z} + \begin{bmatrix} 2 \\ 1 \\ 0 \end{bmatrix} u$$

$$y = [1 \quad -2 \quad -0.5]\mathbf{z}.$$

The third row of the input matrix, $\mathbf{T}^{-1}\mathbf{B}$, is zero and the 'decoupled' state, z_3, is therefore unaffected by the input u. The system S_F is thus uncontrollable.

More conveniently, the $\mathbf{P}$ matrix (Eq. (3.17)) may be calculated:

$$\mathbf{P} = \left[\begin{bmatrix} 1 \\ 0 \\ 2 \end{bmatrix} \quad \begin{bmatrix} -1 & 0 & 0 \\ 0 & 0 & 1 \\ 2 & 0 & -2 \end{bmatrix} \begin{bmatrix} 1 \\ 0 \\ 2 \end{bmatrix} \quad \begin{bmatrix} -1 & 0 & 0 \\ 0 & 0 & 1 \\ 2 & 0 & -2 \end{bmatrix}^2 \begin{bmatrix} 1 \\ 0 \\ 2 \end{bmatrix} \right]$$

$$= \begin{bmatrix} 1 & -1 & 1 \\ 0 & 2 & -2 \\ 2 & -2 & +2 \end{bmatrix}.$$

This has rank 2, (App. A.5), indicating that S_F is uncontrollable.

Heuristically, the zero at $s = -2$ prevents the input u affecting the mode described by the pole at $s = -2$.

It is left to the reader to confirm that postmultiplying $(s\mathbf{I} - \mathbf{A})^{-1}$ by $\mathbf{B}$ results in cancellation of the factor $(s + 2)$ in the characteristic polynomial, $\det(s\mathbf{I} - \mathbf{A})$, to give a 'reduced' transfer function, $\mathbf{F}(s)$, with poles $s = 0, -1$.

(ii) Observability

A system S is *observable* if all the modes affect the output. If the state equations are cast in diagonal form (Sec. 3.3(ii)) and if the ith column of $\mathbf{CT}$ is all zeros, then clearly the corresponding state z_i does not affect the output; that state is unobservable at the output, and the system is said to be *unobservable*.

In the calculation of the system transfer function (Eq. (3.14)), this manifests itself in the cancellation of factors of the characteristic polynomial $\det(s\mathbf{I} - \mathbf{A})$ when $(s\mathbf{I} - \mathbf{A})^{-1}$ is premultiplied by $\mathbf{C}$.

In terms of the state equations (Eq. (3.5)), the equivalent condition is that S is observable if and only if:

$$\mathbf{Q} = (\mathbf{C}^{\mathrm{T}} \quad (\mathbf{CA})^{\mathrm{T}} \quad (\mathbf{CA}^2)^{\mathrm{T}} \quad (\mathbf{CA}^3)^{\mathrm{T}} \quad \ldots \quad (\mathbf{CA}^{m-1})^{\mathrm{T}}) \tag{3.18}$$

has rank m (App. A.5).

EXAMPLE

Consider the system in Fig. 3.7(b).

The single-input, single-output system S_F comprises the two subsystems considered in subsection (i) but connected in reverse order.

Using a similar modeling procedure, state equations for S_F are:

$$\begin{bmatrix} \dot{x}_1 \\ \dot{x}_2 \\ \dot{x}_3 \end{bmatrix} = \begin{bmatrix} 0 & 1 & 0 \\ 0 & -2 & 0 \\ 1 & 0 & -1 \end{bmatrix} \begin{bmatrix} x_1 \\ x_2 \\ x_3 \end{bmatrix} + \begin{bmatrix} 0 \\ 2 \\ 0 \end{bmatrix} u$$

$$y = [1 \quad 0 \quad 1] \begin{bmatrix} x_1 \\ x_2 \\ x_3 \end{bmatrix} + 0u.$$

Casting this in diagonal form (Sec. 3.3(ii)):

$$\dot{\mathbf{z}} = \begin{bmatrix} 0 & 0 & 0 \\ 0 & -1 & 0 \\ 0 & 0 & -2 \end{bmatrix} \mathbf{z} + \begin{bmatrix} 1 \\ -2 \\ 2 \end{bmatrix} u$$

$$y = [1 \quad 1.5 \quad 0]\mathbf{z}.$$

The third column of the output matrix **CT** is zero, so the third 'decoupled' state, z_3, does not affect the output y. The system is therefore unobservable.

More conveniently, the **Q** matrix (Eq. (3.18)) may be calculated:

$$\mathbf{Q} \left[\begin{bmatrix} 1 \\ 0 \\ 1 \end{bmatrix} \begin{bmatrix} 0 & 0 & 1 \\ 1 & -2 & 1 \\ 0 & 0 & -1 \end{bmatrix} \begin{bmatrix} 1 \\ 0 \\ 1 \end{bmatrix} \begin{bmatrix} 0 & 0 & 1 \\ 1 & -2 & 0 \\ 0 & 0 & -1 \end{bmatrix}^2 \begin{bmatrix} 1 \\ 0 \\ 1 \end{bmatrix} \right]$$

$$= \begin{bmatrix} 1 & 1 & -1 \\ 0 & 1 & -1 \\ 1 & -1 & 1 \end{bmatrix}.$$

This has rank 2 (App. A.5), indicating that S_F is unobservable.

Heuristically, the zero at $s = -2$ filters out the mode corresponding to the pole at $s = -2$ before it appears at the output, y.

It is left to the reader to confirm that premultiplying $(s\mathbf{I} - \mathbf{A})^{-1}$ by **C** results in cancellation of the factor $(s+2)$ in the characteristic polynomial, $\det(s\mathbf{I} - \mathbf{A})$, to give a 'reduced' transfer function, $\mathbf{F}(s)$, with poles $s = 0, -1$.

(iii) Stability of LTI Systems

Since the state of a system S can be regarded as the sum of two components, the zero input and zero initial state responses (Eq. 3.10), two distinct types of stability can be identified:

1. S is *asymptotically* stable (a.s.) if and only if the zero input response tends to zero, as t tends to infinity; i.e. $\mathbf{x} \rightarrow \mathbf{0}, t \rightarrow \infty, \mathbf{u} = \mathbf{0}, \mathbf{x}(0) \neq 0$.

2. S is *bounded-input, bounded-output* (b.i.b.o.) stable if and only if the zero initial state response is bounded for every bounded input.
3. Equivalently, S is a.s. if and only if all the eigenvalues of **A** (Eq. (3.5)) have negative real parts. It is b.i.b.o. stable if and only if all the system poles (Eq. (3.14)) have negative real parts – a slightly weaker condition (Sec. 3.3(iii)). Consequently, if S is a.s., this implies b.i.b.o. stability, but the reverse is strictly true only if S is controllable and observable.

EXAMPLE

Consider the system:

$$\dot{\mathbf{x}} = \begin{bmatrix} 0 & 1 & 0 \\ 6 & -1 & 0 \\ -3 & 0 & -1 \end{bmatrix} \mathbf{x} + \begin{bmatrix} 0 \\ 2 \\ 0 \end{bmatrix} u$$

$$y = [1 \quad 0 \quad 1]\mathbf{x}.$$

The characteristic equation (Eq. (3.15)) is:

$$\det \begin{bmatrix} s & -1 & 0 \\ -6 & s+1 & 0 \\ 3 & 0 & s+1 \end{bmatrix} = 0$$

$$(s+1)(s-2)(s+3) = 0.$$

Thus the matrix **A** has eigenvalues at -1, $+2$, -3 and the system is not asymptotically stable. However, the transfer function (Eq. (3.14)) is:

$$\mathbf{F}(s) = [1 \quad 0 \quad 1] \begin{bmatrix} s & -1 & 0 \\ -6 & s+1 & 0 \\ 3 & 0 & s+1 \end{bmatrix}^{-1} \begin{bmatrix} 0 \\ 2 \\ 0 \end{bmatrix}$$

$$= \frac{2(s-2)}{(s+1)(s-2)(s+3)}$$

$$= \frac{2}{(s+1)(s+3)}.$$

Thus the system poles are at -1, -3 and the system is b.i.b.o. stable.

Since there are fewer poles than eigenvalues, the system is either uncontrollable or unobservable; a check reveals that the observability matrix (Eq. (3.18)) is:

$$\mathbf{Q} = \begin{bmatrix} 1 & -3 & 9 \\ 0 & 1 & -4 \\ 1 & -1 & 1 \end{bmatrix}.$$

This has rank 2 (App. A.5) and the system is therefore unobservable.

3.5 REALIZATION OF STATE EQUATIONS

Though it is easy to find the system transfer function from the state equations (Sec. 3.3(iii)), the reverse process is usually more difficult. The simple examples considered in Secs. 3.2, 3.3 have yielded to straightforward treatment, but often a more systematic approach is needed. A considerable variety of methods exists (ref. 3); some useful basic ones are described here.

(i) Single-Input, Single-Output Systems

Consider a single-input, single-output system with transfer function:

$$\frac{Y(s)}{U(s)} = F(s) = \frac{a_q s^q + a_{q-1} s^{q-1} + \cdots + a_0}{s^p + b_{p-1} s^{p-1} + \cdots + b_0}. \tag{3.19}$$

The requirement is to define matrices **A**, **B**, **C**, **D** for the system so that state and output equations (Eqs. (3.5), (3.6)) can be written down.

If $q > p$, $F(s)$ is 'improper', and no state space realization is possible.

If $q = p$, $F(s)$ is 'proper', and $D = F(\infty)$.

Define

$$\mathbf{F}_r(s) = \mathbf{F}(s) - \mathbf{F}(\infty). \tag{3.20}$$

One of the following procedures can now be applied to $\mathbf{F}_r(s)$, which is 'strictly proper'.

1. Without loss of generality, for a strictly proper $F(s)$, since a_i can take any value, including zero, let $q = p - 1$:

$$F(s) = \frac{a_{p-1} s^{p-1} + a_{p-2} s^{p-2} + \cdots + a_0}{s^p + b_{p-1} s^{p-1} + b_{p-2} s^{p-2} + \cdots + b_0}. \tag{3.21}$$

Then state equations can be realized by setting:

$$\begin{aligned}
x_1 &= y \\
x_2 &= \dot{x}_1 + b_{p-1} x_1 - a_{p-1} u \\
x_3 &= \dot{x}_2 + b_{p-2} x_1 - a_{p-2} u \\
&\vdots \\
x_p &= \dot{x}_{p-1} + b_1 x_1 - a_1 u \\
\dot{x}_p &+ b_0 x_1 - a_0 u = 0.
\end{aligned}$$

Rearranging this:

$$\dot{\mathbf{x}} = \begin{bmatrix} -b_{p-1} & 1 & 0 & \cdots & 0 \\ -b_{p-2} & 0 & 1 & \cdots & 0 \\ -b_{p-3} & 0 & 0 & 1 & 0 \\ \vdots & & & & \\ -b_0 & 0 & 0 & \cdots & 0 \end{bmatrix} \mathbf{x} + \begin{bmatrix} a_{p-1} \\ a_{p-2} \\ \cdot \\ \vdots \\ a_0 \end{bmatrix} u \tag{3.22}$$

$$y = [1 \quad 0 \quad \cdots \quad 0]\mathbf{x}. \tag{3.23}$$

There is no guarantee of the controllability or observability of this realization.

2. A 'companion' form which is controllable, and which therefore applies to controllable systems only, is:

$$\dot{\mathbf{x}} = \begin{bmatrix} 0 & 1 & 0 & \cdots & 0 \\ 0 & 0 & 1 & 0\cdots & 0 \\ \cdot & \cdot & & & \vdots \\ \vdots & \vdots & & & 1 \\ -b_0 & -b_1 & -b_2 & \cdots & -b_{p-1} \end{bmatrix} \mathbf{x} + \begin{bmatrix} 0 \\ 0 \\ \vdots \\ 0 \\ 1 \end{bmatrix} u \tag{3.24}$$

$$y = [\, a_0 \quad a_1 \quad a_2 \quad \cdots \quad a_{p-1}]\mathbf{x}.$$

3. A 'companion' form which is observable, and which therefore applies to observable systems only, is:

$$\dot{\mathbf{x}} = \begin{bmatrix} 0 & 0 & \cdots & -b_0 \\ 1 & 0 & \cdots & -b_1 \\ 0 & 1 & \cdots & -b_2 \\ \vdots & & & \vdots \\ 0 & 0 & ..1 & -b_{p-1} \end{bmatrix} \mathbf{x} + \begin{bmatrix} a_0 \\ a_1 \\ a_2 \\ \vdots \\ a_{p-1} \end{bmatrix} u \tag{3.25}$$

$$y = [0 \quad 0 \quad \cdots \quad 1] \quad \mathbf{x}.$$

EXAMPLES

(a) Consider a third-order system governed by the differential equation:

$$\frac{d^3y}{dt^2} + 2\frac{d^2y}{dt^2} + 3\frac{dy}{dt} + 4y = \frac{d^3u}{dt^3} + 4\frac{d^2u}{dt^2} + 5\frac{du}{dt} + 6u.$$

The transfer function of this is:

$$\frac{Y(s)}{U(s)} = F(s) = \frac{s^3 + 4s^2 + 5s + 6}{s^3 + 2s^2 + 3s + 4}.$$

Since $q = p$ (Eq. (3.19)), Eq. (3.20) is appropriate, and

$$\mathbf{D} = \mathbf{F}(s)|_{s=\infty}$$
$$= 1$$
$$\mathbf{F}_r(s) = \frac{2s^2 + 2s + 2}{s^3 + 2s^2 + 3s + 4}$$

From Eqs. (3.22), (3.23), one possible set of state equations is:

$$\dot{\mathbf{x}} = \begin{bmatrix} -2 & 1 & 0 \\ -3 & 0 & 1 \\ -4 & 0 & 0 \end{bmatrix} \mathbf{x} + \begin{bmatrix} 2 \\ 2 \\ 2 \end{bmatrix} u$$

$$y = [\, 1 \quad 0 \quad 0]\mathbf{x} + \; 1u.$$

Realizations which are guaranteed controllable and observable respectively (Eqs. (3.24), (3.25)) are:

$$\dot{\mathbf{x}} = \begin{bmatrix} 0 & 1 & 0 \\ 0 & 0 & 1 \\ -4 & -3 & -2 \end{bmatrix} \mathbf{x} + \begin{bmatrix} 0 \\ 0 \\ 1 \end{bmatrix} u$$

$$y = [2 \quad 2 \quad 2]\mathbf{x} + 1u$$

and:

$$\dot{\mathbf{x}} = \begin{bmatrix} 0 & 0 & -4 \\ 1 & 0 & -3 \\ 0 & 1 & -2 \end{bmatrix} \mathbf{x} + \begin{bmatrix} 2 \\ 2 \\ 2 \end{bmatrix} u$$

$$y = [0 \quad 0 \quad 1]\mathbf{x} + 1u.$$

(b) A reconsideration of the systems in Fig. 3.7 is useful here. The system in Fig. 3.7(a) is uncontrollable – the zero at $s = -2$ filters out any component in the input u which could excite the mode corresponding to the pole at $s = -2$.

A 'complete' transfer function of this system, which admittedly infringes the strict definition of system transfer function (Eq. (3.14)) by not canceling common factors, may be written:

$$F(s) = \frac{2(s+2)}{s(s+1)(s+2)}$$
$$= \frac{2s+2}{s^3 + 3s^2 + 2s.}$$

Since the system is known to be uncontrollable but observable, Eq. (3.25) may be used:

$$\dot{\mathbf{x}} = \begin{bmatrix} 0 & 0 & -0 \\ 1 & 0 & -2 \\ 0 & 1 & -3 \end{bmatrix} \mathbf{x} + \begin{bmatrix} 2 \\ 2 \\ 0 \end{bmatrix} u$$

$$y = [0 \quad 0 \quad 1]\mathbf{x}.$$

The model of Eq. (3.24), since it is controllable, would of course be incorrect for this system.

The system in Fig. 3.7(b) has the same transfer function, but is unobservable; applying similar logic, the controllable model of Eq. (3.24) gives the realization:

$$\dot{\mathbf{x}} = \begin{bmatrix} 0 & 1 & 0 \\ 0 & 0 & 1 \\ -0 & -2 & -3 \end{bmatrix} \mathbf{x} + \begin{bmatrix} 0 \\ 0 \\ 1 \end{bmatrix} u$$

$$y = [2 \quad 2 \quad 0]\mathbf{x}.$$

The model of Eq. (3.25), since it is observable, would be incorrect for this system.

(ii) Multivariable Systems (ref. 3)

(a) A General Formula

A controllable and observable realization due to Ho and Kalman* is as follows.

Consider an $r \times q$ proper transfer function matrix $\mathbf{F}(s)$. This may be expanded:

$$\mathbf{F}(s) = \mathbf{H}_0 + \mathbf{H}_1 s^{-1} + \mathbf{H}_2 s^{-2} + \cdots \tag{3.26}$$

where $\mathbf{H}_0$, $\mathbf{H}_1$, $\cdots$ are constant matrices, $\mathbf{H}_0 = \mathbf{F}(\infty) = \mathbf{D}$

From Eq. (3.14),

$$\begin{aligned}\mathbf{F}(s) &= \mathbf{C}(s\mathbf{I} - \mathbf{A})^{-1}\mathbf{B} + \mathbf{D}\\ &= \mathbf{D} + \mathbf{CB}s^{-1} + \mathbf{CAB}s^{-2} + \mathbf{CA}^2\mathbf{B}s^{-3} + \cdots.\end{aligned} \tag{3.27}$$

In Eqs. (3.26), (3.27), the coefficients of the same powers of s may be equated to give values of **A**, **B**, **C**, **D** by the following procedure:

1. Find $\mathbf{H}_0$, $\mathbf{H}_1, \cdots$ by expanding $\mathbf{F}(s)$ according to Eq. (3.26). Then $\mathbf{D} = \mathbf{H}_0 = \mathbf{F}(\infty)$.
2. If the lowest common denominator of all the elements in $\mathbf{F}(s)$ has degree l, form the $rl \times ql$ matrices:

$$\mathbf{T} = \begin{bmatrix} \mathbf{H}_1 & \mathbf{H}_2 & \mathbf{H}_3 \cdots & \mathbf{H}_l \\ \mathbf{H}_2 & \mathbf{H}_3 & \cdots & \mathbf{H}_{l+1} \\ \vdots & & & \\ \mathbf{H}_l & & & \mathbf{H}_{2l-1} \end{bmatrix}$$

$$\tilde{\mathbf{T}} = \begin{bmatrix} \mathbf{H}_2 & \mathbf{H}_3 & \mathbf{H}_4 & \cdots & \mathbf{H}_{l+1} \\ \vdots & & & & \vdots \\ \mathbf{H}_{l+1} & \mathbf{H}_{l+2} & & & \mathbf{H}_{2l} \end{bmatrix}.$$

3. If the rank of **T** is m (App. A.5), form the $rl \times rl$ matrix **K** and the $ql \times ql$ matrix **L** such that the $rl \times ql$ matrix:

$$\mathbf{KTL} = \begin{bmatrix} \mathbf{I}_m & 0 \\ 0 & 0 \end{bmatrix} = \mathbf{I}_{m,rl}^{\mathrm{T}} \mathbf{I}_{m,ql} \tag{3.28}$$

where $\mathbf{I}_{m,rl}$ is $(m \times rl)$ and $\mathbf{I}_{m,ql}$ is $(m \times ql)$, with ones on the principal diagonal and zeros elsewhere. **KTL** is found by the sequence of row and column operations indicated in the CAD description which follows.

Then:

$$\mathbf{A} = \mathbf{I}_{m,rl} \mathbf{K} \tilde{\mathbf{T}} \mathbf{L} \mathbf{I}_{m,ql}^{\mathrm{T}} \tag{3.29}$$

$$\mathbf{B} = \mathbf{I}_{m,rl} \mathbf{K} \mathbf{T} \mathbf{I}_{q,ql}^{\mathrm{T}} \tag{3.30}$$

*This algorithm in by B.L. Ho and R.E. Kalman, 'Effective construction of linear state variable models from input/output data', *Proc. Third Allerton Conference*, pp. 449–459, 1965.

$$\mathbf{C} = \mathbf{I}_{r,rl}\mathbf{T}\mathbf{L}\mathbf{I}^{\mathrm{T}}_{m,ql} \tag{3.31}$$

$$\mathbf{D} = \mathbf{H}_0 = \mathbf{F}(\infty). \tag{3.32}$$

$\mathbf{I}_{q,ql}$, $\mathbf{I}_{r,rl}$ are $(q \times ql)$ and $(r \times rl)$ matrices respectively constructed in the same way as $\mathbf{I}_{m,rl}$, $\mathbf{I}_{m,ql}$.

This realization is controllable and observable.

CAD Facility

A procedure of this complexity is impractical without CAD; an interactive algorithm is shown in the flow diagram of Fig. 3.8.

The operation of the algorithm is as follows.

BLOCK 1

The details of the transfer function matrix, including the numbers of rows and columns, the numerators and denominators of each element, the lowest

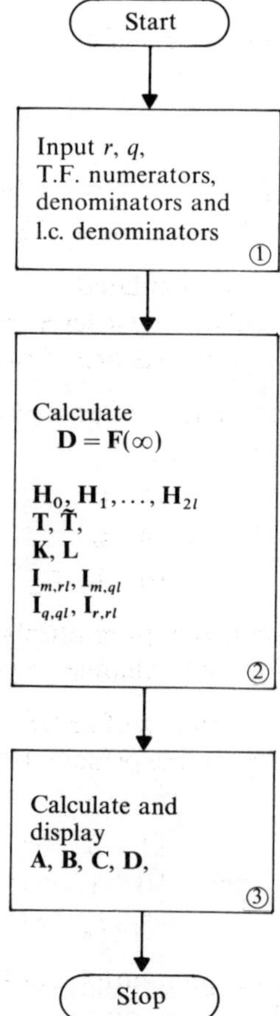

Fig. 3.8 CAD algorithm using Ho and Kalman's method.

common denominator of the elements in polynomial form, and the order of that polynomial, are input, displayed, and corrected if necessary.

BLOCK 2

The matrices $\mathbf{D} = \mathbf{F}(\infty)$ and $\mathbf{F}_r(s) = \mathbf{F}(s) - \mathbf{F}(\infty)$ are calculated.

The matrices $\mathbf{H}_1, \ldots, \mathbf{H}_l$ may now be calculated, where l is the order of the lowest common denominator polynomial, say $h(s)$, as follows:

$$h(s) = s^l + a_1 s^{l-1} + \cdots + a_l.$$

First, $\mathbf{F}_r(s)$ is expanded:

$$\mathbf{F}_r(s) = \frac{1}{h(s)}[\mathbf{R}_1 s^{l-1} + \mathbf{R}_2 s^{l-2} + \cdots + \mathbf{R}_l]$$

where $\mathbf{R}_1, \mathbf{R}_2, \ldots, \mathbf{R}_l$ are real matrices.

This can also be written:

$$\mathbf{F}_r(s) = \mathbf{H}_1 s^{-1} + \mathbf{H}_2 s^{-2} + \cdots.$$

Equating coefficients of equal powers of s:

$$\begin{aligned}
\mathbf{H}_1 &= \mathbf{R}_1 \\
\mathbf{H}_2 &= \mathbf{R}_2 - a_1 \mathbf{H}_1 \\
\mathbf{H}_l &= \mathbf{R}_l - a_1 \mathbf{H}_{l-1} - a_2 \mathbf{H}_{l-2} - \cdots - a_{l-1} \mathbf{H}_1 \\
\mathbf{H}_{l+i} &= -a_1 \mathbf{H}_{l+i-1} - a_2 \mathbf{H}_{l+i-2} - \cdots - a_l \mathbf{H}_i, \qquad i = 0, 1, 2, \ldots l.
\end{aligned} \tag{3.33}$$

The matrices $\mathbf{T}$, $\tilde{\mathbf{T}}$ are now calculated.

$\mathbf{K}$ and $\mathbf{L}$ can now be calculated by a series of 'elementary' operations on the matrix $\mathbf{T}$ (ref. 8). These operations are of three types.

Type I: Let $\mathbf{L}_i$ be the unit matrix of order i with two rows interchanged; e.g. a third-order $\mathbf{L}_i$ matrix is:

$$\mathbf{L}_3 = \begin{bmatrix} 1 & 0 & 0 \\ 0 & 0 & 1 \\ 0 & 1 & 0 \end{bmatrix}.$$

A Type I operation may be defined as premultiplying $\mathbf{T}$ by $\mathbf{L}_{rl}$ or postmultiplying $\mathbf{T}$ by $\mathbf{L}_{ql}$. These operations interchange rows and columns respectively.

Type II: Let $\mathbf{M}_i$ be the unit matrix of order i modified by the introduction of the element l_m in one position off the principal diagonal; e.g. a third-order $\mathbf{M}_i$ matrix is:

$$\mathbf{M}_3 = \begin{bmatrix} 1 & 0 & 0 \\ 0 & 1 & 0 \\ 0 & l_m & 1 \end{bmatrix}.$$

A Type II operation is defined as premultiplying $\mathbf{T}$ by $\mathbf{M}_{rl}$ or postmultiplying $\mathbf{T}$ by $\mathbf{M}_{ql}$. These operations add l_m times one row to another row or l_m times one column to another column respectively.

Type III: Let $\mathbf{N}_i$ be the unit matrix of order i with one diagonal element replaced by l_n; e.g. a third-order $\mathbf{N}_i$ matrix is:

$$\mathbf{N}_3 = \begin{bmatrix} 1 & 0 & 0 \\ 0 & l_n & 0 \\ 0 & 0 & 1 \end{bmatrix}.$$

A Type III operation is defined as premultiplying $\mathbf{T}$ by $\mathbf{N}_{rl}$ or postmultiplying $\mathbf{T}$ by $\mathbf{N}_{ql}$. These operations multiply rows and columns respectively by l_n.

Consider the matrix $\mathbf{T}$. If the top place in the principal diagonal is zero, a nonzero element can be brought to this position by a Type I operation. The element is then reduced to unity by a Type III operation. All the remaining elements in the first row and column can now be reduced to zero by Type II operations. In a similar manner, the second element in the principal diagonal can be reduced to unity, and the remaining elements in the second row and column to zero. This procedure is repeated until the matrix is reduced to the form of Eq. (3.28), and the K, L matrices calculated.

Matrices $\mathbf{I}_{q,ql}$ and $\mathbf{I}_{r,rl}$ are also formulated.

BLOCK 3

The matrices $\mathbf{A}, \mathbf{B}, \mathbf{C}, \mathbf{D}$ are found using Eqs. (3.29), (3.30), (3.31), (3.32) and displayed.

EXAMPLE

Consider the transfer function matrix:

$$\mathbf{F}(s) = \begin{bmatrix} \dfrac{s+1}{s+2} \\ \dfrac{s+3}{(s+2)(s+4)} \end{bmatrix}.$$

The lowest common denominator is:

$$h(s) = (s+2)(s+4) = s^2 + 6s + 8$$

$$\mathbf{D} = \mathbf{F}(\infty) = \begin{bmatrix} 1 \\ 0 \end{bmatrix}$$

$$\mathbf{F}_r(s) = \frac{1}{h(s)}\left[\begin{bmatrix} -1 \\ 1 \end{bmatrix} s + \begin{bmatrix} -4 \\ 3 \end{bmatrix}\right].$$

Also:

$$q = 1, \quad r = 2, \quad l = 2,$$
$$a_1 = 6, \quad a_2 = 8.$$

From Eq. (3.32),

$$\mathbf{H}_0 = \begin{bmatrix} -1 \\ 0 \end{bmatrix}, \quad \mathbf{H}_1 = \begin{bmatrix} -1 \\ 1 \end{bmatrix}, \quad \mathbf{H}_2 = \begin{bmatrix} -4 \\ 3 \end{bmatrix}, \quad \mathbf{H}_3 = \begin{bmatrix} -4 \\ 10 \end{bmatrix}, \quad \mathbf{H}_4 = \begin{bmatrix} 8 \\ 36 \end{bmatrix}$$

$$\mathbf{T}=\begin{bmatrix}1 & -2\\ 1 & -3\\ -2 & -4\\ -3 & 10\end{bmatrix},\quad \tilde{\mathbf{T}}=\begin{bmatrix}-2 & -4\\ -3 & 10\\ -4 & -8\\ 10 & 36\end{bmatrix}.$$

T has rank 2 (App. A.5), so $m=2$.

Applying two Type II operations to **T**,

$$\mathbf{K}=\begin{bmatrix}1 & 0 & 0 & 0\\ 0 & 1 & 0 & 0\\ 0 & 0 & 1 & 0\\ 0 & 0 & 1 & 1\end{bmatrix}\begin{bmatrix}-1 & 0 & 0 & 0\\ 0 & -1 & 0 & 0\\ 1 & 0 & 1 & 0\\ 1 & 0 & 0 & 1\end{bmatrix}$$

$$=\begin{bmatrix}-1 & 0 & 0 & 0\\ 0 & -1 & 0 & 0\\ 2 & 0 & 1 & 0\\ 2 & 0 & 1 & 0\end{bmatrix}$$

$$\mathbf{L}=\begin{bmatrix}3 & 2\\ 1 & 1\end{bmatrix}.$$

From Eqs. (3.29), (3.30), (3.31),

$$\mathbf{A}=\begin{bmatrix}-2 & 0\\ -1 & -4\end{bmatrix},\quad \mathbf{B}=\begin{bmatrix}1\\ -1\end{bmatrix},\quad \mathbf{C}=\begin{bmatrix}-1 & 0\\ 0 & -1\end{bmatrix}$$

$$\mathbf{D}=\begin{bmatrix}1\\ 0\end{bmatrix}.$$

Substitution of these in Eq. (3.14) confirms that this system has the transfer function matrix required.

(b) A Non-general Formula

A controllable and observable realization for the less general, but very common case where $\mathbf{F}(s)$ has m poles with none repeated, can be obtained by the following relatively simple procedure.

1. $\mathbf{D}=\mathbf{F}(\infty)$
2. Find $\mathbf{F}_r(s)=\mathbf{F}(s)-\mathbf{F}(\infty)$ and apply the remaining steps to $\mathbf{F}_r(s)$.
3. Using residues, expand $\mathbf{F}_r(s)$ in the form:

$$\mathbf{F}_r(s)=\mathbf{M}_1\frac{1}{s-\lambda_1}+\mathbf{M}_2\frac{1}{s-\lambda_2}+\cdots+\mathbf{M}_m\frac{1}{s-\lambda_m}$$

   where the poles of $\mathbf{F}_r(s)$, which are known, are $\lambda_1, \lambda_2, \ldots, \lambda_m$.
4. Decompose the matrices $\mathbf{M}_i$ into

$$\mathbf{M}_i=\mathbf{L}_i\mathbf{N}_i$$

   where $\mathbf{L}_i$ is an (r) column vector
   $\mathbf{N}_i$ is a (q) row vector.

This can always be done, since $\mathbf{M}_i$ can be shown to have unity rank.

5. Then:

$$\mathbf{A} = \begin{bmatrix} \lambda_1 & 0 & \cdots & 0 \\ 0 & \lambda_2 & & \\ \vdots & & & \\ 0 & 0 & & \lambda_m \end{bmatrix}$$

$$\mathbf{B} = \begin{bmatrix} \mathbf{N}_1 \\ \mathbf{N}_2 \\ \vdots \\ \mathbf{N}_q \end{bmatrix}$$

$$\mathbf{C} = (\mathbf{L}_1, \mathbf{L}_2, \ldots, \mathbf{L}_r)$$

$$\mathbf{D} = \mathbf{F}(\infty).$$

This realization is controllable and observable.

CAD Facility

As in the previous case, this procedure requires a CAD facility and an interactive algorithm is shown in the flow diagram of Fig. 3.9.

The operation of the algorithm is as follows.

BLOCK 1

The details of the transfer function matrix, including the numbers of rows and columns and the numerators and denominators (in factorized form) of each element are input, displayed, and corrected if necessary.

BLOCK 2

The matrices $\mathbf{F}(\infty)$ and $\mathbf{F}_r(s)$ are calculated.

The matrices $\mathbf{M}_1, \mathbf{M}_2, \ldots, \mathbf{M}_m$ are calculated using $\lambda_1, \lambda_2, \ldots, \lambda_m$ and the residue formula:

$$\mathbf{M}_i = \lim_{s \to \lambda_i} [(s - \lambda_i)\mathbf{F}(s)].$$

BLOCK 3

The vectors

$$\mathbf{L}_i = \begin{bmatrix} l_{i1} \\ l_{i2} \\ \vdots \\ l_{ir} \end{bmatrix}, \quad \mathbf{N}_i = (n_{i1}, n_{i2}, \ldots, n_{iq})$$

are calculated from $\mathbf{L}_i\mathbf{N}_i = \mathbf{M}_i$, which is of unity rank.

BLOCK 4

The matrices $\mathbf{A}, \mathbf{B}, \mathbf{C}, \mathbf{D}$ are found and displayed.

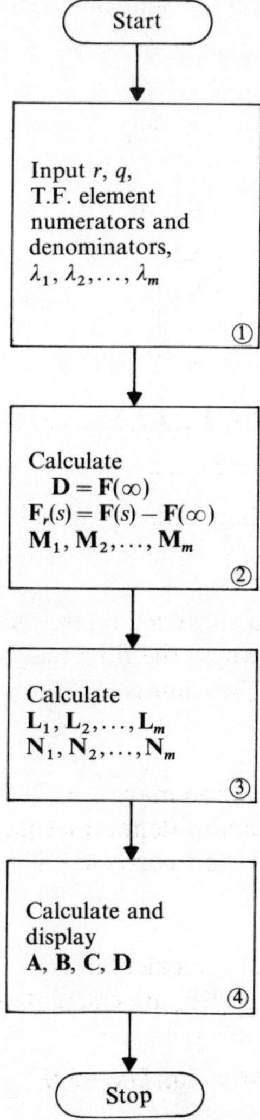

Fig. 3.9 System realization CAD algorithm for system with distinct poles.

EXAMPLE

Consider the transfer function matrix:

$$\mathbf{F}(s) = \begin{bmatrix} \dfrac{s+1}{s+2} \\ \dfrac{s+3}{(s+4)(s+5)} \end{bmatrix}.$$

The poles in this case are at $\lambda_1 = -2$, $\lambda_2 = -4$, $\lambda_3 = -5$.

$$\mathbf{D} = \mathbf{F}(\infty) = \begin{bmatrix} 1 \\ 0 \end{bmatrix}$$

$$\mathbf{F}_r(s) = \begin{bmatrix} \dfrac{-1}{s+2} \\ \dfrac{s+3}{(s+4)(s+5)} \end{bmatrix}$$

$$= \left[\begin{bmatrix} -1 \\ 0 \end{bmatrix} \frac{1}{(s+2)} + \begin{bmatrix} 0 \\ -1 \end{bmatrix} \frac{1}{(s+4)} + \begin{bmatrix} 0 \\ 2 \end{bmatrix} \frac{1}{(s+5)} \right].$$

Hence:

$$\mathbf{L}_1 = \begin{bmatrix} -1 \\ 0 \end{bmatrix}, \quad N_1 = 1$$

$$\mathbf{L}_2 = \begin{bmatrix} 0 \\ -1 \end{bmatrix}, \quad N_2 = 1$$

$$\mathbf{L}_3 = \begin{bmatrix} 0 \\ 2 \end{bmatrix}, \quad N_3 = 1.$$

Also:

$$\mathbf{A} = \begin{bmatrix} -2 & 0 & 0 \\ 0 & -4 & 0 \\ 0 & 0 & -5 \end{bmatrix}$$

$$\mathbf{B} = \begin{bmatrix} 1 \\ 1 \\ 1 \end{bmatrix}$$

$$\mathbf{C} = \begin{bmatrix} -1 & 0 & 0 \\ 0 & -1 & 2 \end{bmatrix}$$

$$\mathbf{D} = \begin{bmatrix} 1 \\ 0 \end{bmatrix}.$$

Substitution of these in Eq. (3.14) confirms that this system has the transfer function matrix required.

3.6 POLE SHIFTING BY STATE FEEDBACK

The eigenvalues (and so the poles) of a controllable LTI system, S, can be 'shifted' by means of state feedback as indicated in Fig. 3.10.

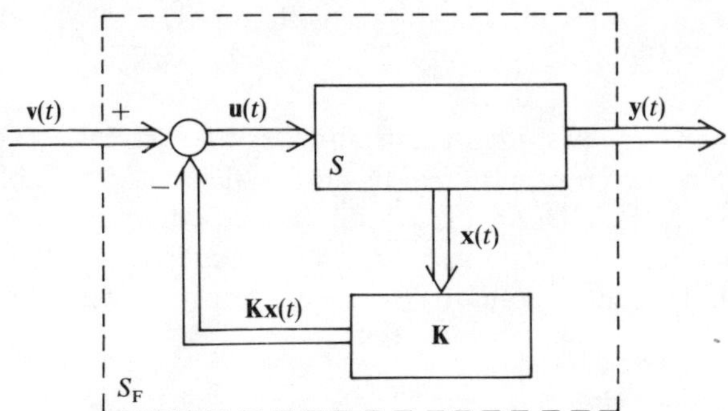

Fig. 3.10 LTI system with state feedback.

The open-loop system S has state equations (Eq. (3.5)):

$$\dot{\mathbf{x}} = \mathbf{Ax} + \mathbf{Bu}$$

and its natural modes are described by the eigenvalues of $\mathbf{A}$ (Sec. 3.3).

The closed-loop system S_F has state equations:

$$\begin{aligned}\dot{\mathbf{x}} &= \mathbf{Ax} + \mathbf{B}(\mathbf{v} - \mathbf{Kx}) \\ &= (\mathbf{A} - \mathbf{BK})\mathbf{x} + \mathbf{Bv}\end{aligned} \tag{3.34}$$

where $\mathbf{v}$ is the input vector $(v_1, v_2, \ldots v_q)^T$, v_i a real variable, $\mathbf{K}$ is a constant $(q \times m)$ state feedback matrix.

The natural modes of S_F are described by the eigenvalues of $(\mathbf{A} - \mathbf{BK})$, and, provided S is controllable, $\mathbf{K}$ can be chosen so that, theoretically at least, these eigenvalues have any desired values. In practice, $\mathbf{K}$ is chosen so that the dominant poles of S_F (i.e. the dominant eigenvalues of $(\mathbf{A} - \mathbf{BK})$) lie on good lines of constant damping ratio as far from the s plane origin as practical constraints allow (Sec. 1.5).

If S is uncontrollable, only eigenvalues which are controllable can be shifted in this manner, and the closed-loop system S_F is also uncontrollable. In this case it is usually possible to manipulate the state equations into two real subsystems, one of which is controllable, using a similarity transformation.

$$\bar{\mathbf{x}} = \mathbf{T}^{-1}\mathbf{x}$$

$\mathbf{T}$ can be found such that:

$$\dot{\bar{\mathbf{x}}} = \begin{bmatrix} \dot{\mathbf{x}}_c \\ \dot{\mathbf{x}}_{\bar{c}} \end{bmatrix} = \begin{bmatrix} \mathbf{A}_c & \mathbf{A}_{c\bar{c}} \\ 0 & \mathbf{A}_{\bar{c}} \end{bmatrix} \begin{bmatrix} \mathbf{x}_c \\ \mathbf{x}_{\bar{c}} \end{bmatrix} + \begin{bmatrix} \mathbf{B}_c \\ 0 \end{bmatrix} \mathbf{u} \tag{3.35}$$

$$\mathbf{y} = [\mathbf{C}_c \quad \mathbf{C}_{\bar{c}}] \begin{bmatrix} \mathbf{x}_c \\ \mathbf{x}_{\bar{c}} \end{bmatrix} = \mathbf{Du}.$$

The subsystem $\dot{\mathbf{x}}_{\bar{c}} = \mathbf{A}_{\bar{c}}\mathbf{x}_{\bar{c}}$ is unaffected by $\mathbf{u}$ or $\mathbf{x}_c$, and is thus uncontrollable.

The remaining controllable system with 'input' $\mathbf{x}_{\bar{c}}$ dropped is:

$$\dot{\mathbf{x}}_c = \mathbf{A}_c\mathbf{x}_c + \mathbf{B}_c\mathbf{u}.$$

The poles of this may be shifted using state feedback as described above, and the whole system then reconstructed by adding the uncontrollable subsystem. A matrix $\mathbf{T}$ with the desired properties can be constructed by taking the independent columns of the controllability matrix $\mathbf{P}$ (Eq. (3.17)) and adding to them further independent columns to make $\mathbf{T}$ nonsingular and square (ref. 3).

(i) Single-input Systems

Where S is single-input, $\mathbf{K}$ is a row matrix which is unique for a set of specified eigenvalues of S_F. $\mathbf{K}$ can be found easily, as demonstrated in the following example. A noteworthy property of these systems is that the zeros of S_F are the same as those of S.

EXAMPLE

Consider the example of Sec. 3.2 (Fig. 3.2) which concerns an amplifier, servomotor, gearbox, and potentiometer and has state equations:

$$\dot{\mathbf{x}} = \begin{bmatrix} 0 & 10 & 0 \\ 0 & 0 & 10 \\ 0 & -6 & -32 \end{bmatrix} \mathbf{x} + \begin{bmatrix} 0 \\ 0 \\ 450 \end{bmatrix} u$$

$$y = [1 \quad 0 \quad 0]\mathbf{x}$$

where

$$x_1 = y, \quad x_2 = 0.1\dot{y}, \quad x_3 = 0.001\ddot{y}.$$

Solving the characteristic equation (Eq. (3.15)) for this reveals that the eigenvalues of the matrix $\mathbf{A}$, and the poles of the system, are at:

$$s = 0, \quad s = -2, \quad s = -30.$$

The dominant poles, $s = 0, s = -2$, are undesirable – indeed the system is unstable – and to shift the poles to, say, $-7 \pm j7$, ($\zeta = 0.707$, Fig. 2.11), -100, a feedback matrix $\mathbf{K} = (k_1 \quad k_2 \quad k_3)$ can be designed and applied as shown in Fig. 3.10.

The state equation of S_F (Eq. (3.34)) is:

$$\dot{\mathbf{x}} = \begin{bmatrix} 0 & 10 & 0 \\ 0 & 0 & 10 \\ -450k_1 & -6-450k_2 & -32-450k_3 \end{bmatrix} \mathbf{x} + \begin{bmatrix} 0 \\ 0 \\ 450 \end{bmatrix} v.$$

The characteristic equation of this (Eq. (3.15)) is:

$$\det \begin{bmatrix} s & -10 & 0 \\ 0 & s & -10 \\ 450k_1 & 6+450k_2 & s+32+450k_3 \end{bmatrix} = 0$$

or:

$$s^3 + s^2(32 + 450k_3) + s(60 + 4500k_2) + 45{,}000k_1 = 0.$$

For the poles (and eigenvalues) to be at $-7 \pm j7$, -100, this must be equal to:

$$(s + 7 + j7)(s + 7 - j7)(s + 100) = 0$$

or:

$$s^3 + 114s^2 + 1498s + 9800 = 0.$$

Equating coefficients of equal powers of s,

$$\begin{aligned}\mathbf{K} &= (k_1 \quad k_2 \quad k_3)\\ &= (0.218 \quad 0.319 \quad 0.182).\end{aligned}$$

The state equations of S_F (Eq. (3.34)) are:

$$\dot{\mathbf{x}} = \begin{bmatrix} 0 & 10 & 0 \\ 0 & 0 & 10 \\ -98.01 & -149.82 & -114 \end{bmatrix} \mathbf{x} + \begin{bmatrix} 0 \\ 0 \\ 450 \end{bmatrix} v$$

$$y = [1 \quad 0 \quad 0]\mathbf{x}.$$

The closed-loop transfer function (Eq. (3.14)) is:

$$\mathbf{F}(s) = [1 \quad 0 \quad 0] \begin{bmatrix} s & -10 & 0 \\ 0 & s & 10 \\ -98.01 & -149.82 & -114 \end{bmatrix}^{-1} \begin{bmatrix} 0 \\ 0 \\ 450 \end{bmatrix} + 0$$

$$= \frac{45{,}000}{s^3 + 114s^2 + 1498s + 9800}.$$

The unit input step function response of this closed-loop system is in Fig. 3.12(a).

While **K** may (in theory) be designed to shift the poles of S to any specified locations, this is not possible in practice. If the new locations are chosen too ambitiously, the elements of **K** are found to be too large or small (or both), or S_F to be badly conditioned in the sense that small parameter variations correspond to gross changes in pole positions. Furthermore, the steady-state gain of S_F, which depends on **K**, can be unrealistically large or small if **K** is not well chosen.

The selection of acceptable 'target' pole positions is thus a matter of experimenting until the properties of S_F are acceptable. A CAD facility is essential for this work.

CAD Facility

The interactive CAD algorithm in Fig. 3.11 uses the basic idea outlined in the example above, after converting S to controllable companion form (Eq. (3.24)). The feedback matrix relevant to that form is found, and the

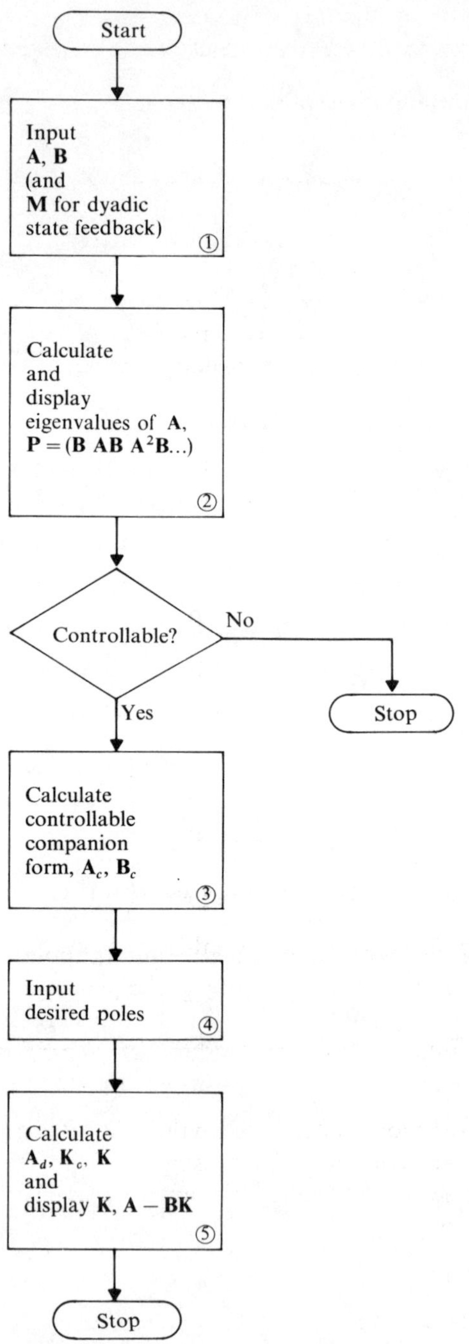

Fig. 3.11 CAD pole-shifting algorithm

algorithm finally converts this to the state feedback matrix **K**. This procedure allows easy extension to the multi-input case (Sec. 3.6(ii)).

The operation of the algorithm is as follows.

BLOCK 1

The open-loop system matrices **A**, **B** are requested, input, displayed, and corrected if necessary.

BLOCK 2

The controllability matrix **P** (Eq. (3.17)) is found and its rank, $\bar{r}$, determined (App. A.5). If $\bar{r} < m$, the system is uncontrollable and the program is halted. **P** is stored. The eigenvalues of **A** are calculated and displayed (App. A.9).

BLOCK 3

The matrices **A, B** are converted to controllable companion form (Eq. (3.24)) using the transformation $\mathbf{x} = \mathbf{Tz}$ to give the state equation:

$$\dot{\mathbf{z}} = \mathbf{A}_c\mathbf{z} + \mathbf{B}_c\mathbf{u}$$

where:

$$\mathbf{A}_c = \mathbf{T}^{-1}\mathbf{A}\mathbf{T}$$

$$\mathbf{B}_c = \mathbf{T}^{-1}\mathbf{B} = (0 \quad 0 \quad \cdots \quad 1)^{\mathrm{T}}.$$

T is an $(m \times m)$ matrix which can be found by several methods, one of which is (ref. 3):

Let $\mathbf{p}_m^{\mathrm{T}}$ be the mth (last) row of $\mathbf{P}^{-1}$.

Then

$$\mathbf{T}^{-1} = \begin{bmatrix} \mathbf{p}_m^{\mathrm{T}} \\ \mathbf{p}_m^{\mathrm{T}}\mathbf{A} \\ \vdots \\ \mathbf{p}_m^{\mathrm{T}}\mathbf{A}^{m-1} \end{bmatrix}.$$

In the controllable form the closed-loop state equations are:

$$\dot{\mathbf{z}} = (\mathbf{A}_c - \mathbf{B}_c\mathbf{K}_c)\mathbf{z} + \mathbf{B}_c\mathbf{u}$$

where $\mathbf{K}_c = (k_{c1} \quad k_{c2} \quad \cdots \quad k_{cm})$ is the (unknown) state feedback $(1 \times m)$ matrix.

BLOCK 4

The desired closed-loop system eigenvalues $\mathrm{P}_1, \mathrm{P}_2, \ldots, \mathrm{P}_m$ are requested, input, displayed, and corrected if necessary.

The desired system characteristic equation coefficients are calculated from the expression:

$$(s - \mathrm{P}_1)(s - \mathrm{P}_2)\cdots(s - \mathrm{P}_m) = s^m + a_{d(m-1)}s^{m-1} + a_{d(m-2)}s^{m-2} + \cdots + a_{d0}.$$

This corresponds to the controllable companion form state equations (Eq. (3.24)):

$$\dot{\mathbf{z}} = \mathbf{A}_d\mathbf{z} + \mathbf{B}_d\mathbf{u}$$

where the mth (last) row of $\mathbf{A}_d$ is:

$$-a_{d0} \quad -a_{d1} \quad \cdots \quad -a_{d(m-1)}.$$

BLOCK 5

By setting $(\mathbf{A}_c - \mathbf{B}_c\mathbf{K}_c) = \mathbf{A}_d$, $\mathbf{B}_c\mathbf{K}_c$ is calculated:

$$\mathbf{B}_c\mathbf{K}_c = (\mathbf{A}_c - \mathbf{A}_d)$$

$\mathbf{B}_c\mathbf{K}_c$ is an $(m \times m)$ matrix of which the mth (last) row is $\mathbf{K}_c$, and all the other rows zero. $\mathbf{K}_c$ is therefore found by this means.

Since:

$$\mathbf{Kx} = \mathbf{KTz} = \mathbf{K}_c\mathbf{z} = \mathbf{K}_c\mathbf{T}^{-1}\mathbf{x}$$
$$\mathbf{K} = \mathbf{K}_c\mathbf{T}^{-1}.$$

$\mathbf{K}$ is calculated and displayed together with the closed-loop system matrix $(\mathbf{A} - \mathbf{BK})$ (Eq. (3.34)).

EXAMPLE

The example already considered in this section may be developed further using this algorithm.

Shifting the poles to:

$$\begin{aligned}
&-10 \pm j10, -30 \text{ requires } \mathbf{K} = (0.133 \quad 0.164 \quad 0.040)\\
&-15 \pm j15, -30 \text{ requires } \mathbf{K} = (0.300 \quad 0.287 \quad 0.062)\\
&-20 \pm j20, -30 \text{ requires } \mathbf{K} = (0.533 \quad 0.431 \quad 0.084).
\end{aligned}$$

The transfer functions of the closed-loop systems can be found using the CAD facility of Sec. 3.2(iii), while the real-time closed-loop behavior can be checked using the facility of Sec. 3.2(i). Closed-loop unit step function responses for these pole positions are in Fig. 3.12.

Comment

The different steady-state gains (Fig. 3.12) of the various closed loops are noteworthy. Any worthwhile design strategy must result in some defined steady-state gain, and this can usually be achieved quite easily by amplification (or attenuation) external to the closed loop. With this proviso, the dynamic system responses in Fig. 3.12 show a distinct improvement on those achieved by classical means for the same open-loop system (Figs. 2.21, 2.26). This example demonstrates the power of the pole-shifting method.

(ii) Multi-input Systems – Dyadic Feedback

If S (Fig. 3.10) is multi-input, $\mathbf{K}$ is not unique for a set of specified poles of S_F, and in general the zeros of S_F differ from those of S. $\mathbf{K}$ may be selected to position some zeros as well as the poles (ref. 4), at the cost of considerably

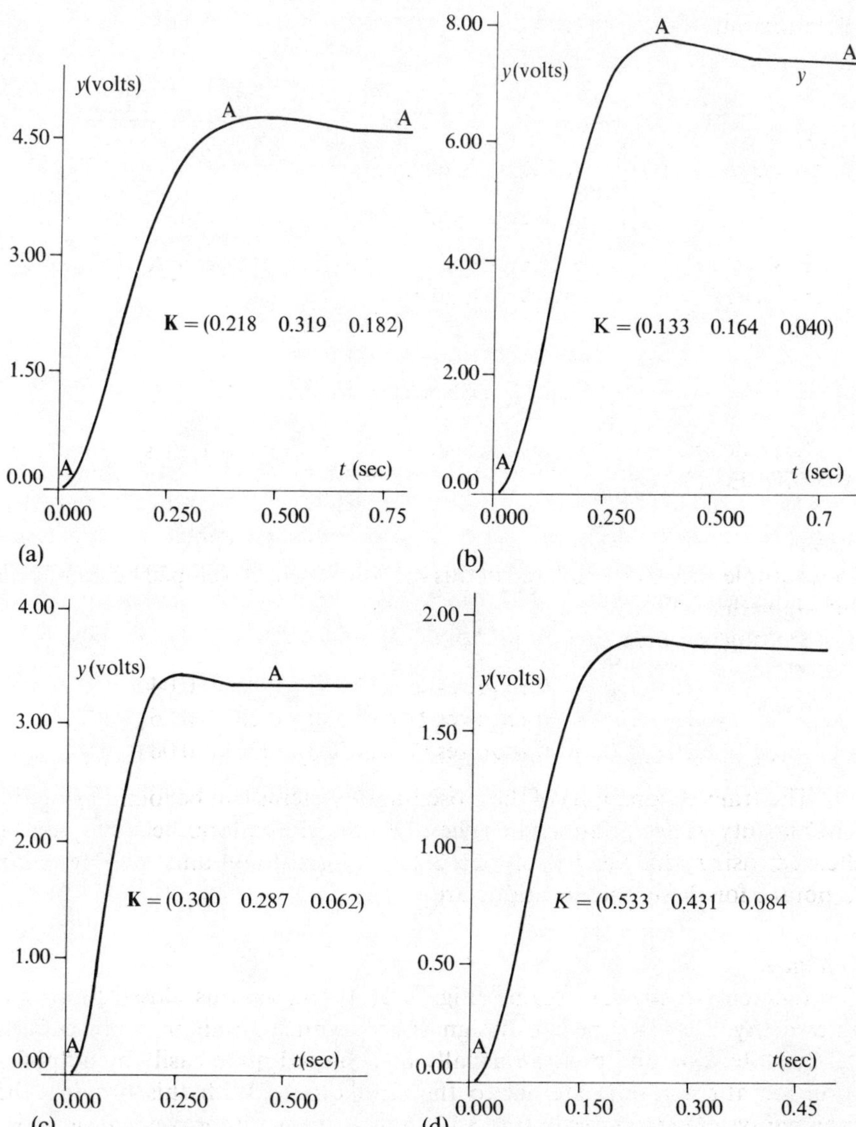

Fig. 3.12 Closed-loop system behavior (examples). (a) Poles at $-7 \pm j7$, -100. (b) Poles at $-10 \pm j10$, -30. (c) Poles at $-15 \pm j15$, -30. (d) Poles at $-20 \pm j20$, -30.

increased computational complexity. A simpler feedback structure which results in an unique **K** is 'dyadic' feedback, illustrated in Fig. 3.13, where **K** is a $(1 \times m)$ matrix, **M** is a $(q \times 1)$ matrix. **MK** is a 'dyadic' pair.

The state equation for S_F is:

$$\dot{\mathbf{x}} = (\mathbf{A} - \mathbf{BMK})\mathbf{x} + \mathbf{Bv}. \tag{3.36}$$

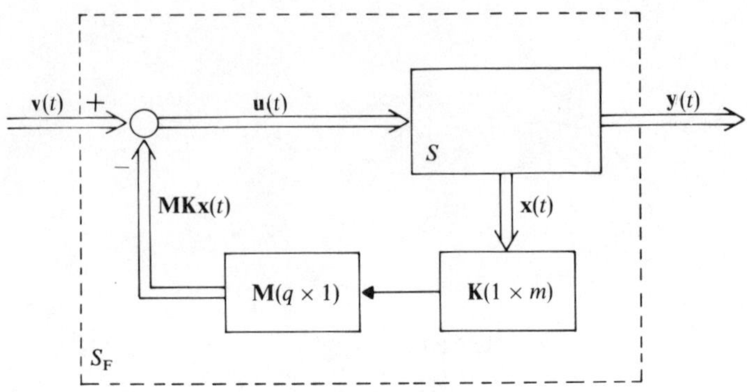

Fig. 3.13 Dyadic state feedback.

The poles of S_F are specified by the eigenvalues of $(\mathbf{A} - \mathbf{BMK})$. Since $(\mathbf{BM})$ is $(m \times 1)$, Eq. (3.36) has the same structure as Eq. (3.34) and provided $\mathbf{M}$ is specified, $\mathbf{K}$ may be designed using the procedures described in Sec. 3.6(i).

The interactive algorithm in Fig. 3.11 applies to dyadic feedback if Block 1 includes input of the matrix $\mathbf{M}$, Block 2 is altered to find the rank on a non-square controllability matrix $\mathbf{P}$ (App. A.5), and if $\mathbf{BM}$, $(\overline{\mathbf{BM}})_c$, replace $\mathbf{B}, \mathbf{B}_c$, respectively in the subsequent calculations.

EXAMPLE

The vertical plane dynamics of a deeply submerged submarine traveling at a forward velocity of $5\,\mathrm{m\,s^{-1}}$ are modeled by the (simplified) equations:

$$\begin{bmatrix} \dot{w} \\ \dot{q} \\ \dot{\theta} \\ \dot{h} \end{bmatrix} = \begin{bmatrix} -0.083 & 0.00115 & 0 & 0 \\ 1 & -0.400 & -0.008 & 0 \\ 0 & 1 & 0 & 0 \\ 1 & 0 & -0.005 & 0 \end{bmatrix} \begin{bmatrix} w \\ q \\ \theta \\ h \end{bmatrix} + \begin{bmatrix} -0.084 & -0.163 \\ 86.000 & -24.000 \\ 0 & 0 \\ 0 & 0 \end{bmatrix} \begin{bmatrix} \delta_b \\ \delta_s \end{bmatrix}$$

where w is the velocity downwards at right angles to the main body axis $(\mathrm{m\,s^{-1}})$,
q is the angular velocity of pitch with respect to the surface (millirad $\mathrm{s^{-1}}$),
θ is the pitch w.r.t the surface (mrad),
h is the depth below the surface (m),
δ_b is the bow control surface angle w.r.t. the body (rad),
δ_s is the stern control surface angle w.r.t the body (rad).

The system is illustrated in Fig. 3.14(a).

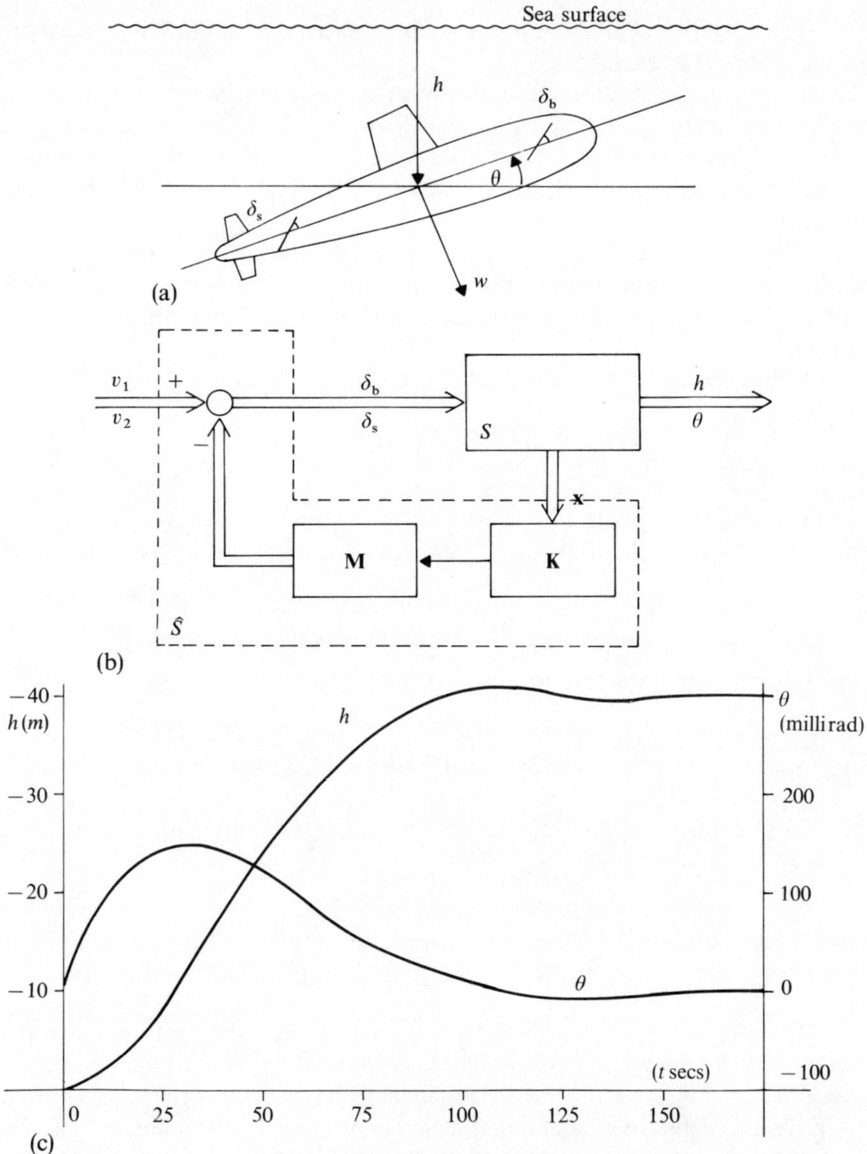

Fig. 3.14 Submarine autopilot design (example). (a) Submarine. (b) Submarine S and autopole Ŝ. (c) System response to step inputs.

The algorithm of Fig. 3.6 reveals the open-loop poles of the submarine (system S, Fig. 3.14(b)) to be:

$$s = 0$$
$$s = -0.383$$
$$s = -0.078$$
$$s = -0.022.$$

The positions of all these poles can be improved considerably; indeed one of them ($s=0$) is unstable.

Traditionally, a submarine autopilot drives the bow and stern control surfaces in symmetrical opposition, i.e. $\delta_b = -\delta_s$. This can be achieved quite easily by specifying the autopilot input, $\mathbf{v} = (h_d \quad -h_d)^T$, where h_d is a suitably scaled demanded depth change, and by setting $\mathbf{M} = (1 \quad -1)^T$ (Fig. 3.13, 3.14(b)).

While the poles can theoretically be shifted to any positions, too ambitious a selection simply results in unrealistic values of $|\theta|$ (over 150 millirad, say), $|\delta_b|$, and $|\delta_s|$ (over 0.5 rad, say) for which the model equations are invalid.

Using the algorithm of Fig. 3.11, it is easy to show that to shift the poles to:

$$s = -0.03 \pm j0.03, \quad -0.08, \quad -0.4$$

with

$$\mathbf{M} = \begin{bmatrix} 1 \\ -1 \end{bmatrix},$$

$$\mathbf{K} = \begin{bmatrix} -0.0105 & +0.0005 & +0.0002 & -0.0013 \\ +0.0105 & -0.0005 & -0.0002 & +0.0013 \end{bmatrix}$$

The closed-loop state equations are (Eq. (3.36)):

$$\begin{bmatrix} \dot{w} \\ \dot{q} \\ \dot{\theta} \\ \dot{h} \end{bmatrix} = \begin{bmatrix} -0.083 & 0.001 & 0 & 0 \\ 2.152 & -0.458 & -0.0275 & 0.1400 \\ 0 & 1 & 0 & 0 \\ 1 & 0 & -0.005 & 0 \end{bmatrix} \begin{bmatrix} w \\ q \\ \theta \\ h \end{bmatrix} + \begin{bmatrix} -0.084 & -0.163 \\ 86.000 & -24.000 \\ 0 & 0 \\ 0 & 0 \end{bmatrix} \begin{bmatrix} h_d \\ -h_d \end{bmatrix}.$$

The algorithm of Fig. 3.4 generates a real-time response to a demanded scaled depth change ($+0.05-0.05$) as shown in Fig. 3.14(c), while the algorithm of Fig. 3.6 confirms that the steady-state gain of the whole closed-loop system (depth over demanded scaled depth) is 800. A strategy of this kind is a promising basis for a submarine autopilot design.

3.7 OPTIMAL CONTROL STRATEGIES

The object of 'optimal' control is to specify an input vector $\mathbf{u}(t)$ which drives a system to a specified target state in such a way that, during the process, a defined 'cost function' is minimized.

A variety of optimal control strategies exists (refs. 2, 5), most notably, perhaps, linear optimal quadratic control, switching curve strategy, and dynamic programming. The first two are briefly described here, while dynamic programming is outlined in Sec. 4.8.

(i) Linear Quadratic Optimal Control

Optimal control is not limited to linear systems, but when a 'linear quadratic' cost function is defined for an LTI system, optimal control strategy devolves into a particular case of pole shifting by state feedback (Sec. 3.6).

Consider an LTI system represented by the familiar state equations (Eq. (3.5)):

$$\dot{\mathbf{x}} = \mathbf{A}\mathbf{x} + \mathbf{B}\mathbf{u}$$
$$\mathbf{y} = \mathbf{C}\mathbf{x}.$$

A linear quadratic cost function is defined:

$$V = \int_{t_0}^{t_1} (\mathbf{x}^{\mathrm{T}}\mathbf{E}\mathbf{x} + \mathbf{u}^{\mathrm{T}}\mathbf{F}\mathbf{u})\,\mathrm{d}t \tag{3.37}$$

where **E** is a positive semidefinite symmetric $(m \times m)$ matrix,
F is a positive definite symmetric $(q \times q)$ matrix (App. A.11).

V describes the cost of the process, $t_0 \leqslant t \leqslant t_1$, and comprises a positive function of the state (typically a description of a desired trajectory between the initial and target states) and a positive function of the input (typically a measure of the input amplitudes).

The design problem is to find an input **u** which drives the state **x** from $\mathbf{x}(t_0)$ to $\mathbf{x}(t_1)$ with minimum V. Without loss of generality, since the states can be redefined appropriately, $\mathbf{x}(t_1)$ can be set at $\mathbf{0}$. If, in addition, $t_1 = \infty$, the solution of the problem, which is found by the calculus of variations, is (refs. 2, 5):

$$\mathbf{u} = -\mathbf{K}\mathbf{x}$$

where

$$\mathbf{K} = \mathbf{F}^{-1}\mathbf{B}^{\mathrm{T}}\mathbf{P} \tag{3.38}$$

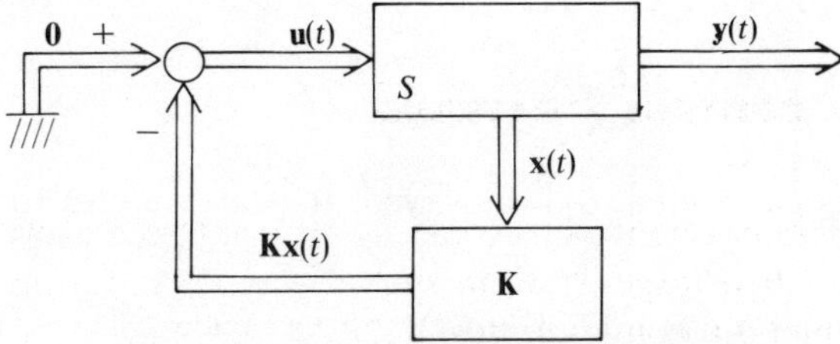

Fig. 3.15 Optimal control by state feedback.

and $\mathbf{P}$ is the $(m \times m)$ matrix solution of the algebraic 'Riccati' equation:

$$\mathbf{E} - \mathbf{PBF}^{-1}\mathbf{B}^{\mathrm{T}}\mathbf{P} + \mathbf{A}^{\mathrm{T}}\mathbf{P} + \mathbf{PA} = 0. \tag{3.39}$$

The controlled system is illustrated in Fig. 3.15.

CAD Facility

The interactive CAD algorithm shown in the flow diagram of Fig. 3.16* solves the algebraic Riccati equation for $\mathbf{P}$, given the matrices $\mathbf{A}$, $\mathbf{B}$, $\mathbf{E}$, $\mathbf{F}$ ($\mathbf{E}$ symmetric positive semi-definite, $\mathbf{F}$ symmetric positive definite) and calculates the required feedback matrix $\mathbf{K}$ according to Eq. (3.38).

The operation of the algorithm is as follows.

BLOCK 1

The system state matrices $\mathbf{A}$, $\mathbf{B}$ are requested, input, displayed, and corrected if necessary.

BLOCK 2

The controllability matrix $\mathbf{P}$ (Eq. (3.17)) is calculated and stored. Its rank is checked (App. A.5) and if less than m, the system is declared uncontrollable.

BLOCK 3

The matrices $\mathbf{E}$, $\mathbf{F}$ are requested, input, and checked to be positive semidefinite symmetric and positive definite symmetric respectively (App. A.11).

BLOCK 4

The matrix $\mathbf{H}$ is calculated:

$$\mathbf{H} = \begin{bmatrix} \mathbf{A} & -\mathbf{BFB}^{\mathrm{T}} \\ -\mathbf{E} & -\mathbf{A}^{\mathrm{T}} \end{bmatrix}.$$

The eigenvectors and eigenvalues of $\mathbf{H}$ are found (App. A.9).

The method assumes that $\mathbf{H}$ has no purely imaginary eigenvalues. m eigenvalues have positive and m eigenvalues have negative real parts. If the eigenvectors, which are $(2m)$ vectors, corresponding to those with positive real parts are written:

$$\mathbf{a}_i = \begin{bmatrix} \mathbf{b}_i \\ \mathbf{c}_i \end{bmatrix}, \quad i = 1, \ldots, m; \quad \mathbf{b}_i,\ \mathbf{c}_i \text{ both } m \text{ vectors,}$$

then the solution of Eq. (3.39) is:

$$\mathbf{P} = (\mathbf{b}_1, \ldots, \mathbf{b}_m)(\mathbf{c}_1, \ldots, \mathbf{c}_m)^{-1}.$$

BLOCK 5

$\mathbf{K}$ (Eq. (3.38)) is calculated and displayed. It is also useful to calculate and display $\mathbf{A} - \mathbf{BK}$.

*This algorithm is due to J.E. Potter, 'Matrix Quadratic Solutions,' *SIAM Journal of Applied Mathematics*, 14 (1966), 496–501.

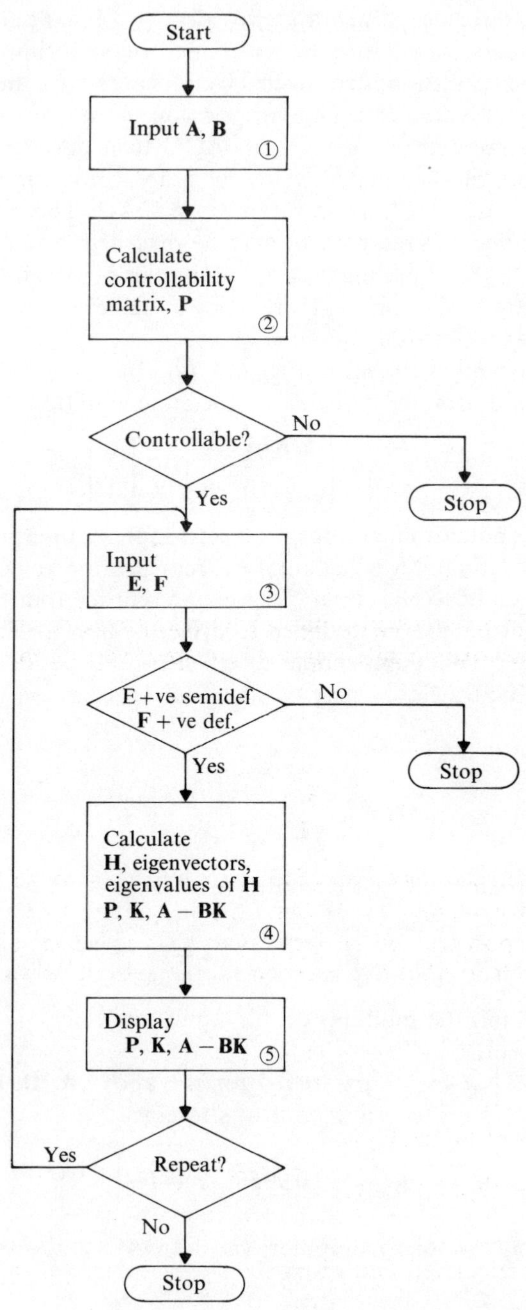

Fig. 3.16 CAD algorithm for matrix Riccati equation solution.

EXAMPLES

(a) A useful laboratory experiment is provided by a balanced vertical-plane mechanical arm driven by a torque servo consisting of a simple flexible steel torsion bar whose deflection is controlled by a local relative angular position, electrically actuated, servo. The angular position of the arm, monitored by a local servo potentiometer, is fed via an ADC to a microcomputer which displays this and the calculated velocity, graphically, on its VDU. The microcomputer drives the torque servo via a DAC. The ADC and DAC sampling rate is high, so the control may be regarded as continuous time.

The object of the experiment is to specify various required state space trajectories, and to design optimal strategies to achieve these, using a CAD facility and the experimental equipment.

The arrangement is shown in Figs. 3.17(a), (b).

The servo and arm are modeled by the state equations:

$$\dot{\mathbf{x}} = \begin{bmatrix} \dot{x}_1 \\ \dot{x}_2 \end{bmatrix} = \begin{bmatrix} -0.01 & 1 \\ 0 & 0 \end{bmatrix} \mathbf{x} + \begin{bmatrix} 0 \\ 1 \end{bmatrix} u$$

where x_1 is the potentiometer output (volts) representing position,
x_2 is the (computer calculated) $\dot{x}_1$, representing velocity,
u is the input to the servo (volts), representing applied torque.

Optimal controls can be designed to drive the arm from various initial states, $\mathbf{x}(0)$, to $\mathbf{x}(t_1) = 0$, along simple specified trajectories.

A straight-line trajectory may be specified, for example:

$$x_1 = -2x_2 \ .$$

Equivalently:

$$(x_1 + x_2)^2 = 0 \, .$$

This is approximated by the control if $\mathbf{x}^{\mathrm{T}}\mathbf{E}\mathbf{x}$ is minimized, where:

$$\mathbf{E} = \begin{bmatrix} 1 & 2 \\ 2 & 4 \end{bmatrix}; \quad \text{i.e.} \quad \mathbf{x}^{\mathrm{T}}\mathbf{E}\mathbf{x} = (x_1 + 2x_2)^2 .$$

$\mathbf{F}$ is selected to limit the modulus of the input, torque, u,
For example, $\mathbf{F} = 1$.

Equations (3.38), (3.39) are solved for the above **A**, **B**, **E**, **F** using the algorithm of Fig. (3.16) to give a control strategy:

$$u = -\mathbf{K}\mathbf{x} = -[0.9955 \quad 2.4477] \begin{bmatrix} x_1 \\ x_2 \end{bmatrix} .$$

This is applied by the microcomputer, via the DAC, and a trajectory for

$$\mathbf{x}[0] = \begin{bmatrix} 1 \\ 0 \end{bmatrix}$$

is shown in Fig. 3.17(c).

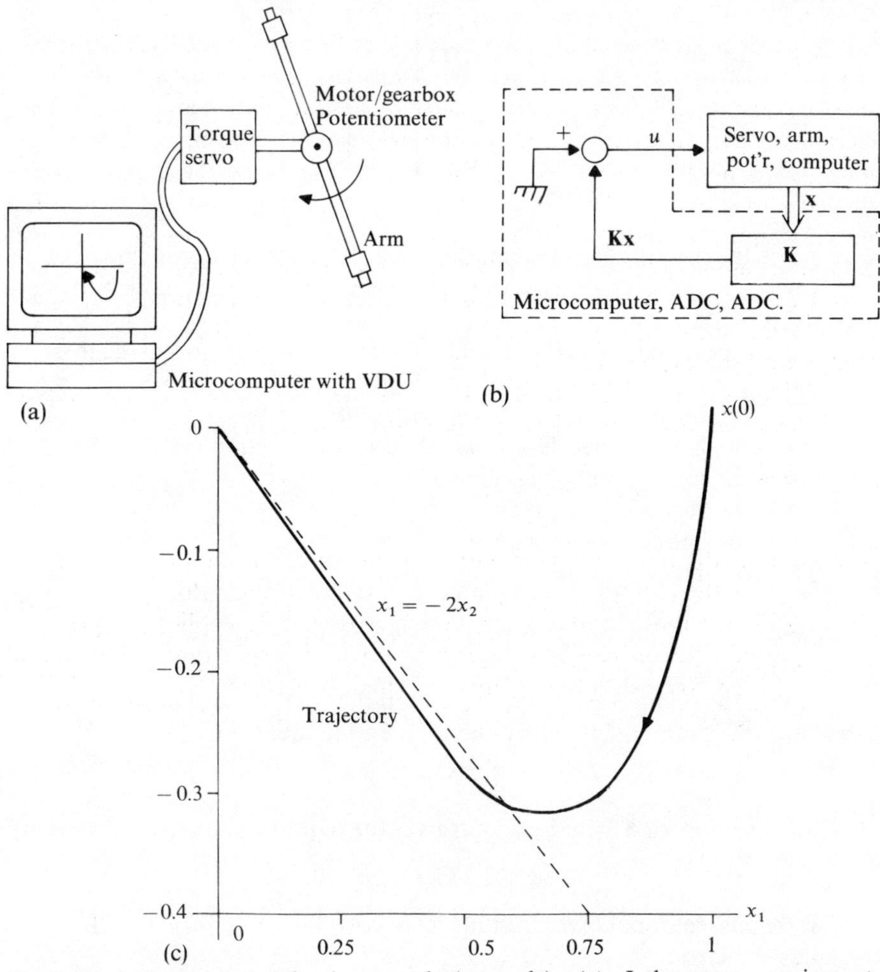

Fig. 3.17 Laboratory optimal control (example). (a) Laboratory equipment. (b) System. (c) Trajectory.

The arm is driven by the control on to the required trajectory in a manner determined mainly by **F**, and follows it, quite accurately, in a manner determined by **E**, until $\mathbf{x} = \mathbf{0}$.

(b) The vertical plane dynamics of an aircraft which is about to land are modeled by the simplified equations:

$$\begin{bmatrix} \dot{u} \\ \dot{w} \\ \dot{q} \\ \dot{\theta} \\ \dot{h} \\ \dot{e} \end{bmatrix} = \begin{bmatrix} -0.058 & 0.065 & 0 & -0.171 & 0 & 1 \\ -0.303 & -0.685 & 1.109 & 0 & 0 & 0 \\ 0.072 & -0.658 & -0.947 & 0 & 0 & 0 \\ 0 & 0 & 1 & 0 & 0 & 0 \\ 0 & -1 & 0 & 1.133 & 0 & 0 \\ 0 & 0 & 0 & 0 & 0 & -0.571 \end{bmatrix} \begin{bmatrix} u \\ w \\ q \\ \theta \\ h \\ e \end{bmatrix}$$

$$+\begin{bmatrix} 0 & 0 & -0.119 \\ -0.054 & 0 & 0.074 \\ -1.117 & 0 & 0.115 \\ 0 & 0 & 0 \\ 0 & 0 & 0 \\ 0 & 0.571 & 0 \end{bmatrix} \begin{bmatrix} \mu \\ \gamma \\ \delta \end{bmatrix} \tag{3.40}$$

where u is the aircraft forward speed along its main body axis (m s^{-1}).
w is the velocity downwards at right angles to the main body axis (m s^{-1}),
q is the angular velocity of pitch with respect to the ground (degrees^{-1}),
θ is the pitch w.r.t. the ground (degrees),
h is the height w.r.t 1 m below the ground (m),
e is the forward acceleration due to throttle action (m s^{-2}),
μ is the elevator angle (degrees),
γ is the throttle value (m s^{-2}),
δ is the spoiler angle (degrees).

The system is illustrated in Fig. 3.18(a).

Consider the design of an input, $(\mu \quad \gamma \quad \delta)^{\mathrm{T}}$ such that the aircraft comes in to land along an exponential path towards a point 1 m below ground level as indicated in Fig. 3.18(b).

Under the (almost correct) assumption that the forward ground speed is constant, this path is defined by the differential equation:

$$\dot{h} + 0.2h = 0.$$

Substituting for $\dot{h}$ from Eq. (3.40) gives the required state space trajectory:

$$-w + 1.133 + 0.2h = 0.$$

It seems reasonable to minimize a cost function (Eq. (3.37)) which includes the term:

$$C^2 = (1.133 - w + 0.2h)^2.$$

Expressing this in the form:

$$[u \quad w \quad q \quad \theta \quad h \quad e]\mathbf{E}[u \quad w \quad q \quad \theta \quad h \quad e]^{\mathrm{T}} = C^2$$

it is easy to show that:

$$\mathbf{E} = \begin{bmatrix} 0 & 0 & 0 & 0 & 0 & 0 \\ 0 & 1 & 0 & -1.133 & -0.2 & 0 \\ 0 & 0 & 0 & 0 & 0 & 0 \\ 0 & -1.133 & 0 & 1.284 & 0.227 & 0 \\ 0 & -0.2 & 0 & 0.227 & 0.04 & 0 \\ 0 & 0 & 0 & 0 & 0 & 0 \end{bmatrix}.$$

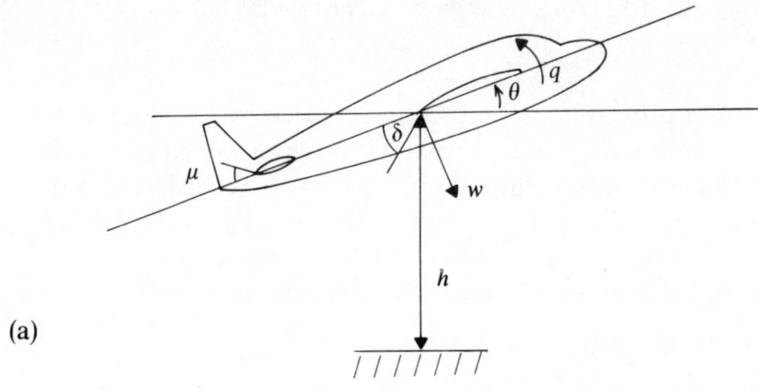

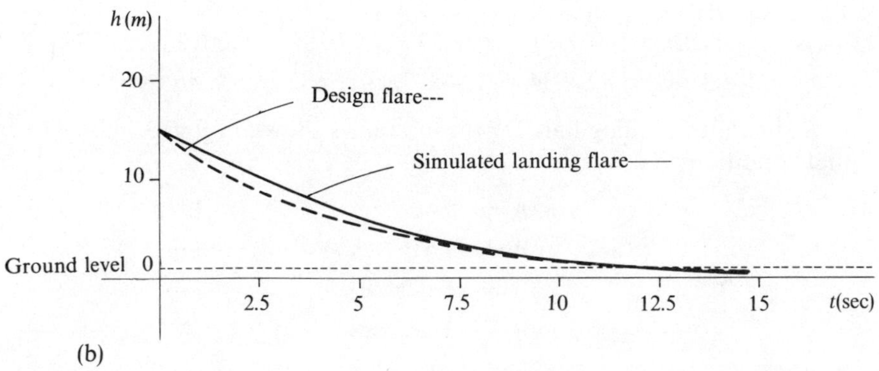

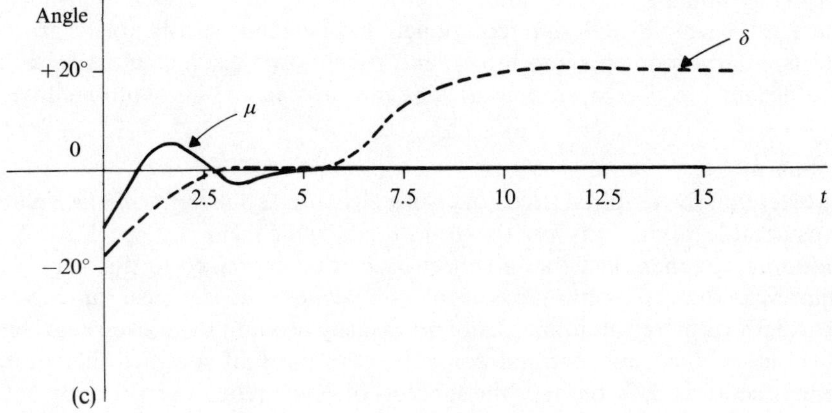

Fig. 3.18 Landing aircraft (example) (a) Aircraft. (b) Design flare and simulated flare. (c) Spoiler and elevator deflections.

Consider the second term in the cost function (Eq. (3.37)):

$$[\mu \quad \gamma \quad \delta]\mathbf{F}[\mu \quad \gamma \quad \delta]^{\mathrm{T}}.$$

The design procedure is first to select a matrix **F** (which determines the maximum excursions of the input variables μ, γ, δ). Equations (3.38), (3.39) are then solved, using the algorithm of Fig. 3.16, to give a feedback matrix **K**, and a simulation run reveals whether any state or input variables exceed practical limits. If so, **F** is reselected and the procedure repeated.

Some trial and error of this kind yields the matrices:

$$\mathbf{F} = \begin{bmatrix} 0.0032 & 0 & 0 \\ 0 & 0.179 & 0 \\ 0 & 0 & 0.0008 \end{bmatrix}$$

$$\mathbf{K} = \begin{bmatrix} -1.7313 & 2.1382 & -2.1793 & -6.4857 & -0.8896 & -0.6598 \\ 0.0200 & -0.0087 & 0.0063 & 0.0182 & 0.0019 & 0.0289 \\ -3.0985 & 30.4834 & -1.8827 & -41.7882 & -6.9286 & -0.9192 \end{bmatrix}.$$

A simulated landing flare for the aircraft is shown in Fig. 3.18(b), where initial conditions are:

$$\begin{aligned} u(0) &= 5\,\mathrm{m\,s^{-1}} \\ w(0) &= -2.5\,\mathrm{m\,s^{-1}} \\ q(0) &= -1\ \text{degree}\,\mathrm{s^{-1}} \\ \theta(0) &= -3\ \text{degrees} \\ h(0) &= 15\,\mathrm{m} \\ e(0) &= 0.5\,\mathrm{m\,s^{-2}}. \end{aligned}$$

The simulated spoiler and elevator angle profiles are in Fig. 3.18(c). Clearly, the simplified design requirement has been met satisfactorily; ground effects, lateral moments and forces, and touchdown distance must be taken into account to develop this basic idea into a practical autoland facility.

Comment

The fact that the required trajectory must be expressed in terms of $\mathbf{x}^{\mathrm{T}}\mathbf{E}\mathbf{x}$ places considerable limitations on the trajectories which can be specified. Furthermore, the mere fact that a trajectory can be expressed in this way is no guarantee that the optimal control can achieve it; the cost function V (Eq. (3.37)) is merely minimized, not necessarily brought to, or even near, zero.

Linear quadratic optimal control is the basis of several other design techniques (refs. 2, 5), perhaps the simplest of which relies on a property of the state equations augmented as follows:

$$\begin{bmatrix} \dot{\mathbf{x}} \\ \dot{\mathbf{p}} \end{bmatrix} = \begin{bmatrix} \mathbf{A} & -\mathbf{B}\mathbf{F}\mathbf{B}^{\mathrm{T}} \\ -\mathbf{E} & -\mathbf{A}^{\mathrm{T}} \end{bmatrix} \begin{bmatrix} \mathbf{x} \\ \mathbf{p} \end{bmatrix}.$$

The eigenvalues of this are symmetrical about 0 and can be represented by $\lambda_1, \lambda_2, \ldots, \lambda_m, -\lambda_1, -\lambda_2, \ldots, -\lambda_m$. $\lambda_1, \ldots, \lambda_m$ are chosen to be in desirable positions (cf. Sec. 3.5) and **E**, **F** can then be calculated. The relevant feedback matrix is then found by solving Eqs. (3.38), (3.39). This strategy places the closed-loop poles in positions which minimize a rather obscurely defined cost function, but trial and error in the selection of $\lambda_1, \lambda_2, \ldots, \lambda_m$ can yield worthwhile results.

(ii) Switching Curve Strategy (ref. 2)

This technique, which is applicable to nonlinear as well as linear systems, depends on constructing a reverse-time model of the system and developing a 'switching curve' from this. Consider a system represented by the familiar state equation (Eq. (3.1)):

$$\dot{\mathbf{x}} = \mathbf{f}(\mathbf{x}, \mathbf{u}, t).$$

A cost function is defined (cf. Eq. 3.37):

$$V(\mathbf{x}, \mathbf{u}, t) = \int_0^{t_1} L(\mathbf{x}, \mathbf{u}, t)\,\mathrm{d}t. \tag{3.41}$$

V describes the cost of the process, $0 \leqslant t \leqslant t_1$ (V and L are scalar functions). The design problem is to find an input $\mathbf{u}$ which drives the state $\mathbf{x}$ from $\mathbf{x}(0)$ to $\mathbf{x}(t_1)$ with minimum V.

A 'Hamiltonian' scalar is formed:

$$\begin{aligned} \mathbf{H} &= L(\mathbf{x}, \mathbf{u}, t) + \dot{\mathbf{x}}^{\mathrm{T}} \cdot \mathbf{p} \\ &= L(\mathbf{x}, \mathbf{u}, t) + \mathbf{f}^{\mathrm{T}}(\mathbf{x}, \mathbf{u}, t) \cdot \mathbf{p}(t) \end{aligned} \tag{3.42}$$

where $\mathbf{p}$ is a 'costate' vector defined by:

$$\dot{\mathbf{p}} = -\frac{\partial H}{\partial \mathbf{x}} = \left[-\frac{\partial H}{\partial x_1} \quad -\frac{\partial H}{\partial x_2} \quad \cdots \quad -\frac{\partial H}{\partial x_m} \right]^{\mathrm{T}}. \tag{3.43}$$

According to Pontryagin's Minimum Principle (refs. 2, 5), the input $\mathbf{u}(t)$ which minimizes V is that which minimizes H w.r.t. $\mathbf{u}$. This is the optimal control, which can be found as a function of $\mathbf{x}, \mathbf{p}, t$, say $\bar{\mathbf{u}}^*(\mathbf{x}, \mathbf{p}, t)$, fairly readily from (Eq. (3.42).

To be useful, however, the optimal input must be found as a function of $\mathbf{x}$, or perhaps $\mathbf{x}$, t, say $\mathbf{u}^*(\mathbf{x}, t)$, and this involves solving Eqs. (3.1), (3.43), $\mathbf{u} = \bar{\mathbf{u}}^*(\mathbf{x}, \mathbf{p}, t)$ given the boundary conditions $\mathbf{x}(0)$, $\mathbf{x}(t_1)$ and using this solution to eliminate $\mathbf{p}$ from $\bar{\mathbf{u}}^*$. Since the boundary conditions are 'mixed' – $\mathbf{x}(0)$ being at the start and $\mathbf{x}(t_1)$ at the end of any attempted simulation run – this is generally very difficult to do.

The 'switching curve' strategy solves this problem for a useful class of cases by considering the behavior of the system in reverse time; $\hat{t} = t_1 - t$, $0 \leqslant \hat{t} \leqslant t_1$, starting from the target state $\mathbf{x}(t_1) = \hat{\mathbf{x}}(\hat{0})$.

The system model is rewritten as a function of $\hat{t}$ and a corresponding Hamiltonian and set of costate equations (all with $\hat{t}$ as the independent variable) are developed:

$$\frac{d\hat{\mathbf{x}}}{d\hat{t}} = \bar{\mathbf{f}}(\hat{\mathbf{x}}, \hat{\mathbf{u}}, \hat{t}) \tag{3.44}$$

$$\bar{H} = \bar{L}(\hat{\mathbf{x}}, \hat{\mathbf{u}}, \hat{t}) + \bar{\mathbf{f}}^{\mathrm{T}}(\hat{\mathbf{x}}, \hat{\mathbf{u}}, \hat{t}).\,\hat{\mathbf{p}}(\hat{t}) \tag{3.45}$$

$$\frac{d\hat{\mathbf{p}}}{d\hat{t}} = -\frac{\partial \bar{H}}{\partial \hat{\mathbf{x}}}. \tag{3.46}$$

As in forward time, the optimal control is found by solving Eqs. (3.44), (3.46) for $\hat{\mathbf{x}}(\hat{t})$, $\hat{\mathbf{p}}(\hat{t})$ and using these to find $\hat{\mathbf{u}}^*(\hat{\mathbf{x}}, \hat{t})$. In this case, however, since the system always starts from the target state $\mathbf{x}(t_1) = \hat{\mathbf{x}}(\hat{0})$, every trajectory and corresponding $\hat{\mathbf{u}}^*$ is of interest. Different trajectories result from choosing different costate initial conditions $\hat{\mathbf{p}}(\hat{0})$.

In a useful class of cases, it turns out that $\hat{\mathbf{u}}^*$, and so $\mathbf{u}^*$, amount to sequences of switching between extreme values of $\mathbf{u}$. In these cases a 'switching curve' can be constructed in the reverse-time state space from a number of reverse-time trajectories generated by an analog (or digital) model of the reverse-time equations. This curve is then used as the basis of an optimal strategy in forward time. The method is perhaps best described by example.

EXAMPLE

A useful laboratory example is provided by an almost balanced pendulum to which a torque may be applied using electrically actuated airjets mounted on the pendulum itself, or, alternatively, by a local torque servo (cf. Fig. 3.17). The arrangement is shown in Fig. 3.19(a).

The angular position and velocity of the pendulum are monitored by a microcomputer (with VDU) via ADC channels and a potentiometer and tachogenerator.

The behavior of the pendulum is modeled by the equation:

$$\ddot{\theta} + 0.01\dot{\theta} + 0.46 \sin\theta = u \tag{3.47}$$

where θ is the angular displacement (rad),
u is the input (rad s^{-2}) provided by the airjets, only one of which operates at any instant ($-0.32 \leqslant u \leqslant +0.32$)

The object of the experiment is to devise a control strategy which brings the pendulum to a stop ($\theta = \dot{\theta} = 0$), starting from any initial position and velocity.

Setting $x_1 = \theta$, $x_2 = \dot{\theta}$, Eq. (3.47) is cast in state equation form (Eq. (3.1)):

$$\dot{x}_1 = x_2 \tag{3.48}$$

$$\dot{x}_2 = -0.01x_2 - 0.46 \sin x_1 + u \tag{3.49}$$

To minimize the time taken by the process, set $L(\mathbf{x}, \mathbf{u}, t) = 1$ so that the cost

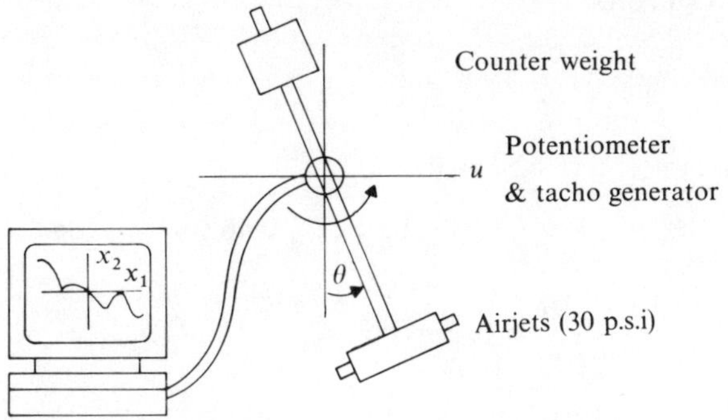

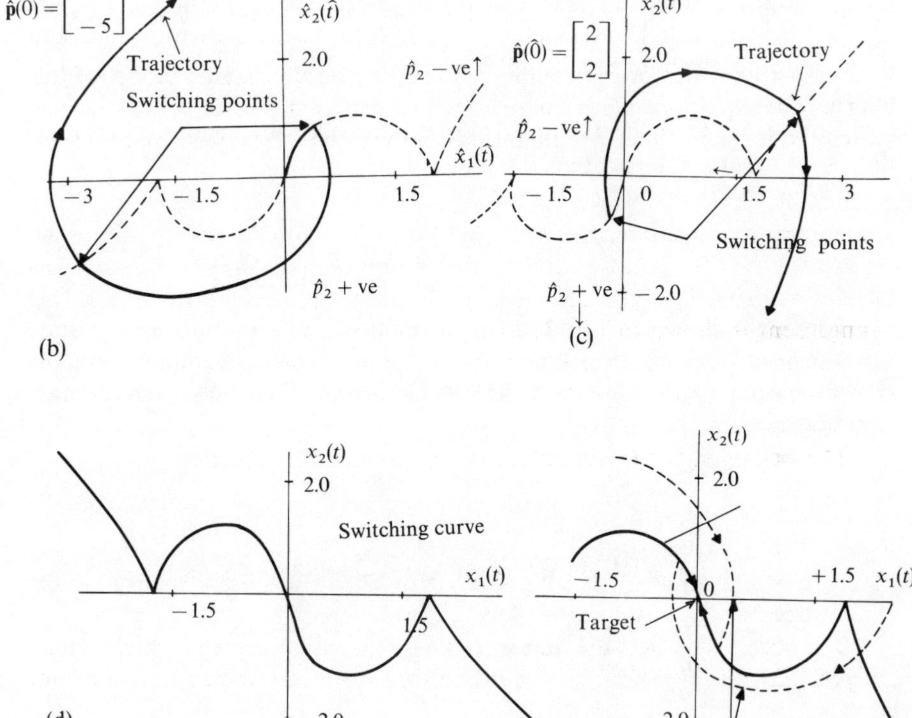

Fig. 3.19 Laboratory pendulum (example). (a) Laboratory pendulum and microcomputer. (b) Reverse-time trajectory. (c) Reverse-time trajectory. (d) Forward-time switching curve. (e) forward-time trajectories.

function (Eq. (3.41)) is:

$$V=\int_0^{t_1} 1\,\mathrm{d}t = t_1 .$$

The Hamiltonian (Eq. (3.42)) is formed:

$$H = 1 + x_2 p_1 + (-0.01x_2 - 0.46\sin x_1 + u)p_2 \tag{3.50}$$

where

$$\dot{p}_1 = -\frac{\partial H}{\partial x_1} = 0.46 p_2 \cos x_1 \tag{3.51}$$

$$\dot{p}_2 = -\frac{\partial H}{\partial x_2} = -p_1 + 0.01 p_2 . \tag{3.52}$$

From Eq. (3.50), it is clear that H is minimized w.r.t. u for

$$u = -0.32\,\mathrm{sign}\,(p_2). \tag{3.53}$$

The optimal control is therefore found in principle by setting $u = -0.32$ sign (p_2), solving Eqs. (3.48) (3.49), (3.51), (3.52) for p_2 and substituting this in Eq. (3.53). Since the boundary conditions are mixed ($\mathbf{x}(0) = \mathbf{x}_0$, $\mathbf{x}(t_1) = \mathbf{0}$), this is impractical.

Following the method outlined above, the process is considered in reverse time starting at the target ($\theta = \dot{\theta} = 0$).

In Eq. (3.47), setting $\hat{t} = t_1 - t$,

$$\frac{\mathrm{d}^2\theta}{\mathrm{d}\hat{t}^2} - 0.01\frac{\mathrm{d}\theta}{\mathrm{d}\hat{t}} + 0.46\sin\theta = u .$$

Casting this in state equations, defining a Hamiltonian and costate equations and substituting the optimal control found by minimizing the Hamiltonian w.r.t. u (cf. Eqs. (3.48) to (3.53)) gives the state and costate equations:

$$\frac{\mathrm{d}\hat{x}_1}{\mathrm{d}\hat{t}} = \hat{x}_2$$

$$\frac{\mathrm{d}\hat{x}_2}{\mathrm{d}\hat{t}} = 0.01\hat{x}_2 - 0.46\sin\hat{x}_1 - 0.32\,\mathrm{sign}\,(\hat{p}_2)$$

$$\frac{\mathrm{d}\hat{p}_1}{\mathrm{d}\hat{t}} = 0.46\hat{p}_2\cos\hat{x}_1$$

$$\frac{\mathrm{d}\hat{p}_2}{\mathrm{d}\hat{t}} = -\hat{p}_1 - 0.01\hat{p}_2 .$$

These equations can be solved using an analog model or Runge–Kutta based simulation (Sec. 1.2) for a variety of initial conditions ($\hat{\mathbf{x}}(\hat{0}) = \mathbf{0}$, $\hat{\mathbf{p}}(\hat{0}) = \hat{\mathbf{p}}_0$). In the two-dimensional state space, several trajectories in reverse time are found and the switching points, where $\hat{p}_2$ changes sign, noted. Two trajectories and their switching points are shown in Figs. 3.19(b), (c). The switching curve is developed by connecting many such switching points smoothly, as indicated by the dotted lines.

The equivalent switching curve for forward time is easily found by substituting:

$$x_1(t) = \theta(t) = \theta(\hat{t}) = \hat{x}_1(\hat{t})$$

$$x_2(t) = \frac{d\theta}{dt} = -\frac{d\theta}{d\hat{t}} = -\hat{x}_2(\hat{t}).$$

This gives the forward-time switching curve shown in Fig. 3.19(d).

The optimal strategy (in forward time) is simply to apply maximum torque to the pendulum in one direction or the other depending on which 'side' of the switching curve the state is, and to reverse that direction when the state 'crosses' the switching curve. Two trajectories are shown in Fig. 3.19(e).

In the example considered here, the switching curve has a 'lobed' shape. A trajectory crosses a lobe only once, and when it reaches the final lobe (nearest the origin), it follows this to the origin. In many cases the switching curve has 'final' lobes only.

In the equipment described here, the switching curve is found by experiments on a reverse-time model and specified to the microcomputer (Fig. 3.19(a)) which displays it on the VDU. A state trajectory is also displayed and developed while a run proceeds. When this trajectory crosses the switching curve, the computer reverses the control jets. The time taken, t_1, to reach $\mathbf{x}(t_1) = \mathbf{0}$ is noted for various initial conditions $\mathbf{x}(0)$, and compared with that taken using a manually operated jet (or torque servo) control.

Comment

Where two states are involved, this strategy is simple and attractive. A three-state system involves a switching surface often of surprising complexity, and finding it can be a major exercise. This problem increases sharply with the number of states.

3.8 STATE ESTIMATORS FOR LTI SYSTEMS

The control strategies described in Secs. 3.6 and 3.7 depend on the system states being available as transducer or other output signals. If this is not the case the states can sometimes be estimated by a 'state estimator' or 'observer', which is basically a continuously corrected analog model of the system running in parallel with it.

Consider an observable LTI system, S, with dynamical equations: (Eqs. (3.5), (3.6)):

$$\dot{\mathbf{x}} = \mathbf{Ax} + \mathbf{Bu}$$

$$\mathbf{y} = \mathbf{Cx} + \mathbf{Du}.$$

A state estimator, $\hat{S}$, estimates or models the state $\mathbf{x}$, given the system input $\mathbf{u}$ and output $\mathbf{y}$.

Two types of estimator are common.

(i) The Asymptotic State Estimator

An asymptotic state estimator, $\hat{S}$, is shown in schematic form in Fig. 3.20.

The estimator dynamical equations, which are the system equations with an added corrective term, are:

$$\begin{aligned}\dot{\hat{\mathbf{x}}} &= \mathbf{A}\hat{\mathbf{x}} + \mathbf{B}\mathbf{u} + \mathbf{M}(\mathbf{y} - \mathbf{C}\hat{\mathbf{x}} - \mathbf{D}\mathbf{u}) \\ &= (\mathbf{A} - \mathbf{M}\mathbf{C})\hat{\mathbf{x}} + \mathbf{B}\mathbf{u} + \mathbf{M}\mathbf{y} - \mathbf{M}\mathbf{D}\mathbf{u}\end{aligned} \tag{3.54}$$

where $\mathbf{M}$ is an $(m \times r)$ matrix.

From Eqs. (3.5), (3.54), the estimator 'error,' $(\mathbf{x} - \hat{\mathbf{x}})$, behaves according to the equation:

$$(\dot{\mathbf{x}} - \dot{\hat{\mathbf{x}}}) = (\mathbf{A} - \mathbf{M}\mathbf{C})(\mathbf{x} - \hat{\mathbf{x}}) \cdot \tag{3.55}$$

If S is observable, $\mathbf{M}$ can be selected to ensure that the eigenvalues of $(\mathbf{A} - \mathbf{M}\mathbf{C})$ correspond to a rapid, well-behaved decay of any estimation error (cf. Sec. 3.6).

If S is unobservable, only the modes which are observable can be estimated in this manner. As in the dual case of shifting the poles of uncontrollable systems (Sec. 3.6), it is usually possible to manipulate the unobservable state equations into two real subsystems, one observable and one unobservable, using a similarity transformation (ref. 3):

$$\bar{\mathbf{x}} = \mathbf{T}\mathbf{x} .$$

$\mathbf{T}$ can be found such that (cf. Eq. (3.35)):

$$\dot{\bar{\mathbf{x}}} = \begin{bmatrix} \dot{\mathbf{x}}_o \\ \dot{\mathbf{x}}_{\bar{o}} \end{bmatrix} = \begin{bmatrix} \mathbf{A}_o & 0 \\ \mathbf{A}_{o\bar{o}} & \mathbf{A}_{\bar{o}} \end{bmatrix} \begin{bmatrix} \mathbf{x}_o \\ \mathbf{x}_{\bar{o}} \end{bmatrix} + \begin{bmatrix} \mathbf{B}_o \\ \mathbf{B}_{\bar{o}} \end{bmatrix} \mathbf{u}$$

$$\mathbf{y} = (\mathbf{C}_o \quad 0) \begin{bmatrix} \mathbf{x}_o \\ \mathbf{x}_{\bar{o}} \end{bmatrix}$$

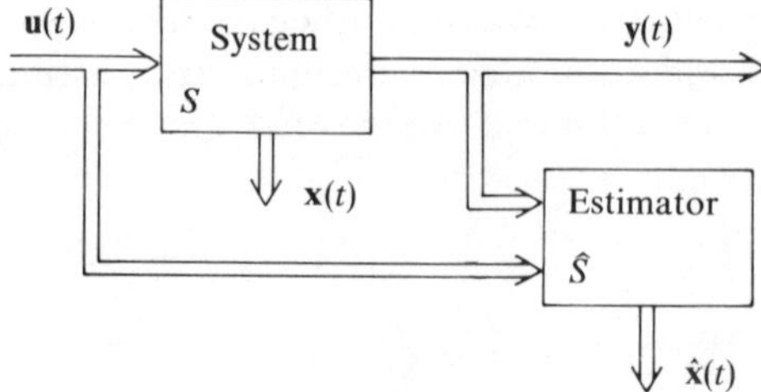

Fig. 3.20 Asymptotic state estimator.

The states of the observable subsystem

$$\dot{\mathbf{x}} = \mathbf{A}_o\mathbf{x}_o + \mathbf{B}_o\mathbf{u}$$

may be estimated by means of an observer.

As with Eq. (3.35), the determination of **T** is not particularly easy. A matrix **T** with the desired properties can be constructed by taking independent rows of the observability matrix **Q** (Eq. (3.18)) and adding to them further independent rows to make **T** a nonsingular square matrix (ref. 3).

If S is multi-output, the matrix **M** is not unique for a specified set of eigenvalues of $(\mathbf{A} - \mathbf{MC})$, and one simple way of ensuring uniqueness (cf. Sec. 3.6(ii)) is to make **M** dyadic; i.e. $\mathbf{M} = \mathbf{NR}$ where **N** is $(m \times 1)$, $\mathrm{R}(1 \times r)$. In this case Eq. (3.55) becomes:

$$(\dot{\mathbf{x}} - \dot{\hat{\mathbf{x}}}) = (\mathbf{A} - \mathbf{NRC})(\mathbf{x} - \hat{\mathbf{x}}).$$

Provided **R** is specified, **N**, and so **M**, may be designed uniquely. **R** is chosen to 'apportion' the system outputs taken into account by the estimator. To allow all the eigenvalues of $(\mathbf{A} - \mathbf{MC})$ to be assigned, **A**, **MC** must satisfy the observability criterion of Eq. (3.18).

CAD Facility

The algorithm of Fig. 3.11 finds a $(1 \times m)$ vector **K** such that $(\mathbf{A} - \mathbf{BMK})$ has specified eigenvalues, **A**, **B**, and **M** being defined by the user.

Since $(\mathbf{A} - \mathbf{NRC})^{\mathrm{T}} = \mathbf{A}^{\mathrm{T}} - \mathbf{C}^{\mathrm{T}}\mathbf{R}^{\mathrm{T}}\mathbf{N}^{\mathrm{T}}$, the same algorithm can be used to find a $(1 \times r)$ vector $\mathbf{N}^{\mathrm{T}}$ such that $(\mathbf{A} - \mathbf{NRC})^{\mathrm{T}}$, and so $(\mathbf{A} - \mathbf{NRC})$, has specified eigenvalues, $\mathbf{A}^{\mathrm{T}}$, $\mathbf{C}^{\mathrm{T}}$, with $\mathbf{R}^{\mathrm{T}}$ being defined by the user.

A CAD algorithm for asymptotic state estimator design is therefore simply that in Fig. 3.11 with **A**, **B**, **M** (Block 1) replaced with $\mathbf{A}^{\mathrm{T}}$, $\mathbf{C}^{\mathrm{T}}$, $\mathbf{R}^{\mathrm{T}}$ respectively. This automatically checks system observability (Block 2) and replaces the generated gain and state matrices **K**, $(\mathbf{A} - \mathbf{BMK})$ (Block 5) with the transposed estimator gain and state matrices $\mathbf{R}^{\mathrm{T}}$, $(\mathbf{A} - \mathbf{NRC})^{\mathrm{T}}$.

(ii) The Reduced-order Estimator (Luenberger Observer) (ref. 3)

If the observable system S has r outputs, **y**, and m states, **x**, r 'similar system' states (Sec. 3.3(ii)) may in principle be deduced algebraically from **y**, a 'reduced order' estimator built to estimate the remaining $(m - r)$ 'similar system' states, and an algebraic transformation performed to generate estimates of the original states. This arrangement is shown in Fig. 3.21.

The design procedure is as follows:

1. Define a diagonal $(m - r) \times (m - r)$ matrix **L** with elements (and so eigenvalues) 'faster' than, and in no case equal to, the eigenvalues of **A**.
2. Define an $((m - r) \times r)$ matrix **H** and solve the equation:

$$\mathbf{TA} - \mathbf{LT} = \mathbf{HC}$$

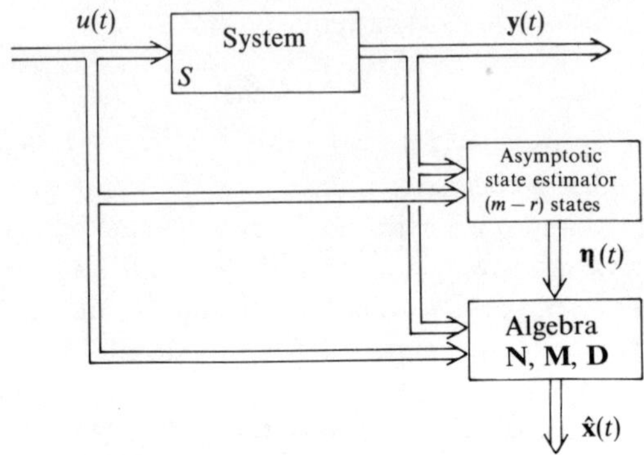

Fig. 3.21 Luenberger observer.

for the $(m-r) \times m$ matrix $\mathbf{T}$ which must be such that the $(m \times m)$ matrix

$$\mathbf{R} = \begin{bmatrix} \mathbf{C} \\ \mathbf{T} \end{bmatrix} \quad \text{is not singular.}$$

If $\mathbf{R}$ turns out to be singular, redefine $\mathbf{H}$.

3. Calculate $\mathbf{G} = \mathbf{TB}$ and

$$(\mathbf{MN}) = \mathbf{R}^{-1}$$

where $\mathbf{M}$ is $(m \times r)$, $\mathbf{N}$ is $m \times (m - r)$.

4. The estimator equations are:

$$\dot{\boldsymbol{\eta}} = \mathbf{L}\boldsymbol{\eta} + (\mathbf{G} - \mathbf{HD})\mathbf{u} + \mathbf{Hy} \tag{3.56}$$

$$\hat{\mathbf{x}} = \mathbf{N}\boldsymbol{\eta} - \mathbf{MDu} + \mathbf{My} \tag{3.57}$$

where $\boldsymbol{\eta}$ is an $(m - r)$ vector representing $(m - r)$ states.

CAD Facility

The interactive algorithm shown in Fig. 3.22 follows the design procedure outlined above.

The operation of the algorithm is as follows.

BLOCK 1

The system parameters **A**, **B**, **C**, **D** are requested, input, displayed, and corrected if necessary.

BLOCK 2

The observability matrix **Q** (Eq. (3.18)) is calculated. If the rank of $\mathbf{Q} < m$ (App. A.5), the system is unobservable and the program is halted.

BLOCK 3

The eigenvalues of **A** are calculated (App. A.9) stored, and displayed.

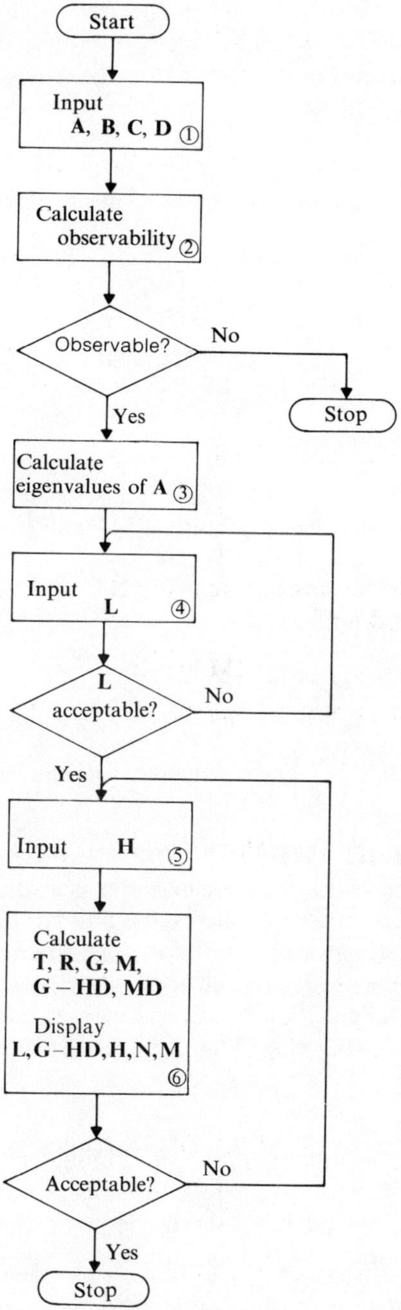

Fig. 3.22 CAD Algorithm for Luenberger observer design.

BLOCK 4
An $((m-r)\times(m-r))$ diagonal matrix $\mathbf{L}$ is requested, input and checked to ensure that its elements (and so eigenvalues) have negative real parts and that no element is an eigenvalue of $\mathbf{A}$.

BLOCK 5
The $((m-r)\times r)$ matrix $\mathbf{H}$ is requested, input, displayed, and corrected if necessary.

BLOCK 6
The equation

$$\mathbf{TA}-\mathbf{LT}=\mathbf{HC}$$

can be solved for $\mathbf{T}$ if $\mathbf{A}$, $\mathbf{L}$ have no common eigenvalues.
The jth row of $\mathbf{T}$ has elements:

$$\begin{bmatrix} t_{j1} \\ \vdots \\ t_{jm} \end{bmatrix} = (\mathbf{A}^{\mathrm{T}} - l_{jj}\mathbf{I})^{-1} \begin{bmatrix} \sum_{i=1}^{r} h_{ji}C_{i1} \\ \vdots \\ \sum_{i=1}^{r} h_{ji}C_{im} \end{bmatrix}.$$

The matrix $\mathbf{T}$ is thus calculated, and the square matrix $\begin{bmatrix}\mathbf{C}\\ \mathbf{T}\end{bmatrix}$ constructed.

If $\det\begin{bmatrix}\mathbf{C}\\ \mathbf{T}\end{bmatrix} \leqslant 0.1$, $\begin{bmatrix}\mathbf{C}\\ \mathbf{T}\end{bmatrix}$ is considered singular, and a new $\mathbf{H}$ matrix is requested.

The matrices $\mathbf{L}$, $(\mathbf{G}-\mathbf{HD})$, $\mathbf{H}$, $\mathbf{N}$, $\mathbf{MD}$, $\mathbf{M}$, are constructed and displayed.

Estimated states $\hat{\mathbf{x}}$ may be used in feedback strategies directly in place of the actual states, $\mathbf{x}$, with acceptable results. This is a feature of the 'separation property' of such estimators; the poles of the estimator do not affect those of the system and vice versa. It must be appreciated, of course, that an estimator introduces additional poles into the overall system, but if these are 'fast' (i.e. the eigenvalues of $(\mathbf{A}-\mathbf{MC})$ (Eq. 3.54) are 'fast'), their effect can usually be neglected.

EXAMPLE
The power of these estimation and pole shifting techniques is demonstrated well by simple laboratory equipment consisting of a flywheel driven by a local hydraulic (or electric) position servo via a flexible steel torsion bar. A potentiometer monitors the flywheel angular position.

The system is illustrated in Fig. 3.23(a).

The behavior of the system is described by the differential equation:

$$\frac{\mathrm{d}^2 y}{\mathrm{d}t^2} + 0.46\frac{\mathrm{d}y}{\mathrm{d}t} + 210y = 210u$$

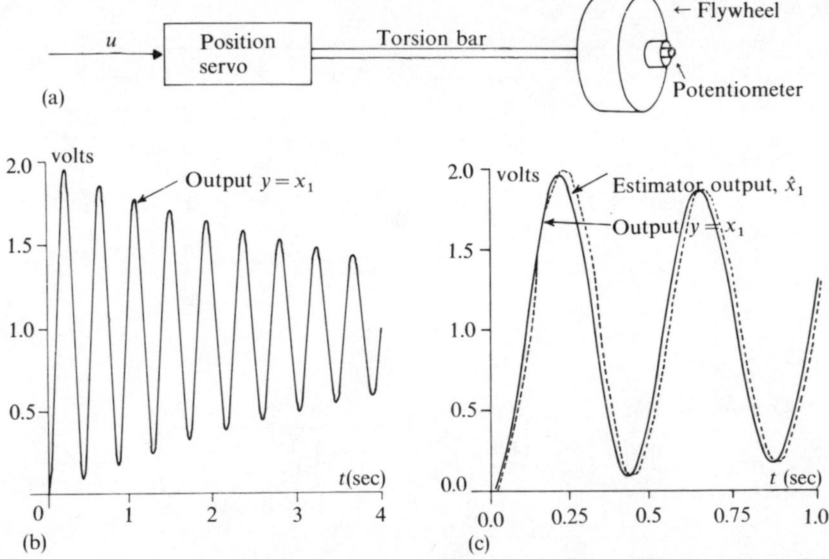

Fig. 3.23 Flywheel torsion bar, step response and estimator behavior (example). (a) Torsion bar and flywheel. (b) Unit step function response. (c) Step response and estimator output.

where u is the input (volts) to the drive servo, y is the output (volts) from the flywheel potentiometer.

Equivalently, the transfer function is:

$$\frac{Y(s)}{U(s)}=\frac{210}{s^2+0.46s+210} \qquad \text{(poles at } s=-0.23\pm j14.49\text{).}$$

This may be cast as state equations (Eqs. (3.22), (3.23)):

$$\dot{\mathbf{x}}=\begin{bmatrix}-0.46 & 1\\ -210 & 0\end{bmatrix}\mathbf{x}+\begin{bmatrix}0\\ 210\end{bmatrix}u$$
$$y=[1 \quad 0]\mathbf{x}$$

where $x_1=y; x_2=\dot{y}+0.46y=\dot{x}_1+0.46x_1$.

An asymptotic state estimator with eigenvalues of $(\mathbf{A}-\mathbf{MC})$ at $s=-10$, -10 (Sec. 3.7(i)) gives (Fig. 3.21):

$$\mathbf{M}=\begin{bmatrix}19.0\\ -110.0\end{bmatrix}$$
$$\dot{\hat{\mathbf{x}}}=\begin{bmatrix}-20 & 1\\ -100 & 0\end{bmatrix}\hat{\mathbf{x}}+\begin{bmatrix}19.0\\ -110.0\end{bmatrix}y+\begin{bmatrix}0\\ 210\end{bmatrix}u.$$

The system behavior is highly undesirable, as shown by the unit step function response of Fig. 3.23(b). The estimator follows this response accurately (Fig. 3.23(c)) after a small transient error lasting under 0.05 sec.

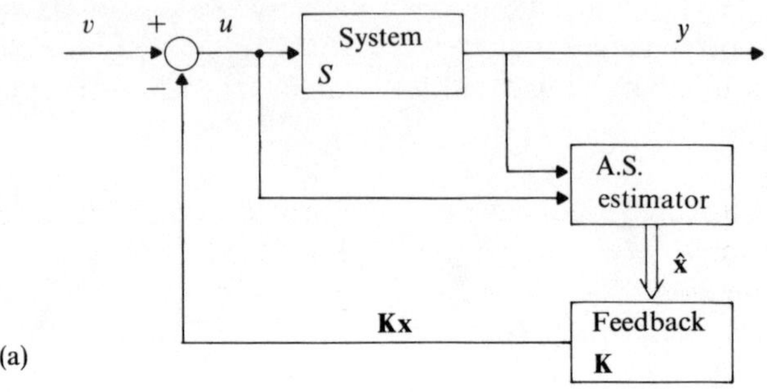

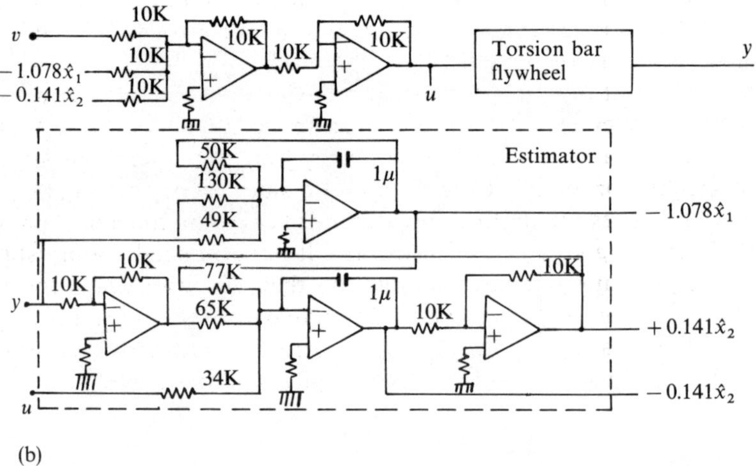

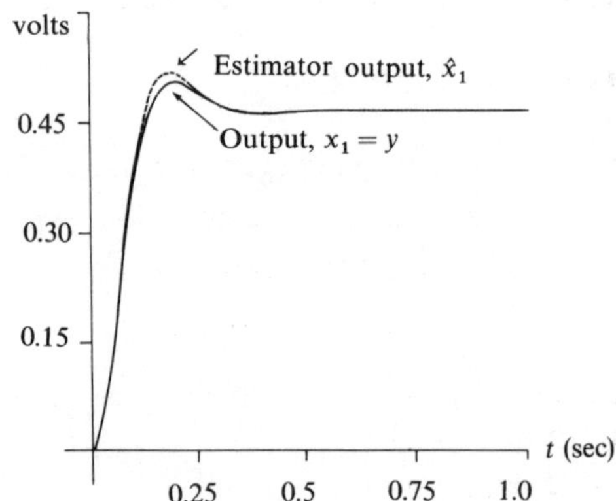

Fig. 3.24 Estimator, state feedback and system behavior (example). (a) Control strategy. (b) Estimator and feedback mechanization. (c) Unit step function response ($v = 1$).

To shift the very undesirable open-loop poles at $s = -0.23 \pm j14.49$ to better closed-loop positions, say $s = -15 \pm j15$, the algorithm of Sec. 3.6, used under the not quite correct assumption that $\hat{\mathbf{x}} = \mathbf{x}$, yields a feedback matrix:

$$\mathbf{K} = [1.078 \quad 0.141].$$

The estimator and feedback strategies are shown in block form in Fig. 3.24(a), an estimator mechanization in Fig. 3.24(b), and a step function response for the closed-loop system in Fig. 3.24(c).

Comment

Though the eigenvalues of $(\mathbf{A} - \mathbf{MC})$ (asymptotic estimator) and $\mathbf{L}$ (Luenberger observer) may theoretically be chosen with any values, in practice only those resulting in well-behaved, nonsaturating estimators can properly be selected. This implies a certain amount of trial and error which renders CAD facilities essential for these design technqiues.

A comment on the comparative behavior of asymptotic estimators and Luenberger observers is appropriate here. In the latter case, the equivalent of r state estimates are calculated algebraically from the system output, without the corrective property of the asymptotic state estimator (Eq. (3.54)). The quality of these estimates therefore depends heavily on the model parameters $\mathbf{A}$, $\mathbf{B}$, $\mathbf{C}$, $\mathbf{D}$ being correct, which can result in surprisingly poor estimates if these are only approximate – a situation that can arise where for example the state equations have been derived by a linearization procedure (Eqs. (3.7), (3.8)). In contrast, with its inherent corrective property, an asymptotic estimator can give excellent results in similar circumstances.

3.9 MODELING AND BEHAVIOR OF STOCHASTIC SYSTEMS (REFS. 1, 6)

(i) Systems Affected by White Noise

A stochastic, or 'noisy', linear system may be modeled by an extension of Eqs. (3.3), (3.4):

$$\dot{\mathbf{x}} = \mathbf{A}(t)\mathbf{x} + \mathbf{B}(t)\mathbf{u} + \mathbf{w}(t) \tag{3.58}$$

$$\mathbf{y} = \mathbf{C}(t)\mathbf{x} + \mathbf{D}(t)\mathbf{u} + \mathbf{v}(t) \tag{3.59}$$

$\mathbf{w}(t)$ and $\mathbf{v}(t)$ are usually vectors of Gaussian white noise (Sec. 1.21) with known means and covariances (App. B).

$$\mathscr{E}(\mathbf{w}(t)) = \mathbf{0}$$

$$\text{cov}(\mathbf{w}(t)) = \mathscr{E}[\mathbf{w}(t)\mathbf{w}^{\mathrm{T}}(t+\tau)] = \mathbf{Q}\delta(\tau)$$

where $\delta(\tau)$ is a Dirac delta function (App. C.1).

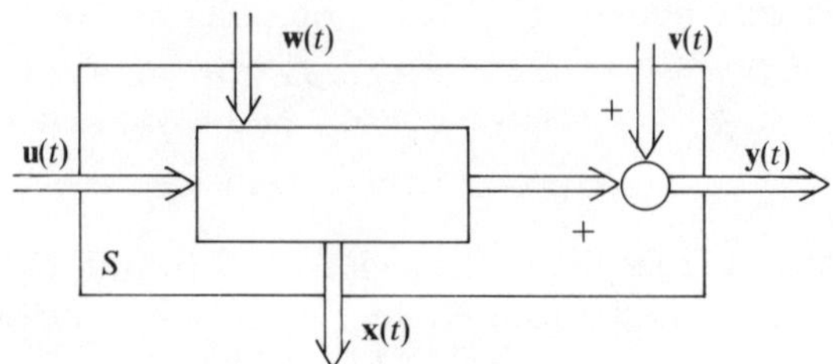

Fig. 3.25 State representation of linear stochastic system S

Also,
$$\mathscr{E}(\mathbf{v}(t)) = \mathbf{0}$$
$$\operatorname{cov}(\mathbf{v}(t)) = \mathscr{E}[\mathbf{v}(t)\mathbf{v}^{\mathrm{T}}(t+\tau)] = \mathbf{R}\delta(\tau)\,.$$

The model of Eqs. (3.58), (3.59) is, strictly speaking, inadequate, owing to the mathematical properties of the white noise **w**. A rigorous analysis requires a better model using stochastic differential equations (refs. 1, 6). Equations (3.58), (3.59), however, do amount to a useful representation which with certain limitations allows correct results to be deduced. This model is therefore accepted in this section, though some of the results presented cannot be derived directly from it.

A block diagram of the stochastic system is shown in Fig. 3.25.

$\mathbf{w}(t)$ represents a noise forcing function which disturbs the system S – e.g. wave motion affecting the motion of a submarine near the surface, while $\mathbf{v}(t)$ represents measurement noise – e.g. noise on the detected depth of the submarine caused by the waves above it. (Admittedly, neither noise source in this case can be regarded as white, but the example remains helpful.)

Noisy LTI systems may be modeled by the corresponding extensions of Eqs. (3.5), (3.6):

$$\mathbf{x} = \mathbf{A}\mathbf{x} + \mathbf{B}\mathbf{u} + \mathbf{w} \tag{3.60}$$
$$\mathbf{y} = \mathbf{C}\mathbf{x} + \mathbf{D}\mathbf{u} + \mathbf{v} \tag{3.61}$$

where **A**, **B**, **C**, **D** are constant matrices and **w**, **v** are usually Gaussian white noise vectors.

(ii) Systems Affected by 'Colored' Noise

In the above models, **w**, **v** are taken as Gaussian white noise; but where this is inappropriate, as it very often is, nonwhite, or 'colored' noise can be modeled by augmenting the state equations as indicated in Fig. 3.26.

The colored noise forcing function on S is $\mathbf{y}_{\mathrm{w}}$ (colored), which is generated by Gaussian white noise, **w**, via the filter S_{w} (Sec. 1.21).

S_{w} has state equations:

$$\dot{\mathbf{x}}_{\mathrm{w}} = \mathbf{A}_{\mathrm{w}}\mathbf{x}_{\mathrm{w}} + \mathbf{B}_{\mathrm{w}}\mathbf{w}$$
$$\mathbf{y}_{\mathrm{w}} = \mathbf{C}_{\mathrm{w}}\mathbf{x}_{\mathrm{w}} + \mathbf{D}_{\mathrm{w}}\mathbf{w}$$

where $\mathbf{A}_{\mathrm{w}}$, $\mathbf{B}_{\mathrm{w}}$, $\mathbf{C}_{\mathrm{w}}$, $\mathbf{D}_{\mathrm{w}}$ are constant matrices.

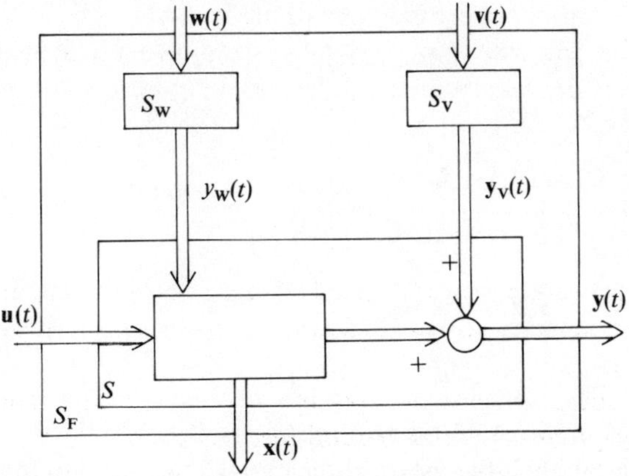

Fig. 3.26 Augmented system modeling colored noise.

The output of S is corrupted with colored noise $\mathbf{y}_v$ which is generated by Gaussian white noise, $\mathbf{v}$, via the filter S_v.

S_v has state equations:

$$\dot{\mathbf{x}}_v = \mathbf{A}_v\mathbf{x}_v + \mathbf{B}_v\mathbf{v}$$
$$\mathbf{y}_v = \mathbf{C}_v\mathbf{x}_v + \mathbf{D}_v\mathbf{v}.$$

Combining Eqs. (3.60), (3.61) with these gives the augmented system state equations for S_F:

$$\begin{bmatrix}\dot{\mathbf{x}}\\ \dot{\mathbf{x}}_w\\ \dot{\mathbf{x}}_v\end{bmatrix} = \begin{bmatrix}\mathbf{A} & \mathbf{C}_w & 0\\ 0 & \mathbf{A}_w & 0\\ 0 & 0 & \mathbf{A}_v\end{bmatrix}\begin{bmatrix}\dot{\mathbf{x}}\\ \mathbf{x}_w\\ \mathbf{x}_v\end{bmatrix} + \begin{bmatrix}\mathbf{D}_w\\ \mathbf{B}_w\\ 0\end{bmatrix}\mathbf{w} + \begin{bmatrix}0\\ 0\\ \mathbf{B}_v\end{bmatrix}\mathbf{v} + \begin{bmatrix}\mathbf{B}\\ 0\\ 0\end{bmatrix}\mathbf{u} \tag{3.62}$$

$$\mathbf{y} = [\mathbf{C} \quad 0 \quad \mathbf{C}_v]\begin{bmatrix}\mathbf{x}\\ \mathbf{x}_w\\ \mathbf{x}_v\end{bmatrix} + \mathbf{D}\mathbf{u} + \mathbf{D}_v\mathbf{v} \tag{3.63}$$

These equations are of the form Eqs. (3.60), (3.61), with the noteworthy feature that the forcing function (white) noise and measurement (white) noise are not independent.

(iii) The Mean and Covariance of the State and Output

Since Eqs. (3.60), (3.61) involve stochastic variables $\mathbf{w}$, $\mathbf{v}$, the state $\mathbf{x}$ and output $\mathbf{y}$ are also stochastic variables which are rather coarsely described by (Sec. 1.21):

$$\begin{aligned}\mathbf{m}_x(t) &= \mathscr{E}(\mathbf{x}(t))\\ \bar{\mathbf{R}}_x(t, t+\tau) &= \operatorname{cov}(\mathbf{x}(t))\\ &= \mathscr{E}[(\mathbf{x}(t) - \mathbf{m}_x(t))(\mathbf{x}(t+\tau) - \mathbf{m}_x(t+\tau))^{\mathrm{T}}].\end{aligned}$$

Slightly fuller symbols are used here than in Sec. 1.21, in the interests of clarity. It can be shown, though not by straightforward means, from Eqs. (3.60), (3.61), that the deterministic differential equations governing $\mathbf{m}_x(t)$ and $\bar{\mathbf{R}}_x(t, t)$ are, for $\mathbf{u}(t) = 0$:

$$\dot{\mathbf{m}}_x(t) = \mathbf{A}(t)\mathbf{m}_x(t), \tag{3.64}$$

$$\dot{\bar{\mathbf{R}}}_x(t) = \mathbf{A}(t)\bar{\mathbf{R}}_x(t) + \bar{\mathbf{R}}_x(t)\mathbf{A}^{\mathrm{T}}(t) + \mathbf{Q} \tag{3.65}$$

where

$$\bar{\mathbf{R}}_x(t) = \bar{\mathbf{R}}_x(t, t).$$

Also:

$$\bar{\mathbf{R}}_x(t, t+\tau) = \boldsymbol{\phi}(t+\tau, t)\bar{\mathbf{R}}_x(t) \tag{3.66}$$

where $\boldsymbol{\phi}(t+\tau, t)$ is the state transition matrix (Eq. (3.9)).

In the case of an LTI system, Eqs. (3.64), (3.65), (3.66) may be solved to give:

$$\mathbf{m}_x(t) = e^{\mathbf{A}t}\mathbf{m}_x(0) \tag{3.67}$$

$$\bar{\mathbf{R}}_x(t) = e^{\mathbf{A}t}\bar{\mathbf{R}}_x(0)(e^{\mathbf{A}t})^{\mathrm{T}} + \int_0^t e^{\mathbf{A}(t-\lambda)}\mathbf{Q}(e^{\mathbf{A}(t-\lambda)})^{\mathrm{T}}\,\mathrm{d}\lambda \tag{3.68}$$

$$\bar{\mathbf{R}}_x(t, t+\tau) = e^{\mathbf{A}\tau}\bar{\mathbf{R}}_x(t). \tag{3.69}$$

It is obvious from these equations that the values of $\mathbf{m}_x(t)$, $\bar{\mathbf{R}}_x(t)$, and $\bar{\mathbf{R}}_x(t, t+\tau)$ develop from arbitrary initial conditions (perhaps at switch-on) in a manner roughly comparable to the value of $\mathbf{x}(t)$ in a deterministic system (Eq. (3.10)), and that they converge to steady-state, stationary, values if the system is stable (the eigenvalues of $\mathbf{A}$ have negative real parts), in a manner influenced by state feedback. A simple CAD facility can be constructed to find these. The design of this is very similar to that of the algorithm in Fig. 3.4, with slightly more complex algebra to accommodate the differences between Eqs. (3.67), (3.68), (3.69) and Eq. (3.10).

3.10 MULTIVARIABLE SYSTEM DESIGN METHODS (REFS. 7, 9)

In contrast to the state feedback methods outlined in Secs. 3.6 to 3.9, there exists a number of output feedback techniques which extend the classical ideas of Chap. 2 to LTI multivariable systems. These seem particularly relevant to certain types of process control problem.

One such technique is described here in a form suited to systems with equal numbers of inputs and outputs and strictly proper transfer function elements (Sec. 1.5). More complicated cases, which are less usual, can be dealt with by elaborations of these ideas (ref. 7).

The design task is to control an r-input, r-output system with $(r \times r)$ transfer function matrix $\mathbf{G}(s)$ (Eq. (3.14)), using the strategy outlined in the block diagram of Fig. 3.27.

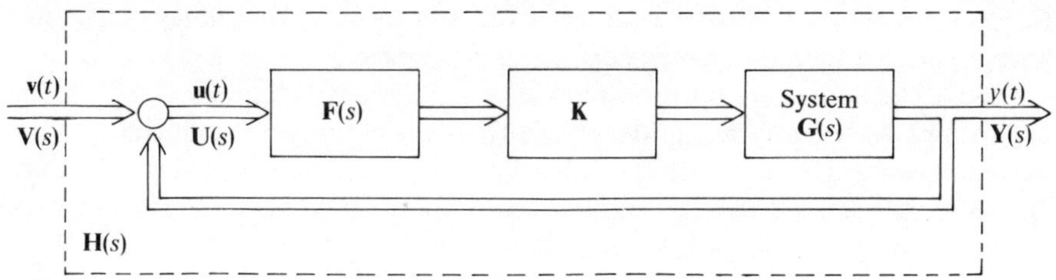

Fig. 3.27 Multivariable system control strategy.

This entails designing a constant 'partial decoupling' matrix **K**, and a diagonal shaping network matrix **F**(*s*) (cf. Fig. 2.19).

The closed-loop transfer function matrix is:

$$\mathbf{H}(s) = (\mathbf{I} + \mathbf{G}(s)\mathbf{K}\mathbf{F}(s))^{-1}\mathbf{G}(s)\mathbf{K}\mathbf{F}(s) \tag{3.70}$$

and the inverse closed-loop transfer function matrix is:

$$\hat{\mathbf{H}}(s) = \hat{\mathbf{F}}(s)\hat{\mathbf{K}}\hat{\mathbf{G}}(s) + \mathbf{I} \tag{3.71}$$

where, using the nomenclature usual in this context,

$$\hat{\mathbf{F}}(s) = (\mathbf{F}(s))^{-1}, \quad \hat{\mathbf{K}} = \mathbf{K}^{-1}, \quad \hat{\mathbf{G}}(s) = (\mathbf{G}(s))^{-1}.$$

Extensions of the Nyquist and inverse Nyquist diagram design methods (Secs. 2.4, 2.6, 2.7) can be used to design **K**, **F**(*s*) so that the system **H**(*s*) is well behaved. For a number of reasons, including the easy relationship between the open- and closed-loop systems (Eq. (3.71)), the inverse Nyquist method is the more popular, and this is described here.

The first task is to find a matrix, $\hat{\mathbf{K}}$, which makes the matrix $\hat{\mathbf{K}}\hat{\mathbf{G}}(j\omega)$ 'row diagonally dominant', $0 \leqslant \omega \leqslant \infty$.

$$\hat{\mathbf{Q}}(j\omega) = \hat{\mathbf{K}}\hat{\mathbf{G}}(j\omega)$$

$$= \begin{bmatrix} \hat{q}_{11}(j\omega) & \hat{q}_{12}(j\omega) & \cdots & \hat{q}_{rr}(j\omega) \\ \hat{q}_{21}(j\omega) & & \cdots & \\ \vdots & & & \\ \hat{q}_{r1}(j\omega) & & \cdots & \hat{q}_{rr}(j\omega) \end{bmatrix}.$$

$\hat{Q}(j\omega)$ is row diagonally dominant if:

$$|\hat{q}_{ii}(j\omega)| > \sum_{\substack{h=1 \\ i \neq h}}^{r} |\hat{q}_{ih}(j\omega)|, \qquad \begin{array}{l} 0 \leqslant \omega \leqslant \infty, \\ i = 1, 2, \ldots, r. \end{array} \tag{3.72}$$

An inverse Nyquist diagram is plotted for each (already 'inverse') element $\hat{q}_{ii}(j\omega)$, $0 \leqslant \omega \leqslant \infty$ (Sec. 2.6) under the initial assumption that $\mathbf{K} = \mathbf{I}$, i.e. $\hat{\mathbf{Q}}(j\omega) = \hat{\mathbf{G}}(j\omega)$. *Gershgorin* circles, with centres on $\hat{q}_{ii}(j\omega)$ and radii equal to the sum of the off-diagonal elements (as in Eq. (3.72)) are superimposed on each diagram for various values of ω. An envelope of these circles, known as a

Gershgorin band, is also drawn on each diagram. r such diagrams, one corresponding to each element $\hat{q}_{ii}(j\omega)$, are prepared.

$\hat{\mathbf{K}}\hat{\mathbf{G}}(j\omega)$ is diagonally dominant if no Gershgorin band, $i = 1, 2, \ldots, r$, encloses $(0 + j0)$. If any band does enclose $(0 + j0)$, a decoupling matrix $\hat{\mathbf{K}} \neq \mathbf{I}$ is designed using, if necessary, a special design algorithm for the purpose. With $\hat{\mathbf{K}}$ designed, the diagrams are redrawn, and these may now be regarded as representing $\hat{\mathbf{F}}(j\omega)\hat{\mathbf{K}}\hat{\mathbf{G}}(j\omega)$, $\hat{\mathbf{F}}(j\omega) = \mathbf{I}$.

The second task is to determine if the closed-loop system $\mathbf{H}(s)$ is stable and well behaved, and if not, to design $\hat{\mathbf{F}}(j\omega) \neq \mathbf{I}$ to make it so. The design criteria and procedure for this are as follows.

If $\hat{\mathbf{G}}(s)$ has no right half-plain zeros ($\mathbf{G}(s)$ has no r.h.p. poles), $\mathbf{H}(s)$ is stable (with $\mathbf{F}(s) = \mathbf{I}$) if no Gershgorin band encircles, even partially, the point $(-1 + j0)$ on the right, $0 \leqslant \omega \leqslant \infty$. If $\hat{\mathbf{G}}(s)$ does have r.h.p. zeros, the criterion for stability of $\mathbf{H}(s)$ is significantly more complicated (refs. 7, 9), but this, fortunately, is unusual.

For $\mathbf{H}(s)$ to be *guaranteed* well behaved, $\mathbf{F}(s) = \mathbf{I}$, all the Gershgorin bands should avoid $(-1 + j0)$ by margins corresponding to the rules for single-input, single-output inverse Nyquist diagram design (Secs. 2.6, 2.7). However, it is quite possible for the bands to be very close to, or even include $(-1 + j0)$ and for the closed-loop system to be stable and well-behaved. Confirmation of the closed-loop behavior by simulation seems an essential adjunct to this method. Should this behavior be unsatisfactory at this stage, a diagonal shaping network matrix, $\mathbf{F}(s)$ is designed:

$$\mathbf{F}(s) = \begin{bmatrix} f_{11}(s) & 0 & \cdots & 0 \\ 0 & f_{22}(s) & \cdots & 0 \\ \vdots & & & \vdots \\ 0 & & \cdots & f_{rr}(s) \end{bmatrix}.$$

The method allows each element $f_{ii}(s)$ to be designed independently as if it affected only the corresponding diagonal element of $\mathbf{G}(s)\mathbf{K}$ (which is not of course true). In practice, the inverse transform, $s = j\omega$, of the shaping network is designed:

$$\hat{\mathbf{F}}(j\omega) = \begin{bmatrix} \hat{f}_{11}(j\omega) & 0 & \cdots & 0 \\ 0 & \hat{f}_{22}(j\omega) & \cdots & 0 \\ \vdots & & & \vdots \\ 0 & & \cdots & \hat{f}_{rr}(j\omega) \end{bmatrix}.$$

Each element of this, $\hat{f}_{ii}(j\omega)$, is designed using the inverse Nyquist diagram, including its Gershgorin bands, of the corresponding system element, $\hat{q}_{ii}(j\omega)$. $\hat{f}_{ii}(j\omega)$ is chosen so that the Gershgorin band on $[\hat{f}_{ii}(j\omega)\hat{q}_{ii}(j\omega)]$, which has circle radii (Eq. 3.72)

$$\sum_{\substack{h=1 \\ h \neq i}}^{r} |f_{ii}(j\omega) q_{ih}(j\omega)|,$$

is an improvement on the uncompensated band. Strictly, this can be achieved by ensuring that the band is tangential to a good M circle (Secs. 2.6, 2.7), but this is a very severe requirement, generally, and narrower *Ostrowski* bands can be defined and calculated (ref. 7) to allow more room for maneuver. Alternatively, simulation runs can be interspersed with attempts to improve the Gershgorin band configurations to the point where satisfactory performance is achieved.

In summary, **K** is designed to achieve row diagonal dominance of $\hat{\mathbf{K}}\hat{\mathbf{G}}(j\omega)$. Gershgorin bands are then used on inverse Nyquist diagrams to indicate the attainment of this, and to suggest suitable shaping elements $\hat{f}_{ii}(j\omega)$. A diagonal matrix $\mathbf{F}(s)$ of shaping network elements is compiled from these, and the design confirmed by simulation.

CAD Facility

The interactive algorithm shown in Fig. 3.28 allows the procedure described above to be implemented.

The operation of the algorithm is as follows.

BLOCK 1

The system transfer function matrix $\mathbf{G}(s)$ is requested, input, displayed, and corrected if necessary. This is perhaps best organized by specifying the number of elements and then each element in the form of a gain and a number of poles and zeros.

BLOCKS 2, 3

The design of a 'partial decoupling' matrix, **K**, is commenced by specifying a frequency, ω, and a row, i, to be considered.

BLOCK 4

The ith row vector of the matrix $\hat{\mathbf{K}}$ is calculated and displayed by the following procedure (ref. 9).

For convenience, let the elements of $\hat{\mathbf{G}}(j\omega)$ be designated:

$$\hat{g}_{hl}(j\omega) = \alpha_{hl} + j\beta_{hl}.$$

The values of α_{hl}, β_{hl} are calculated and a symmetric square matrix, $\mathbf{A}_i$, found whose elements $a_{hn}^{(i)}$ are:

$$a_{hn}^{(i)} = \sum_{\substack{l=1 \\ l \neq i}}^{r} (\alpha_{hl}\alpha_{nl} + \beta_{hl}\beta_{nl}), \qquad \begin{aligned} h &= 1, 2, \ldots, r \\ n &= 1, 2, \ldots, r. \end{aligned}$$

The eigenvectors and eigenvalues of $\mathbf{A}_i$ (which is real and symmetric) are now found (App. A.9), and the vector $\hat{\mathbf{k}}_i$ corresponding to the smallest eigenvalue is selected, stored, and displayed. $\hat{\mathbf{k}}_i$ is the ith row of the matrix $\hat{\mathbf{K}}$.

$$\hat{\mathbf{K}} = (\hat{\mathbf{k}}_1, \hat{\mathbf{k}}_2, \ldots, \hat{\mathbf{k}}_r)^{\mathrm{T}}.$$

This process is repeated, $i = 1, 2, \ldots, r$, until the matrix $\hat{\mathbf{K}}$ is assembled.

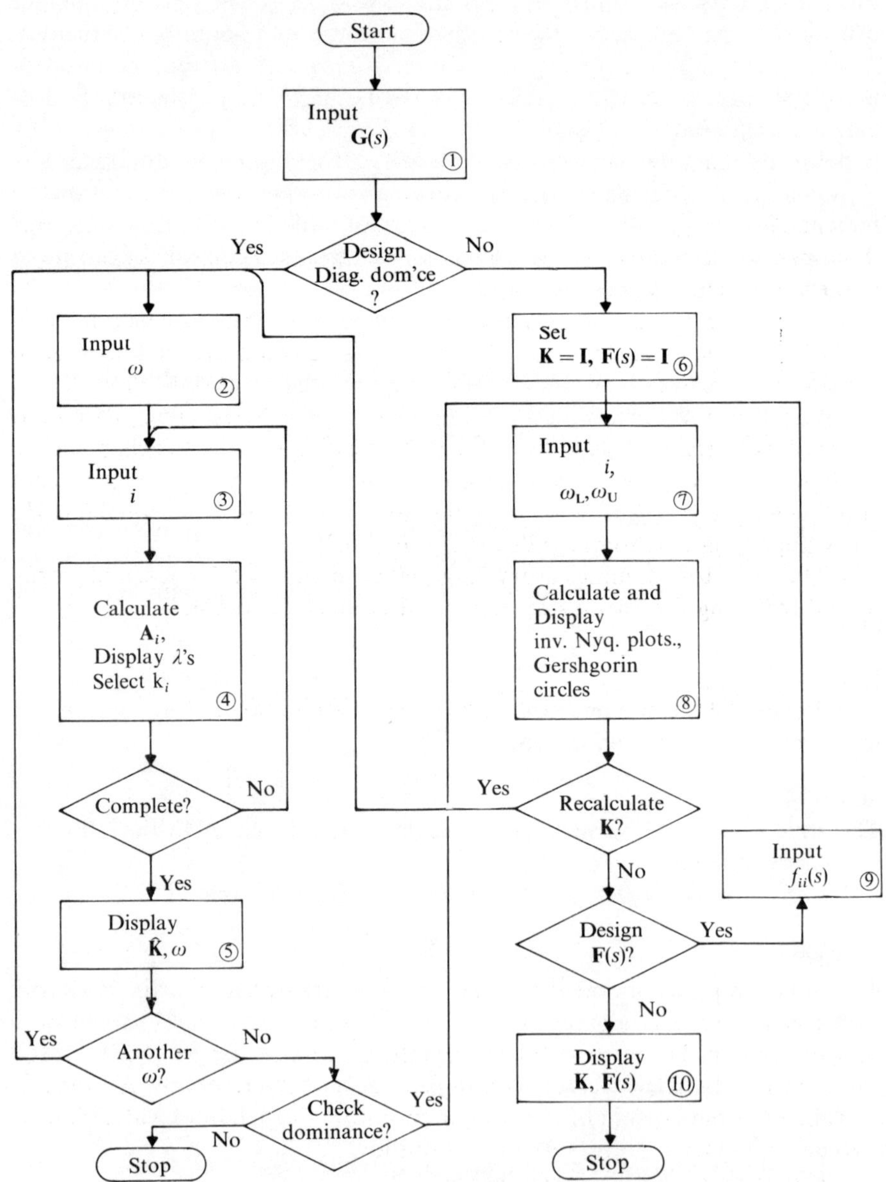

Fig. 3.28 Multivariable system design algorithm.

BLOCK 5

$\hat{\mathbf{K}}$ and ω are displayed, and a decision taken by the operator whether to repeat the entire process for a different frequency, or to check the diagonal dominance of $\hat{\mathbf{K}}\hat{\mathbf{G}}(s)$ for a range of frequencies by means of Gershgorin bands.

BLOCK 6

The matrices $\hat{\mathbf{F}}(s)$ and $\hat{\mathbf{K}}$ are set $\hat{\mathbf{F}}(s) = \mathbf{I}$, $\hat{\mathbf{K}} = \mathbf{I}$, later to be overwritten by matrices designed in the appropriate blocks if these are called up.

BLOCKS 7, 8, 9

The operator selects a diagonal element, ii, of the $(r \times r)$ matrix $\hat{\mathbf{K}}\hat{\mathbf{G}}(s)$, and a range of frequencies $\omega_L \leqslant \omega \leqslant \omega_U$.

An inverse Nyquist diagram for this diagonal element is generated and displayed on the VDU (in graphics mode or its equivalent) and a number of Gershgorin circles are calculated and superimposed on this display.

The operator now decides whether the system is diagonally dominant at all frequencies (band does not include $(0 + j0)$), and whether the shape of the band indicates stability (band does not enclose $(-1 + j0)$, $\omega_L \leqslant \omega \leqslant \omega_U$) and good behavior (margin between band and $(-1 + j0)$ is reasonable). He takes appropriate action as indicated in Fig. 3.28.

The operator examines every diagonal element, $i = 1, 2, \ldots, r$, and works to and fro using the various options until the task is complete.

BLOCK 10

The final matrices $\mathbf{F}(s)$, $\mathbf{K}$ are calculated and displayed:

$$\mathbf{F}(s) = \begin{bmatrix} \hat{f}_{11}^{-1}(s) & 0 & \cdots & 0 \\ 0 & \hat{f}_{22}^{-1}(s) & \cdots & 0 \\ \vdots & & & \vdots \\ 0 & & \cdots & \hat{f}_{rr}^{-1}(s) \end{bmatrix}$$

$$\mathbf{K} = \hat{\mathbf{K}}^{-1}.$$

EXAMPLE

A useful example is provided by laboratory equipment in which a reservoir tank feeds water via a precisor-driven valve and short delay channel to a perspex column. The water is taken from the bottom of this via another valve to a sump tank, from which it is pumped back to the reservoir. A pressure transducer at the bottom of the column monitors the height of water while a turbine flowmeter monitors its rate of outflow.

The arrangement is shown in Fig. 3.29.

The valve precisors are driven via electropneumatic transducers and the flow and pressure meters feed suitable amplifiers.

u_1 is the input to the upper valve (volts),
u_2 is the input to the lower valve (volts),

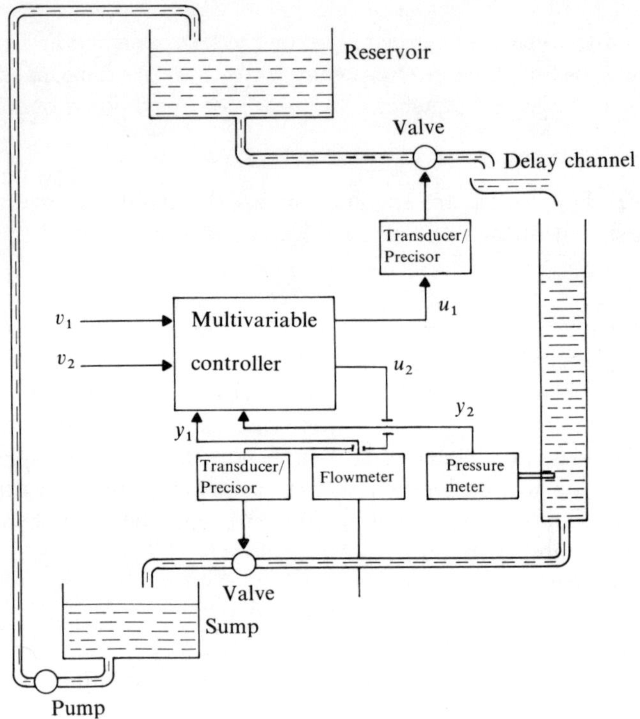

Fig. 3.29 Laboratory multivariable system (example).

y_1 is the flow transducer amplifier output (volts),
y_2 is the pressure transducer amplifer output (volts).

A transfer function analyzer can be used to develop Bode plots for the various transfer function matrix elements, and from these a transfer function matrix is constructed:

$$\mathbf{Y}(s) = \mathbf{G}(s)\mathbf{U}(s), \text{ where:}$$

$$\mathbf{G}(s) = \begin{bmatrix} \dfrac{1.5e^{-0.2s}}{(s+3)} & \dfrac{1.5s}{(s+1)(s+3)} \\ \dfrac{3e^{-0.2s}}{(s+3)} & \dfrac{-3}{(s+3)} \end{bmatrix}.$$

Inverting this:

$$\hat{\mathbf{G}}(s) = (s+3) \begin{bmatrix} \dfrac{(s+1)e^{0.2s}}{3(s+0.5)} & \dfrac{se^{0.2s}}{6(s+0.5)} \\ \dfrac{(s+1)}{3(s+0.5)} & \dfrac{-1(s+1)}{6(s+0.5)} \end{bmatrix}$$

As with classical process systems (Sec. 2.15), it is sometimes helpful to model

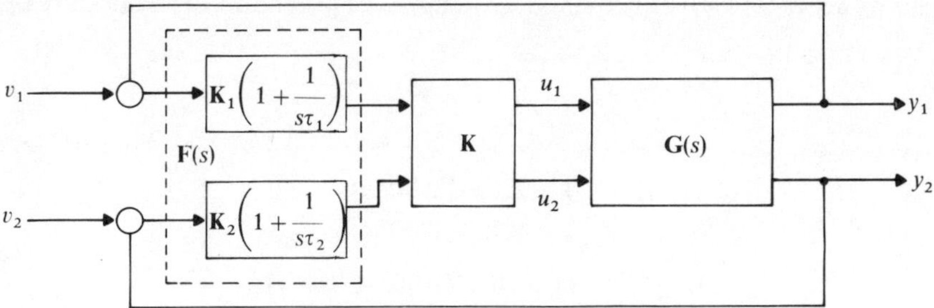

Fig. 3.30 Multivariable control strategy (example).

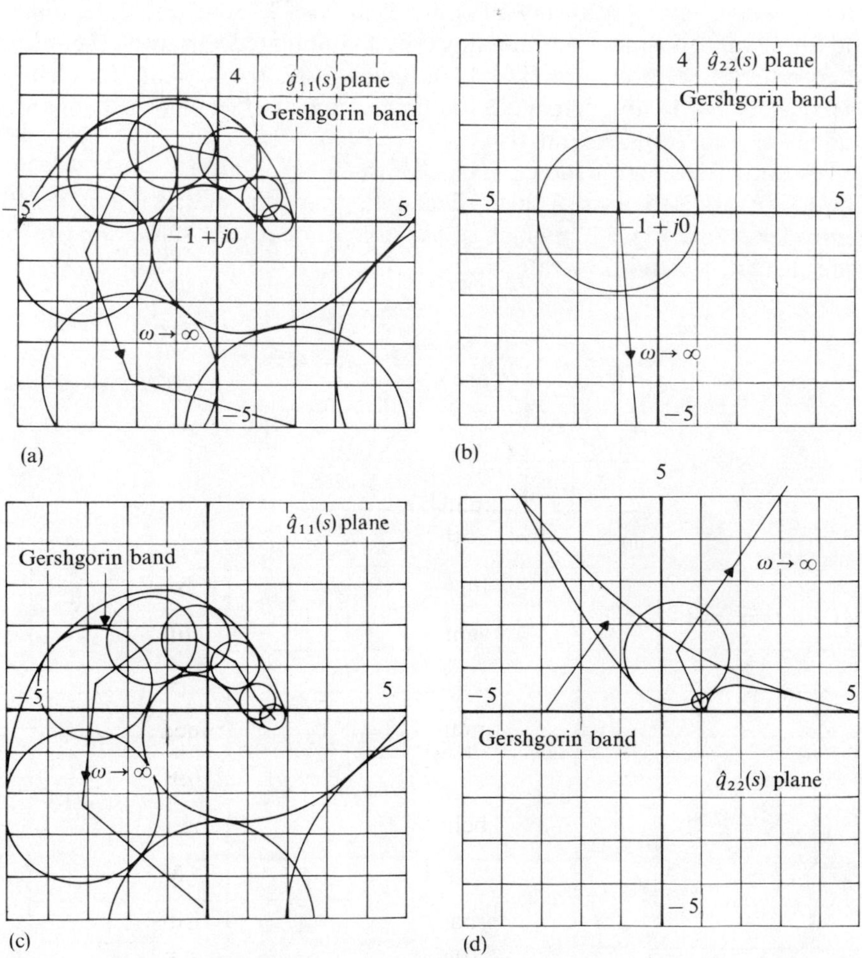

(a) (b)

(c) (d)

Fig. 3.31 Inverse Nyquist diagrams with Gershgorin bands (examples). (a) Diagram for $\hat{g}_{11}(j\omega)$. (b) Diagram for $\hat{g}_{22}(j\omega)$. (c) Diagram for $\hat{q}_{11}(j\omega)$. (d) Diagram for $\hat{q}_{22}(j\omega)$.

transport lag (Eq. (1.8)) approximately in terms of poles and zeros. In the case above:

$$e^{-0.2s} = \frac{e^{-0.1s}}{e^{0.1s}}$$

$$\approx \frac{1 - 0.1s + 0.005s^2}{1 + 0.1s + 0.005s^2}$$

$$= \frac{(s - 10 + j10)(s - 10 - j10)}{(s + 10 + j10)(s + 10 - j10)}.$$

It is a simple matter to control either the column output flow rate y_1 or the column water height, y_2, using PI controllers (Sec. 2.15) in which the upper and lower valve positions are the respective manipulated variables. However, when both these loops are closed simultaneously, the system is violently unstable: $\hat{\mathbf{G}}(j\omega)$ is not diagonally dominant and its Gershgorin bands include $(-1 + j0)$ (Figs. 3.31(a), (b)).

A control strategy is shown in block form in Fig. 3.30 (cf. Figs. 3.27, 3.29). The design task is to devise a partial decoupling constant matrix $\mathbf{K}$ and two PI controllers ($\mathbf{F}(s)$, Fig. 3.27) which allow independent and stable control of water height, y_2, and flow rate, y_1.

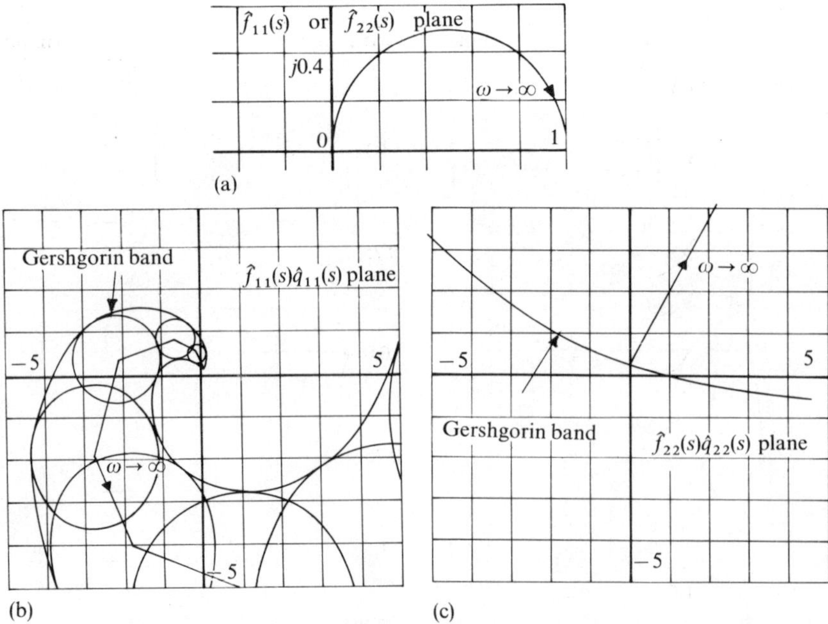

Fig. 3.32 Inverse Nyquist diagrams with Gershgorin bands (example). (a) Diagram for $\hat{f}_{11}(j\omega)$, $\hat{f}_{22}(j\omega)$. (b) Diagram for $\hat{f}_{11}(j\omega)\hat{q}_{11}(j\omega)$. (c) Diagram for $\hat{f}_{22}(j\omega)\hat{q}_{22}(j\omega)$.

In this case, $\hat{\mathbf{G}}(s)$ is simple enough to allow a matrix $\hat{\mathbf{K}}$ to be designed by inspection, so that $\hat{\mathbf{K}}\hat{\mathbf{G}}(j\omega)$ is row diagonally dominant at least over a useful frequency range.

$$\hat{\mathbf{K}} = \begin{bmatrix} 1 & 0 \\ 1 & -1 \end{bmatrix} \quad \text{or} \quad \mathbf{K} = (\hat{\mathbf{K}})^{-1} = \begin{bmatrix} 1 & 0 \\ 1 & -1 \end{bmatrix}.$$

The row diagonal dominance of $\hat{\mathbf{K}}\hat{\mathbf{G}}(j\omega)$ is demonstrated by the CAD algorithm of Fig. 3.28; the inverse Nyquist diagrams, and Gershgorin bands, for its diagonal elements are shown in Figs. 3.31(c), (d), where the diagonal

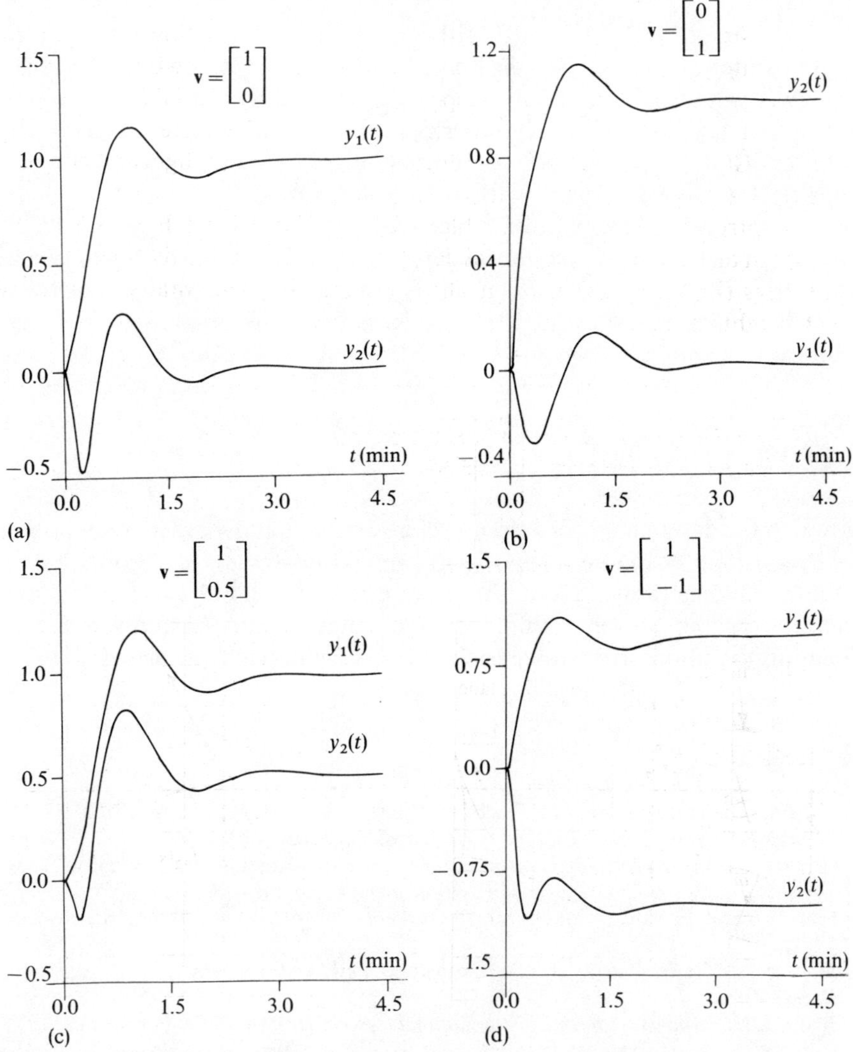

Fig. 3.33 Closed-loop system behavior (example). (a) Setpoint change to flow only. (b) Setpoint change to height only. (c) Setpoint changes to flow and height. (d) Setpoint changes to flow and height.

elements of $(\hat{\mathbf{K}}, \hat{\mathbf{G}}(j\omega))$ are designated $(\hat{\mathbf{K}}\hat{\mathbf{G}}(j\omega))_{11} = \hat{q}_{11}(j\omega)$, $(\hat{\mathbf{K}}\hat{\mathbf{G}}(j\omega))_{22} = \hat{q}_{22}(j\omega)$ respectively.

From Figs. 3.31(c), (d), it is clear that $\hat{\mathbf{K}}\hat{\mathbf{G}}(j\omega) = \hat{\mathbf{Q}}(j\omega)$ is diagonally dominant and that the system would be stable if the loops were simply closed $(\mathbf{F}(s) = \mathbf{I})$. This would give poor performance, however, since good inverse Nyquist criteria (Secs. 2.6, 2.7) are not met, and the lack of any integral control would allow steady state errors.

PI controllers (Sec. 2.15) can easily be included to bring the Gershgorin bands, especially that of Fig. 3.31(d), near good M circles.

After some experimentation, including some simulation runs, PI controllers (Sec. 2.15) are devised as follows:

$$\hat{f}_{11}(s) = \hat{f}_{22}(s) = \frac{s}{s+4}$$

or

$$f_{11}(s) = f_{22}(s) = 1\left(1 + \frac{4}{s}\right)(PI \text{ controller, } PB = 100\%,\ \tau_i = 0.25).$$

The inverse Nyquist diagrams for $\hat{f}_{11}(j\omega)$, $\hat{f}_{22}(j\omega)$ is in Fig. 3.32(a), while those for $\hat{f}_{11}(j\omega)\cdot\hat{q}_{11}(j\omega)$ and $\hat{f}_{22}(j\omega)\hat{q}_{22}(j\omega)$ are in Fig. 3.32(b), (c) respectively.

The behavior of the closed-loop system for various combinations of setpoint step demands is shown in Fig. 3.33. Although the Gershgorin band of Fig. 3.32(b) virtually overlaps $(-1 + j0)$, and diagonal dominance can be shown not to exist for all frequencies, the system behavior is reasonably satisfactory.

Comment

Although the design method described here does not allow exact prediction of the closed-loop behavior owing to the breadth of the Gershgorin bands (leading to uncertainty about the attainment of a good inverse Nyquist diagram), as the above example demonstrates, it can certainly assist in designing useful control strategies for these very difficult problems.

REFERENCES

1. Astrom, K.J., *Introduction to Stochastic Control Theory.* Academic Press, 1970.
2. Athans, M., and Falb, P.L., *Optimal Control.* McGraw-Hill, 1966.
3. Chen, C.T., *Linear System Theory and Design.* Holt, Rinehart and Winston, 1984.
4. Fallside, F., *Control System Design by Pole Zero Assignment.* Academic Press, 1977.
5. Kwakernaak, H., and Sivan, R., *Linear Optimal Control Systems.* Wiley Interscience, 1972.
6. Maybeck, P.S., *Stochastic Models, Estimation, and Control,* Vols. I, II, III. Academic Press, 1982.
7. Munro, N., *Modern Approaches to Control System Design.* Peter Peregrinus, 1979.
8. Ralston, A., and Rabinowitz, P. *A First Course in Numerical Analysis.* McGraw-Hill, 1978.
9. Rosenbrock, H.H., *Computer Aided Control System Design.* Academic Press, 1974.

4 DISCRETE TIME STATE SPACE SYSTEM MODELS

4.1 INTRODUCTION

Since the basic analysis of discrete time systems closely parallels that of continuous time systems, the following sections are brief and depend on cross-reference to Chap. 3 for explanation of much of the detail.

4.2 STATE EQUATIONS (CF. SEC. 3.2) (REFS. 1, 2, 3, 4)

Most discrete systems can be modeled by a set of first-order difference equations of the form:

$$\begin{aligned}\mathbf{x}(n+1) &= \mathbf{f}(\mathbf{x}(n), \mathbf{u}(n), n) \\ \mathbf{y}(n) &= \mathbf{g}(\mathbf{x}(n), \mathbf{u}(n), n)\end{aligned} \tag{4.1}$$

where $\mathbf{x}(n)$ is the state vector $(x_1(n) \quad x_2(n) \quad \ldots \quad x_m(n))^{\mathrm{T}}$, $x_i(n)$ a real variable.
$\mathbf{u}(n)$ is the input vector $(u_1(n) \quad u_2(n) \quad \ldots \quad u_q(n))^{\mathrm{T}}$, $u_i(n)$ a real variable.
$\mathbf{y}(n)$ is the output vector $(y_1(n) \quad y_2(n) \quad \ldots \quad y_r(n))^{\mathrm{T}}$, $y_i(n)$ a real variable.
n is the independent variable, discrete time $(-\infty, \ldots, -1, 0, 1, 2, \ldots + \infty)$.

This is shown schematically in Fig. 4.1.

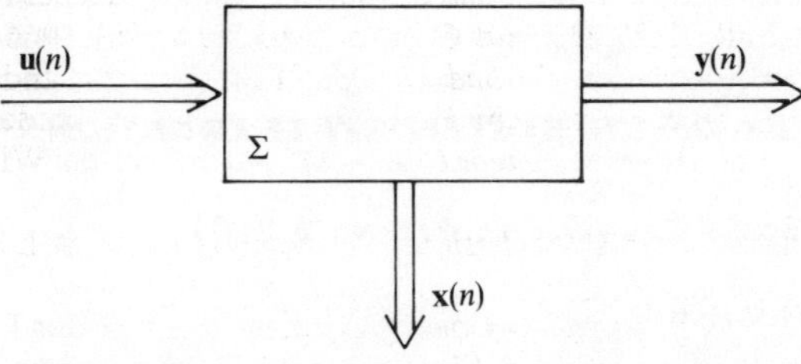

Fig. 4.1 State space representation of a discrete time system Σ.

As with continuous time systems, linear discrete systems may be represented by a subset of these equations:

$$\mathbf{x}(n+1) = \boldsymbol{\phi}(n)\mathbf{x}(n) + \boldsymbol{\psi}(n)\mathbf{u}(n) \tag{4.2}$$

$$\mathbf{y}(n) = \mathbf{C}(n)\mathbf{x}(n) + \mathbf{D}(n)\mathbf{u}(n) \tag{4.3}$$

where the matrices $\boldsymbol{\phi}(n)$, $\boldsymbol{\psi}(n)$, $\mathbf{C}(n)$, $\mathbf{D}(n)$ vary with time, $\mathbf{n}$.

Linear time invariant (LTI) systems may be represented by equations with constant coefficients:

$$\mathbf{x}(n+1) = \boldsymbol{\phi}\mathbf{x}(n) + \boldsymbol{\psi}\mathbf{u}(n) \tag{4.4}$$

$$\mathbf{y}(n) = \mathbf{C}\mathbf{x}(n) + \mathbf{D}\mathbf{u}(n) \tag{4.5}$$

where $\boldsymbol{\phi}$ (the 'state transition matrix'), $\boldsymbol{\psi}$, $\mathbf{C}$, $\mathbf{D}$ are all constant matrices.

EXAMPLE

A discrete system is governed by the difference equation:

$$y(n+3) - 2.4y(n+2) + 1.85y(n+1) - 0.45y(n) = 400u(n) \tag{4.6}$$

where $u(n)$ is the input to the system,
$y(n)$ is the output.

Let:

$$x_1(n) = y(n)$$
$$x_2(n) = y(n+1)$$
$$x_3(n) = y(n+2).$$

Equation (4.6) can now be written:

$$\begin{bmatrix} x_1(n+1) \\ x_2(n+1) \\ x_3(n+1) \end{bmatrix} = \begin{bmatrix} 0 & 1 & 0 \\ 0 & 0 & 1 \\ 0.45 & -1.85 & 2.4 \end{bmatrix} \begin{bmatrix} x_1(n) \\ x_2(n) \\ x_3(n) \end{bmatrix} + \begin{bmatrix} 0 \\ 0 \\ 400 \end{bmatrix} u(n)$$

$$y(n) = [1 \quad 0 \quad 0] \begin{bmatrix} x_1(n) \\ x_2(n) \\ x_3(n) \end{bmatrix}.$$

This is in the form of Eqs. (4.4), (4.5).

4.3 PROPERTIES OF LTI SYSTEM STATE EQUATIONS (CF. SEC. 3.3)

(i) Solution of State Equations (cf. Sec. 3.3(i))

The solution of Eq. 4.4 is:

$$\mathbf{x}(n) = \boldsymbol{\phi}^n\mathbf{x}(0) + \sum_{i=0}^{n-1} \boldsymbol{\phi}^i\boldsymbol{\psi}\mathbf{u}(n-1-i) \tag{4.7}$$

$\boldsymbol{\phi}^n\mathbf{x}(0)$ is the zero input response, i.e. the natural modes of the system, while

$$\sum_{i=0}^{n-1} \boldsymbol{\phi}^i \boldsymbol{\psi} u(n-1-i)$$

is the zero initial state response, i.e. the response of the system Σ to the input, $\mathbf{u}(n)$, assuming zero initial state, $\mathbf{x}(0) = \mathbf{0}$.

The behavior of the system Σ is easily simulated on a digital computer and this is useful in the form of a CAD facility.

CAD Facility

The interactive CAD algorithm shown in the flow diagram of Fig. 4.2 calculates the states of a discrete LTI system for step function inputs of specified amplitudes and specified initial conditions.

The operation of the algorithm is as follows.

BLOCK 1

The system parameter matrices $\boldsymbol{\phi}$, $\boldsymbol{\psi}$, $\mathbf{C}$, $\mathbf{D}$ are requested, input, displayed, and corrected if necessary. $\mathbf{x}(0)$ is also input.

BLOCK 2

The number of interations required, N, the input step amplitudes, $\mathbf{u}(n)$, and a 'printout ratio', R, are specified. R is the ratio of the number of calculations of $\mathbf{x}(n)$, $\mathbf{y}(n)$ performed to the number presented as output (Block 4).

BLOCKS 3, 4

$\mathbf{x}(n)$, $\mathbf{y}(n)$ are calculated, $n = 1, 2, \ldots, N$, given $\mathbf{x}(0)$, from Eqs. (4.4), (4.5), and the values appropriate to the printout ratio are output.

BLOCK 5

Facilities for increasing the number of values calculated are useful.

EXAMPLE

Consider the computer algorithm (Fig. 1.17):

$$y(n) = u(n-1) + 0.5u(n-2) + y(n-1) - 0.5y(n-2). \tag{4.8}$$

Let

$$x_1(n) = y(n)$$
$$x_2(n) = x_1(n+1) - x_1(n) - u(n).$$

Using these, Eq. (4.8) may be cast as state equations (Sec. 4.5(i)):

$$\mathbf{x}(n+1) = \begin{bmatrix} 1 & 1 \\ -0.5 & 0 \end{bmatrix} \mathbf{x}(n) + \begin{bmatrix} 1 \\ 0.5 \end{bmatrix} u(n)$$

$$y(n) = [1 \quad 0]\mathbf{x}(n)$$

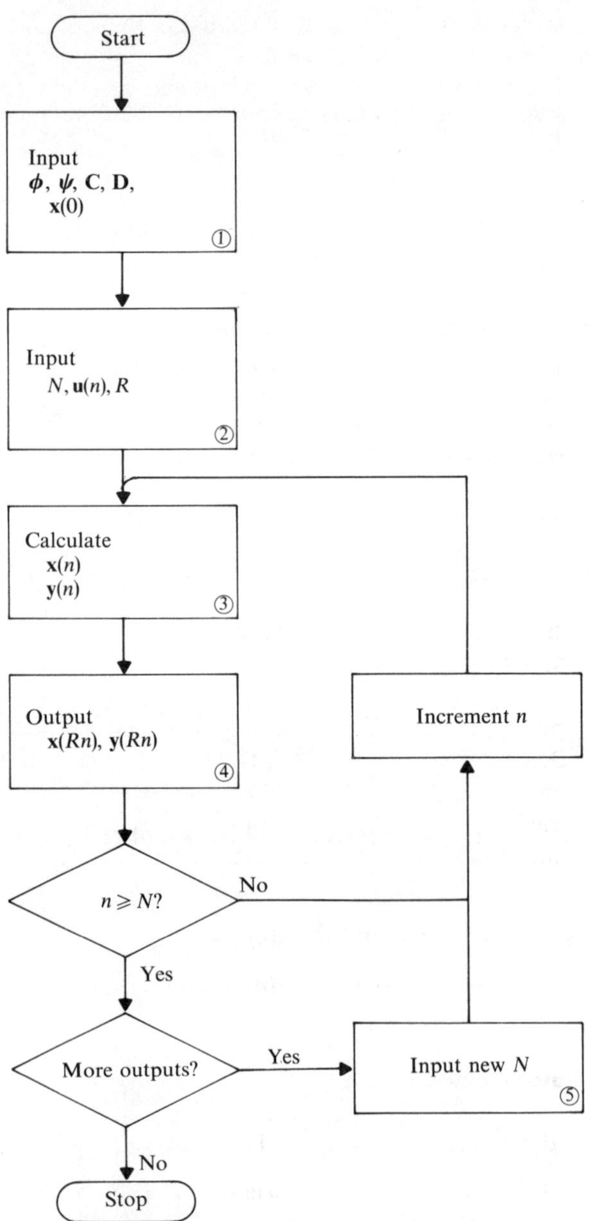

Fig. 4.2 Step function response CAD program.

where

$$\mathbf{x}(n) = [x_1(n) \quad x_2(n)]^{\mathrm{T}}.$$

The response of this system to a unit step input, $u(n) = 1$, $n = 0, 1, \ldots, \infty$, $\mathbf{x}(0) = \mathbf{0}$, easily found by the above CAD algorithm, is shown in Fig. 4.3, for $N = 20$. (For ease of presentation, adjacent values of $x_1(n)$, $x_2(n)$, are connected with straight lines.)

Figure 4.3 agrees with the result obtained differently in Sec. 1.10 (Fig. 1.18).

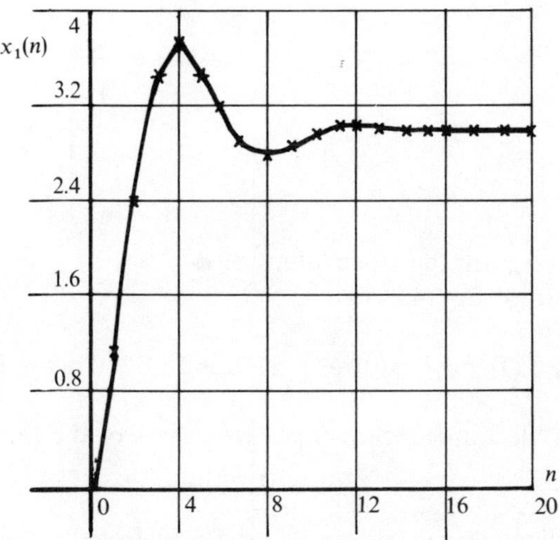

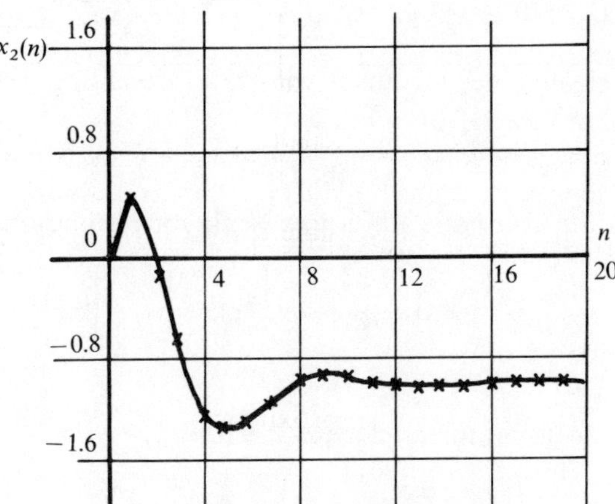

Fig. 4.3 System response (example).

(ii) Similar Systems (cf. Sec. 3.3(ii))

The definition of system states is not unique (ref. 1).

Let
$$\mathbf{x}(n) = \mathbf{T}\mathbf{z}(n)$$

where $\mathbf{T}$ is a square matrix. Eqs. (4.4), (4.5) can be written:

$$\begin{aligned}\mathbf{z}(n+1) &= \mathbf{T}^{-1}\boldsymbol{\phi}\mathbf{T}\mathbf{z}(n) + \mathbf{T}^{-1}\boldsymbol{\psi}\mathbf{u}(n) \\ \mathbf{y}(n) &= \mathbf{C}\mathbf{T}\mathbf{z}(n) + \mathbf{D}\mathbf{u}(n).\end{aligned} \tag{4.9}$$

If $\boldsymbol{\phi}$ has m district eigenvectors and $\mathbf{T}$ consists of these eigenvectors (App. A.9), $\mathbf{T}^{-1}\boldsymbol{\phi}\mathbf{T}$ is diagonal:

$$\mathbf{T}^{-1}\boldsymbol{\phi}\mathbf{T} = \begin{bmatrix} \lambda_1 & \cdot & \cdot & \cdot & 0 \\ 0 & \lambda_2 & \cdot & \cdot & 0 \\ \cdot & & & & \cdot \\ \cdot & & & & \cdot \\ 0 & \cdot & \cdot & \cdot & \lambda_m \end{bmatrix}$$

where $\lambda_1, \lambda_2, \ldots, \lambda_m$ are the eigenvalues of $\boldsymbol{\phi}$.

The solution of Eq. (4.9) is:

$$\mathbf{z}(n) = (\mathbf{T}^{-1}\boldsymbol{\phi}\mathbf{T})^n\mathbf{z}(0) + \sum_{i=0}^{n-1} (\mathbf{T}^{-1}\boldsymbol{\phi}\mathbf{T})^i\mathbf{T}^{-1}\boldsymbol{\psi}\mathbf{u}(n-1-i).$$

This yields 'decoupled' zero-input responses of the form:

$$z_i(n) = (\lambda_i)^n z_i(0).$$

If $\boldsymbol{\phi}$ has fewer than m eigenvectors, $\mathbf{T}^{-1}\boldsymbol{\phi}\mathbf{T}$ may be expressed in Jordan form (App. A.10), and the zero-input responses are of the form:

$$z_i(n) = (\lambda_i)^n z_i(0) + n(\lambda_i)^{n-1} z_{i+1}(0) + \frac{n(n-1)}{1.2}(\lambda_i)^{n-2} z_{i+2}(0) + \cdots.$$

Complex eigenvalues occur in pairs, representing responses with a sinusoidal content (Sec. 1.10).

EXAMPLE

Consider again the computer algorithm with state equations:

$$\mathbf{x}(n+1) = \begin{bmatrix} 1 & 1 \\ -0.5 & 0 \end{bmatrix}\mathbf{x}(n) + \begin{bmatrix} 1 \\ 0.5 \end{bmatrix}u(n)$$

$$y(n) = [1 \quad 0]\mathbf{x}(n).$$

Casting this in diagonal form (cf. Sec. 3.3(ii)):

$$\mathbf{z}(n+1) = \begin{bmatrix} 0.5 + j0.5 & 0 \\ 0 & 0.5 - j0.5 \end{bmatrix}\mathbf{z}(n) + \begin{bmatrix} 0.5 & -j \\ 0.5 & +j \end{bmatrix}u(n)$$

$$y(n) = [1 \quad 1]\mathbf{z}(n).$$

The zero-input response is:

$$z_1(n) = (0.5 + j0.5)^n z_1(0) = 0.707 e^{j\pi n/4} z_1(0)$$
$$z_2(n) = (0.5 - j0.5)^n z_2(0) = 0.707 e^{-j\pi n/4} z_2(0)$$
$$y(n) = (0.707)^n 2\cos(\pi n/4).$$

This agrees with the result illustrated in Fig. 4.3.

(iii) Solution of State Equations by z Transforms (cf. Sec. 3.3(iii))

Taking z transforms (Sec. 1.8) of Eqs. (4.4), (4.5) gives, for $\mathbf{x}(0) = \mathbf{0}$:

$$z\bar{\mathbf{X}}(z) = \boldsymbol{\phi}\bar{\mathbf{X}}(z) + \boldsymbol{\psi}\bar{\mathbf{U}}(z)$$
$$\bar{\mathbf{Y}}(z) = \mathbf{C}\bar{\mathbf{X}}(z) + \mathbf{D}\bar{\mathbf{U}}(z).$$

Eliminating $\bar{\mathbf{X}}(z)$ yields the transfer function matrix:

$$\bar{\mathbf{F}}(z) = \mathbf{C}(z\mathbf{I} - \boldsymbol{\phi})^{-1}\boldsymbol{\psi} + \mathbf{D} \tag{4.10}$$

where

$$\bar{\mathbf{Y}}(z) = \bar{\mathbf{F}}(z)\bar{\mathbf{U}}(z).$$

This follows the algebra of the continuous time transfer function development exactly (with z replacing s), and consequently the same properties may be defined.

Briefly, the system poles are included in the roots of the characteristic equation:

$$\det(z\mathbf{I} - \boldsymbol{\phi}) = 0. \tag{4.11}$$

The poles describe the natural modes of the system. From Eqs. (4.10), (4.11), it is clear that the poles of the system lie at eigenvalues of $\boldsymbol{\phi}$, namely λ_1, $\lambda_2, \ldots, \lambda_m$.

The natural modes are of the form:

$$(\lambda_i)^n z_i(0).$$

Where multiple poles (and repeated eigenvalues and eigenvectors) exist, the natural modes are of the form:

$$(\lambda_i)^n z_i(0) + n(\lambda_i)^{n-1} z_{i+1}(0) + \frac{n(n-1)}{1.2}(\lambda_i)^{n-2} z_{i+2}(0) + \cdots.$$

EXAMPLE

Consider the example of the previous two subsections. From Eq. (4.10):

$$\bar{F}(z) = [1 \quad 0]\begin{bmatrix} z-1 & -1 \\ 0.5 & z \end{bmatrix}^{-1}\begin{bmatrix} 1 \\ 0.5 \end{bmatrix} + 0$$
$$= \frac{z + 0.5}{z^2 - z + 0.5}.$$

The characteristic equation is (Eq. (4.11)):

$$z^2 - z + 0.5 = 0.$$

This has roots at $z = 0.5 \pm j0.5$, or $z = 0.707e^{\pm j\pi/4}$

Considered together, these describe a mode (Sec. 1.10):

$$(0.707)^n \cos(\pi n/4).$$

This agrees with the previous analysis of this system (Sec. 4.3(ii)).

CAD Facility

Since Eqs. (4.4), (4.5), and (4.10) are algebraically identical to Eqs. (3.5), (3.6) and (3.14), the CAD algorithm for deriving the continuous time transfer function (Fig.3.6) serves the discrete time case equally well.

4.4 PROPERTIES OF LTI SYSTEMS (CF. SEC. 3.4)

(i) Reachability (or Controllability) (ref. 1)

The property of controllability, or 'reachability' as it is frequently called for discrete systems, is defined in the same way as for continuous time systems (Sec. 3.4(i)), and the same algebraic formulas apply, with $\boldsymbol{\phi}$, $\boldsymbol{\psi}$, z replacing $\mathbf{A}$, $\mathbf{B}$, s.

(ii) Observability

The property of observability is also defined in the same way as for continuous time systems (Sec. 3.4(ii)), and the same algebraic formulas apply, with $\boldsymbol{\phi}$, $\mathbf{C}$, z replacing $\mathbf{A}$, $\mathbf{C}$, s.

(iii) Stability of LTI Systems

The time domain definitions of asymptotic and bounded-input, bounded-output stability for continuous time systems apply also to discrete systems (Sec. 3.4(iii)).

The equivalent frequency domain statements (Sec. 3.4(iii)) refer of course to the z plane, i.e. Σ (Fig. 4.1) is a.s. if and only if all the eigenvalues of $\boldsymbol{\phi}$ lie inside the unit circle $|\mathbf{z}| = 1$, and b.i.b.o. stable if and only if all the poles of Σ lie inside the unit circle.

4.5 REALIZATION OF STATE EQUATIONS (CF. SEC. 3.5)

Since the algebras of the continuous and discrete time state equations and transfer functions are identical, with **A**, **B**, **C**, **D**, s replaced with $\boldsymbol{\phi}$, $\boldsymbol{\psi}$, **C**, **D**, z, the methods of realizing state equations from transfer function matrices are also identical.

For illustration, the most straightforward methods are described briefly here, with cross-references to Sec. 3.5 where fuller explanations are to be found.

(i) Single-input, Single-output Systems

A transfer function may be expressed in the form:

$$\frac{\bar{Y}(z)}{\bar{U}(z)} = \bar{F}(z) = \frac{a_q z^q + a_{q-1} z^{q-1} + \cdots + a_0}{z^p + b_{p-1} z^{p-1} + \cdots + b_0}.$$

As in Sec. 3.5, the requirement is to generate a set of state equations from $\bar{F}(z)$.

If $q > p$, $\bar{F}(z)$ is 'improper', and no state realization is possible.

If $q = p$, $\bar{F}(z)$ is 'proper', and $D = \bar{F}(\infty)$

Define $\bar{F}_r(z) = \bar{F}(z) - \bar{F}(\infty)$, $\bar{F}_r(z)$ being 'strictly proper', and one of the following procedures can be applied to $\bar{F}_r(z)$.

A strictly proper transfer function, without loss of generality, can be expressed as (cf. Eq. 3.21):

$$\bar{F}(z) = \frac{\bar{Y}(z)}{\bar{U}(z)} = \frac{a_{p-1} z^{p-1} + a_{p-2} z^{p-2} + \cdots + a_0}{z^p + b_{p-1} z^{p-1} + \cdots + b_0}$$

where $a_{p-1}, a_{p-2}, \ldots, a_0, b_{p-1}, b_{p-2}, \ldots, b_0$ are constants.

1. Using the rules outlined in Sec. 3.5(i), the following state equations are realized:

$$\mathbf{x}(n+1) = \begin{bmatrix} -b_{p-1} & 1 & 0 & \cdots & 0 \\ -b_{p-2} & 0 & 1 & \cdots & 0 \\ \vdots & & & & \vdots \\ -b_0 & . & . & . & 0 \end{bmatrix} \mathbf{x}(n) + \begin{bmatrix} a_{p-1} \\ a_{p-2} \\ \vdots \\ a_0 \end{bmatrix} u(n) \qquad (4.12)$$

$$y(n) = [1 \quad 0 \ldots 0]\mathbf{x}(n).$$

2. A reachable companion form, which applies only to reachable systems, is:

$$\mathbf{x}(n+1) = \begin{bmatrix} 0 & 1 & 0 & \cdots & 0 \\ 0 & 0 & 1 & \cdots & 0 \\ \vdots & \vdots & & & \vdots \\ -b_0 & -b_1 & -b_2 & & -b_{p-1} \end{bmatrix} \mathbf{x}(n) + \begin{bmatrix} 0 \\ 0 \\ \vdots \\ 1 \end{bmatrix} u(n) \quad (4.13)$$

$$y(n) = [a_0 \quad a_1 \ldots a_{p-1}]\mathbf{x}(n).$$

3. An observable companion form, which applies only to observable systems, is:

$$\mathbf{x}(n+1) = \begin{bmatrix} 0 & 0 & \cdots & -b_0 \\ 1 & 0 & \cdots & -b_1 \\ 0 & 1 & \cdots & -b_2 \\ \vdots & \vdots & & \vdots \\ 0 & 0 & \cdots & -b_{p-1} \end{bmatrix} \mathbf{x}(n) + \begin{bmatrix} a_0 \\ a_1 \\ \vdots \\ a_{p-1} \end{bmatrix} u(n) \tag{4.14}$$

$$y(n) = [0 \quad 0 \quad \cdots \quad 1]\mathbf{x}(n).$$

EXAMPLE

Consider the third-order algorithm governed by the difference equation:

$$y(n+3) - 1.8y(n+2) + 1.07y(n+1) - 0.21y(n)$$
$$= u(n+3) + 3u(n+2) + 3u(n+1) + u(n)\,.$$

Taking z transforms of this yields the transfer function:

$$\frac{\bar{Y}(x)}{\bar{U}(z)} = \bar{F}(z) = \frac{z^3 + 3z^2 + 3z + 1}{z^3 - 1.8z^2 + 1.07z - 0.21}$$

$$D = \bar{F}(\infty) = 1.$$

From this and Eq. (4.12), a state realization is:

$$\mathbf{x}(n+1) = \begin{bmatrix} 1.8 & 1 & 0 \\ -1.07 & 0 & 1 \\ +0.21 & 0 & 0 \end{bmatrix} \mathbf{x}(n) + \begin{bmatrix} 4.8 \\ 1.93 \\ 1.21 \end{bmatrix} \mathbf{u}(n)$$

$$y(n) = [1 \quad 0 \quad 0]\mathbf{x}(n) + 1u(n).$$

Reachable and observable realizations are easily found from Eqs. (4.13), (4.14) respectively.

(ii) Multivariable Systems

Since the same algebra governs the continuous and discrete time systems the algorithms described in Sec. 3.5(ii)(a), (b) can be applied directly to the discrete time case, with of course **A**, **B**, **C**, **D**, s replaced with $\boldsymbol{\phi}$, $\boldsymbol{\psi}$, **C**, **D**, z.

4.6 SAMPLED DATA SYSTEMS

One type of discrete system of particular interest is the sampled data system (cf. Sec. 1.11). This is a continuous time system 'buffered' with analog-to-digital

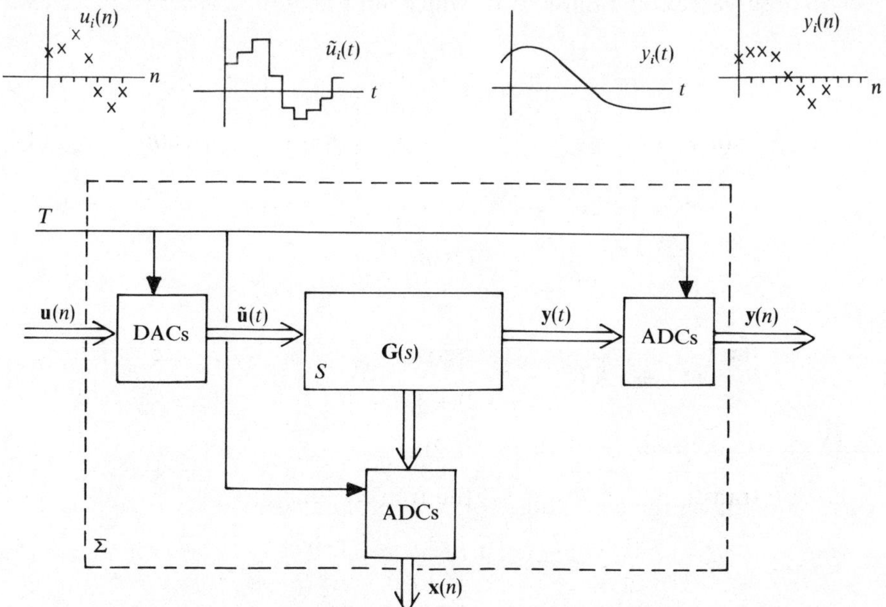

Fig. 4.4 Sampled data system.

and digital-to-analog coverters (ADCs and DACs) as shown in Fig. 4.4.

A DAC generates a 'staircase' continuous time function from a discrete sequence of numbers as illustrated in Fig. 4.4. A vector of such functions $\tilde{\mathbf{u}}(t)$ is generated by a set of DACs from a vector $\mathbf{u}(n)$ of discrete functions.

An ADC generates a discrete sequence of numbers from a continuous time function sampled at intervals T, as illustrated in Fig. 4.4. A vector of such numbers, $\mathbf{x}(n)$ or $\mathbf{y}(n)$, is generated from continuous time vectors $\mathbf{x}(t)$, $\mathbf{y}(t)$ by a set of ADCs.

In Fig. 4.4, the DACs and ADCs are clocked simultaneously at intervals of duration T. In practice a relatively small delay (compared with T) may be necessary between the ADC and DAC clocks – but this is either negligible or can be accounted for easily by adding an equivalent delay term to the transfer function $\mathbf{G}(s)$ (Sec. 1.14).

If the state equations of S (Eqs. (3.4), (3.5)) are known and the DACs and ADCs are clocked simultaneously, discrete state equations for Σ may readily be found from the solution of the continuous time equations for the time $((n+1)T)$, taking $\mathbf{x}(nT)$ as initial conditions and $\mathbf{u}(nT)$ as the input (Eq. 3.10):

$$\mathbf{x}((n+1)T) = e^{\mathbf{A}T}\mathbf{x}(nT) + \int_{nT}^{(n+1)T} e^{\mathbf{A}\tau}\mathbf{B}\mathbf{u}(nT)\,\mathrm{d}\tau$$

$$\mathbf{y}(nT) = \mathbf{C}\mathbf{x}(nT) + \mathbf{D}\mathbf{u}(nT).$$

Writing this in standard form (Eqs. (4.4), (4.5)):

$$\mathbf{x}(n+1) = \boldsymbol{\phi}\mathbf{x}(n) + \boldsymbol{\psi}\mathbf{u}(n)$$
$$\mathbf{y}(n) = \mathbf{C}\mathbf{x}(n) + \mathbf{D}\mathbf{u}(n)$$

where

$$\boldsymbol{\phi} = e^{\mathbf{A}T} = \mathbf{I} + \mathbf{A}T + \frac{1}{2!}\mathbf{A}^2T^2 + \frac{1}{3!}\mathbf{A}^3T^3 + \cdots \tag{4.15}$$

$$\boldsymbol{\psi} = \int_0^T e^{\mathbf{A}\tau}\,\mathrm{d}\tau\mathbf{B} \quad (\text{since } \mathbf{u}(nT) \text{ is constant, } nT < \tau \leqslant (n+1)T)$$

$$= [\mathbf{I}T + \tfrac{1}{2}\mathbf{A}T^2 + \frac{1}{3!}\mathbf{A}^2T^3 + \cdots]\mathbf{B}. \tag{4.16}$$

C, **D** are unchanged.

If T is small compared with the time constants and cycle periods corresponding to the eigenvalues of **A**, approximations may be sufficient:

$$\boldsymbol{\phi} \approx \mathbf{I} + \mathbf{A}T; \qquad \boldsymbol{\psi} \approx T\mathbf{B}. \tag{4.17}$$

CAD Facility

Equations (4.15), (4.16) cannot be evaluated in practice without a CAD facility, and the flow diagram of a simple interactive algorithm, which closely parallels that of Fig. 3.4, is shown in Fig. 4.5.

The operation of the algorithm is as follows.

BLOCK 1

The continuous time system parameters **A**, **B**, **C**, **D** and the sampling period, T, are requested, input, displayed, and corrected if necessary.

BLOCK 2

The matrices $e^{\mathbf{A}T}$ and $\int_0^T e^{\mathbf{A}\tau}\,\mathrm{d}\tau$ are calculated from the formulas:

$$e^{\mathbf{A}T} = \mathbf{I} + \mathbf{A}T + \frac{1}{2!}\mathbf{A}^2T^2 + \frac{1}{3!}\mathbf{A}^3T^3 + \cdots$$

and

$$\int_0^T e^{\mathbf{A}\tau}\,\mathrm{d}\tau = \mathbf{I}T + \frac{1}{2!}\mathbf{A}T^2 + \frac{1}{3!}\mathbf{A}^2T^3 + \cdots.$$

The numbers of terms taken in these series may be limited by taking an upper bound, say 10^{-4}, on any element in the last term taken into account. If, using this criterion, the number of terms exceeds, say, 20, the program declares a calculation failure and stops.

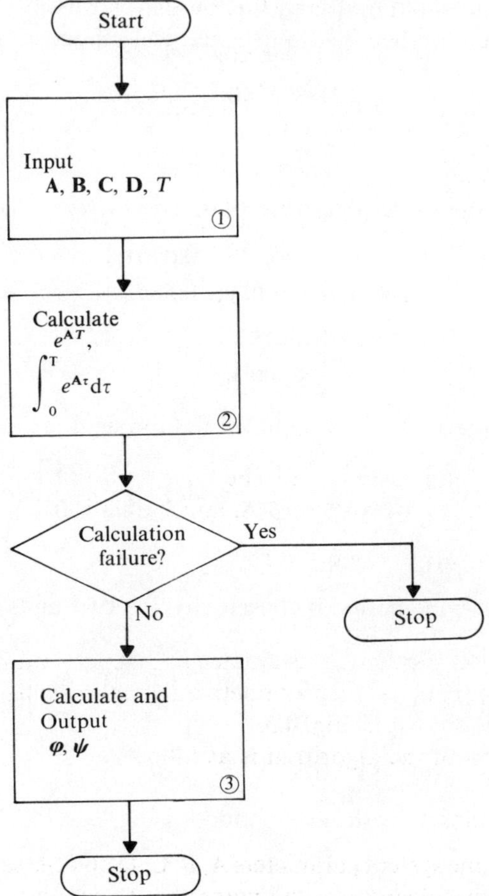

Fig. 4.5 Algorithm for evaluating sampled data system parameters.

BLOCK 3

Using the results of Block 2, $\boldsymbol{\phi}$, $\boldsymbol{\psi}$ are calculated and output, where:

$$\boldsymbol{\phi} = e^{\mathbf{A}T}, \qquad \boldsymbol{\psi} = \int_0^T e^{\mathbf{A}\tau} \, d\tau \, \mathbf{B}.$$

EXAMPLE

Consider the continuous time system of Sec. 3.3, buffered with DACs and ADCs as in Fig. 4.4.

$$\dot{\mathbf{x}} = \begin{bmatrix} -2 & 1 \\ -101 & 0 \end{bmatrix} \mathbf{x} + \begin{bmatrix} 0 \\ 101 \end{bmatrix} u$$

$$y = [1 \quad 0]\mathbf{x}.$$

This has poles at $s = -1 \pm j10$ and the longest permissible sampling period which avoids aliassing (Sec. 1.12) of the natural mode is:

$$T = \frac{2\pi}{2 \times 10} = 0.314 \text{ units}.$$

Let $T = 0.1$ units.

Then, using the CAD algorithm outlined above:

$$\boldsymbol{\phi} = \begin{bmatrix} 0.413 & 0.0761 \\ -7.689 & 0.565 \end{bmatrix}$$

$$\boldsymbol{\psi} = \begin{bmatrix} 0.435 \\ 8.560 \end{bmatrix}.$$

The state equations of the sampled data system are:

$$\mathbf{x}(n+1) = \begin{bmatrix} 0.413 & 0.0761 \\ -7.689 & 0.565 \end{bmatrix} \mathbf{x}(n) + \begin{bmatrix} 0.435 \\ 8.560 \end{bmatrix} u(n)$$

$$y(n) = [1 \quad 0]\mathbf{x}(n).$$

If a short sampling period is chosen, say $T = 0.01$ units, the discrete state equations are:

$$\mathbf{x}(n+1) = \begin{bmatrix} 0.980 & 0.01 \\ -1.010 & 1 \end{bmatrix} \mathbf{x}(n) + \begin{bmatrix} 0 \\ 1.01 \end{bmatrix} u(n)$$

$$y(n) = [1 \quad 0]\mathbf{x}(n).$$

This agrees closely with results found using Eqs. (4.17).

4.7 POLE SHIFTING BY STATE FEEDBACK (CF. SEC. 3.6)

Since the algebras of the continuous and discrete time state equations and transfer functions are identical (with **A**, **B**, **C**, **D**, s replaced with $\boldsymbol{\phi}$, $\boldsymbol{\psi}$, **C**, **D**, z), the methods of pole shifting are also identical, and the same CAD program (Fig. 3.11) is directly applicable.

A closed-loop system with state feedback is illustrated in Fig. 4.6.

The state equations of the closed-loop system Σ_F are (cf. Eq. (3.34)):

$$\mathbf{x}(n+1) = (\boldsymbol{\phi} - \boldsymbol{\psi}\mathbf{K})\mathbf{x}(n) + \boldsymbol{\psi}\mathbf{v}(n).$$

The natural modes of Σ_F are described by the eigenvalues of $(\boldsymbol{\phi} - \boldsymbol{\psi}\mathbf{K})$, and **K** may be chosen so that, theoretically at least, these have any desired values.

In practice **K** is chosen so that the dominant poles of Σ_F, i.e. the dominant eigenvalues of $(\boldsymbol{\phi} - \boldsymbol{\psi}\mathbf{K})$, lie on good lines of constant damping ratio (Fig.

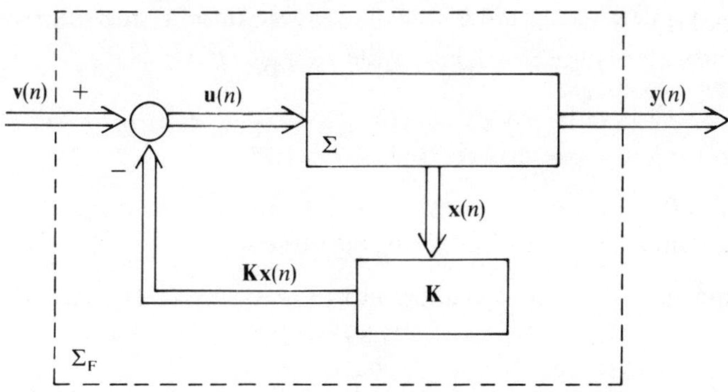

Fig. 4.6 Discrete LTI system with state feedback.

2.35). The poles of Σ_F must be selected so that the elements of $\mathbf{K}$ have reasonable values and so that Σ_F is well-conditioned in the sense that small parameter variations in Σ do not give rise to gross changes in the behavior of Σ_F. These constraints are essentially the same as those governing the design of continuous time systems (Sec. 3.6), with an added difficulty introduced by the sampling; generally, the longer the sampling period, T, the more constrained are the practical closed-loop pole positions. At the other extreme, the selection of too short a sampling period means a heavy computing load and can even result in resolution problems in the control algorithm (refs. 2, 3).

As with continuous time systems, only those poles which are reachable (controllable) may be shifted in this manner, and if any poles of Σ are in fact unreachable, Σ_F is also unreachable (cf. Eq. (3.35)).

EXAMPLES

(a) Single-input system

Consider the example of Sec. 3.6, which concerns a system with state equations:

$$\dot{\mathbf{x}} = \begin{bmatrix} 0 & 10 & 0 \\ 0 & 0 & 10 \\ 0 & -6 & -32 \end{bmatrix} \mathbf{x} + \begin{bmatrix} 0 \\ 0 \\ 450 \end{bmatrix} u$$

$$y = [1 \quad 0 \quad 0]\mathbf{x}.$$

If this system is buffered with DACs and ADCs (Fig. 4.4), $T = 0.020$, to give a system Σ, the state equations for Σ using the CAD program in Fig. 4.5, are easily found:

$$\mathbf{x}(n+1) = \begin{bmatrix} 1.000 & 0.199 & 0.0163 \\ 0 & 0.990 & 0.147 \\ 0 & -0.0882 & 0.519 \end{bmatrix} \mathbf{x}(n) + \begin{bmatrix} 0.0515 \\ 0.734 \\ 6.621 \end{bmatrix} u(n)$$

$$y(n) = [1 \quad 0 \quad 0]\mathbf{x}(n).$$

Using the CAD program of Fig. 3.6, this is found to have eigenvalues at:

$$z = 1,\ 0.9606,\ 0.5484.$$

For the dominant poles of Σ_F on the $\zeta = 0.707$ lines (Fig. 2.35), $z = 0.22 \pm j0.4$, and $z = 0.3$, the program of Fig. 3.11 yields:

$$\mathbf{K} = [2.0327 \quad 0.9623 \ . \ 0.1449].$$

A unit step function response of the closed-loop system Σ_F is found using the CAD program of Fig. 4.2, and this is illustrated in Fig. 4.7.

(b) Multi-input systems–dyadic feedback
Consider the state equations of the submarine vertical plane dynamics in the example of Sec. 3.6(ii) (Fig. 3.14):

$$\mathbf{x} = \begin{bmatrix} -0.083 & 0.00115 & 0 & 0 \\ 1 & -0.400 & -0.008 & 0 \\ 0 & 1 & 0 & 0 \\ 1 & 0 & -0.005 & 0 \end{bmatrix} \mathbf{x} + \begin{bmatrix} -0.084 & -0.163 \\ 86.000 & -24.00 \\ 0 & 0 \\ 0 & 0 \end{bmatrix} \mathbf{u}$$

$$\mathbf{y} = \begin{bmatrix} 0 & 0 & 1 & 0 \\ 0 & 0 & 0 & 1 \end{bmatrix} \mathbf{x}$$

where

$$\mathbf{x} = (w \quad q \quad \theta \quad h)^{\mathrm{T}}$$
$$\mathbf{u} = (\delta_b \quad \delta_s)^{\mathrm{T}}$$
$$\mathbf{y} = (\theta \quad h)^{\mathrm{T}}.$$

For sampling period $T = 1$ second, the equivalent discrete equations are (Fig. 4.5):

$$\mathbf{x}(n+1) = \begin{bmatrix} 0.921 & 0.00091 & 0 & 0 \\ 0.788 & 0.668 & -0.0066 & 0 \\ 0.427 & 0.823 & 0.996 & 0 \\ 0.959 & -0.0017 & -0.0005 & 1 \end{bmatrix} \mathbf{x}(\mathbf{n}) + \begin{bmatrix} -0.038 & -0.168 \\ 70.764 & -19.827 \\ 37.763 & -10.566 \\ -0.091 & -0.065 \end{bmatrix} \mathbf{u}(n)$$

$$\mathbf{y}(n) = \begin{bmatrix} 0 & 0 & 1 & 0 \\ 0 & 0 & 0 & 1 \end{bmatrix} \mathbf{x}(n).$$

This has eigenvalues at:

$$z = 1,\ 0.978,\ 0.925,\ 0.682.$$

A digital autopilot for controlling θ, h is illustrated schematically in Fig. 4.8 (cf. Fig. 3.13).

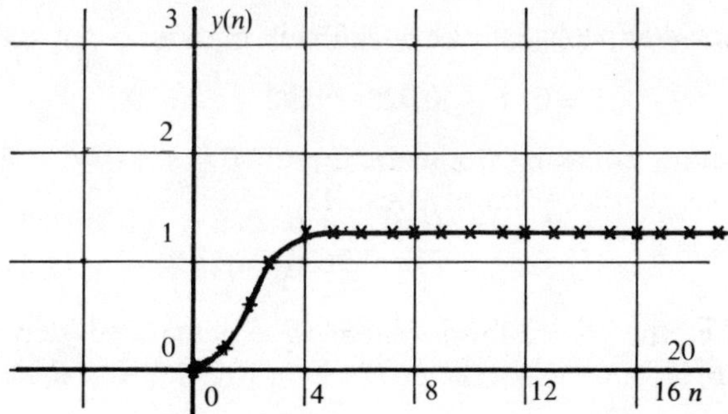

Fig. 4.7 System response to a step function (example).

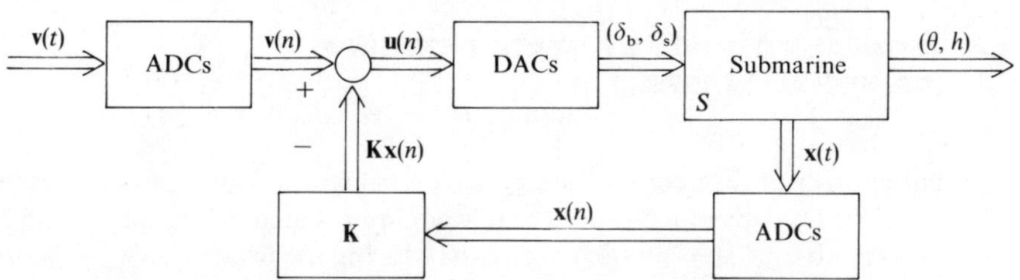

Fig. 4.8 Submarine digital autopilot design (example).

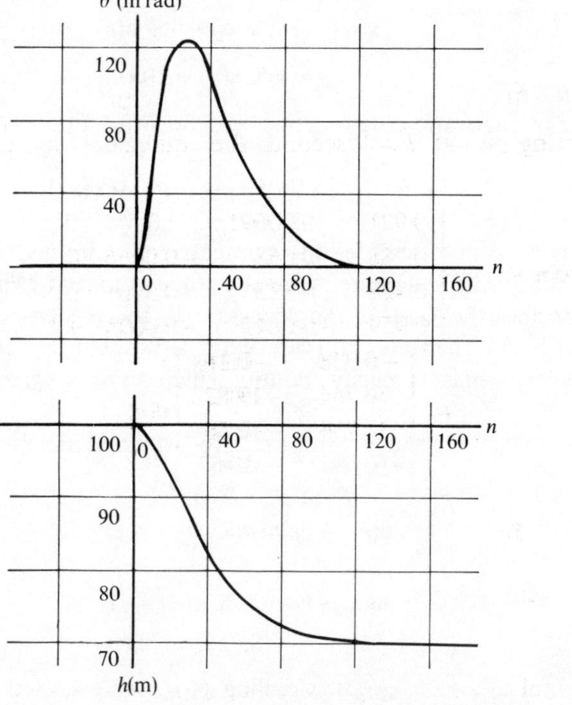

Fig. 4.9 Response of submarine with autopilot (example).

The autopilot consists basically of a feedback matrix **K**. To shift the poles to:

$$z = 0.96 \pm j0.028, \quad 0.923, \quad 0.670,$$

the matrix **K** is found, using the procedure of Fig. 3.11:

$$\mathbf{K} = \begin{bmatrix} 0.259 & +0.001 & +0.0003 & -0.0218 \\ +0.259 & -0.001 & -0.0003 & +0.0218 \end{bmatrix}.$$

The response of the closed-loop system to a demanded step depth change (100 m to 70 m) is found using the program of Fig. 4.5, and this is illustrated in Fig. 4.9.

4.8 OPTIMAL CONTROL BY STATE FEEDBACK (CF. SEC. 3.7) (REFS. 2, 5)

Optimal control of discrete LTI systems closely parallels that of continuous time systems. The object is to specify an input, $\mathbf{u}(n)$, which drives the system $\boldsymbol{\Sigma}$ to a target state in N steps in such a way that during the process a defined cost function is minimized. Naturally in these circumstances $\mathbf{u}(n)$ is a function of the state $\mathbf{x}(n)$.

Consider the system described by state equations (Eqs. (4.4), (4.5)):

$$\mathbf{x}(n+1) = \boldsymbol{\phi}\mathbf{x}(n) + \boldsymbol{\psi}\mathbf{u}(n)$$
$$\mathbf{y}(n) = \mathbf{C}\mathbf{x}(n) + \mathbf{D}\mathbf{u}(n).$$

A linear quadratic cost function is defined (cf. Eq. 3.37):

$$V = \sum_{n=n_0}^{n_0+N} [\mathbf{x}^T(n)\mathbf{E}\mathbf{x}(n) + \mathbf{u}^T(n)\mathbf{F}\mathbf{u}(n)] \tag{4.18}$$

where **E** is a positive semidefinite symmetric $(m \times m)$ matrix,
F is a positive definite symmetric $(q \times q)$ matrix (App. A.11).

V describes the 'cost' of the process over the N steps $n_0 < n \leqslant (n_0 + N)$.

As in Sec. 3.7, the states $\mathbf{x}(n)$ can be defined so that the target state is $\mathbf{0}$, and the design problem is to specify an input which drives $\mathbf{x}(n_0)$ to $\mathbf{0}$ in N steps with minimum V.

One strategy, due to Bryson and Ho, which achieves this is (ref. 2):

$$\begin{aligned} \mathbf{u}(n_0) &= -\mathbf{K}_0\mathbf{x}(n_0) \\ \mathbf{u}(n_0+1) &= -\mathbf{K}_1\mathbf{x}(n_0+1) \\ &\vdots \\ \mathbf{u}(n_0+r) &= -\mathbf{K}_r\mathbf{x}(n_0+r) \\ &\vdots \\ \mathbf{u}(n_0+N) &= -\mathbf{K}_N\mathbf{x}(n_0+N) \end{aligned}$$

where

$$\mathbf{K}_{r-1} = (\mathbf{F} + \boldsymbol{\psi}^{\mathrm{T}}\mathbf{P}_r\boldsymbol{\psi})^{-1}\boldsymbol{\psi}^{\mathrm{T}}\mathbf{P}_r\boldsymbol{\phi} \tag{4.19}$$

$$\mathbf{M}_r = \mathbf{P}_r - \mathbf{P}_r\boldsymbol{\psi}(\mathbf{F} + \boldsymbol{\psi}^{\mathrm{T}}\mathbf{P}_r\boldsymbol{\psi})^{-1}\boldsymbol{\psi}^{\mathrm{T}}\mathbf{P}_r \tag{4.20}$$

$$\mathbf{P}_{r-1} = \boldsymbol{\phi}^{\mathrm{T}}\mathbf{M}_r\boldsymbol{\phi} + \mathbf{E} \tag{4.21}$$

and

$$\mathbf{P}_N = \mathbf{E}, \quad \mathbf{K}_N = \mathbf{0}. \tag{4.22}$$

The cost incurred during the N steps is:

$$V = \tfrac{1}{2}\mathbf{x}^{\mathrm{T}}(n_0)\mathbf{P}_0\mathbf{x}(n_0).$$

The whole strategy is calculated in advance, in 'reverse' sequence, starting with $\mathbf{P}_N = \mathbf{E}$, $\mathbf{K}_N = \mathbf{0}$, and applied in real time to the system, regardless of the values of $\mathbf{x}(n_0)$ and N.

There is of course no guarantee that the strategy will achieve $\mathbf{x}(n_0 + N) = \mathbf{0}$ in N steps; it merely minimizes the cost function V over this sequence, and as in the case of continuous time optimal control (cf. Sec. 3.7), V can often be chosen so that certain fairly simple trajectories can be reasonably approximated. The desired trajectory is described by $\mathbf{E}$, while $\mathbf{F}$ determines the maximum modulus of the system input, $\mathbf{u}(n)$.

The sequence length, N, is perhaps best selected by a process of trial and error, depending on $\mathbf{x}(n_0)$; too short a sequence gives a poor approximation to the required trajectory, while too long a sequence is unnecessarily slow. This implicit dependence of N on $x(n_0)$, of course, places limitations on the straightforward application of this control strategy.

EXAMPLE

A useful laboratory example is provided by the equipment described in example (a) of Sec. 3.7(i) (Fig. 3.17), with the microcomputer applying a control sequence, $-\mathbf{K}_r\mathbf{x}(n_0 + r)$ to drive the arm from the (digitized) state $\mathbf{x}(n_0)$ to $\mathbf{0}$ in a sequence of N steps. N is specified by the operator and $\mathbf{K}_r$ is calculated in advance (in reverse sequence) by the microcomputer (Eqs. (4.19) to (4.22)). The system is shown in schematic form in Fig. 4.10(a).

The servo and arm, seen through DAC and ADC (cf. Fig. 4.4), are modeled by discrete equations derived from the continuous time model using the algorithm of Fig. 4.5, sampling time $T = 0.1$ sec (in practice, the ADC sampling time is much faster to allow reliable computer calculation of x_2):

$$\mathbf{x}(n+1) = \begin{bmatrix} 0.999 & 0.099 \\ 0 & 0 \end{bmatrix}\mathbf{x}(n) + \begin{bmatrix} 0.005 \\ 0.100 \end{bmatrix}u(n)$$

where $x_1(n)$ is the potentiometer output (volts),
$x_2(n)$ is the (computer calculated) rate of change of potentiometer output.
$u(n)$ is the input to the servo (volts).

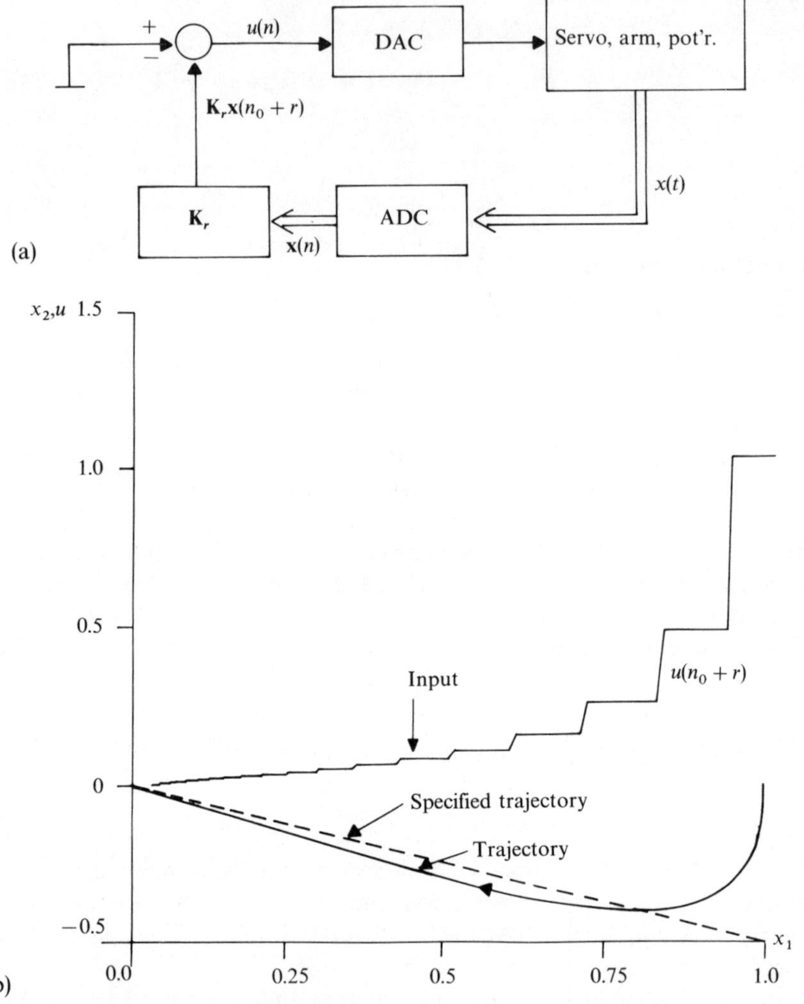

Fig. 4.10 Laboratory optimal control (example). (a) Closed-loop system. (b) Trajectory and input.

A straight-line trajectory may be specified:

$$x_1(n) = -2x_2(n)$$

i.e.

$$(x_1(n) + 2x_2(n))^2 = 0.$$

This is approximated by the control if $\mathbf{x}^{\mathrm{T}}\mathbf{E}\mathbf{x}$ is minimized, where:

$$\mathbf{E} = \begin{bmatrix} 1 & 2 \\ 2 & 4 \end{bmatrix}.$$

F is selected to limit the modulus of $u(n)$. For example, $F = 100$.

Equations (4.19), (4.20), (4.21), (4.22) are solved for these values of **E**, **F**, and $N = 20$, to give the sequence $\mathbf{K}_r$, $r = 0, 1, 2, \ldots, 20$.

This is applied by the microcomputer, via the DAC, and a trajectory for $\mathbf{x}(n_0) = \begin{bmatrix} 1 \\ 0 \end{bmatrix}$ is shown in Fig. 4.10(b) together with the sequence $\mathbf{u}(n_0 + r)$, $r = 0, 1, \ldots, 20$.

The arm is driven successfully on to the required trajectory and follows it quite accurately almost to $\mathbf{x}(n_0 + 20) = \mathbf{0}$.

Clearly, a whole variety of strategies and initial conditions can be tried.

Comment

Discrete linear optimal quadratic control has the apparent advantage over the continuous time strategy that it can drive the state to zero in a finite time. However, this is not guaranteed; the control merely minimizes the cost function V (Eq. (4.18)) without necessarily attaining the required trajectory. Consequently, the wrong choice of sequence length, N, can result in poor performance, and since the right choice is largely a matter of trial and error (depending perhaps on 'zones' of initial conditions $\mathbf{x}(n_0)$), a practical strategy of this kind can be difficult to formulate; success may well depend on selecting a strategy to suit the initial conditions.

4.9 STATE ESTIMATORS FOR LTI SYSTEMS (CF. SEC. 3.8) (REFS. 1, 2)

Since the algebras of the continuous and discrete time systems are identical (with **A**, **B**, **C**, **D**, s replaced with $\boldsymbol{\phi}$, $\boldsymbol{\psi}$, **C**, **D**, z), the methods of state estimator design are identical and the same CAD programs (Sec. 3.8) apply. These methods are briefly reviewed here, this time in the discrete system context.

Consider an observable LTI system Σ with state equations (Eqs. (4.4), (4.5)):

$$\mathbf{x}(n+1) = \boldsymbol{\phi}\mathbf{x}(n) + \boldsymbol{\psi}\mathbf{u}(n)$$

$$\mathbf{y}(n) = \mathbf{C}\mathbf{x}(n) + \mathbf{D}\mathbf{u}(n).$$

(i) The Asymptotic State Estimator

The estimator dynamical equations (cf. Eq. (3.54)) are:

$$\hat{\mathbf{x}}(n+1) = (\boldsymbol{\phi} - \mathbf{MC})\hat{\mathbf{x}}(n) + \boldsymbol{\psi}\mathbf{u}(n) + \mathbf{My}(n) - \mathbf{MDu}(n). \tag{4.23}$$

The estimator 'error' behaves according to the relationship (cf. Eq. (3.55)):

$$(\mathbf{x}(n+1) - \hat{\mathbf{x}}(n+1)) = (\boldsymbol{\phi} - \mathbf{MC})(\mathbf{x}(n) - \hat{\mathbf{x}}(n)). \tag{4.24}$$

As with the continuous time estimator, if Σ is observable, **M** may be designed to ensure that the eigenvalues of $(\boldsymbol{\phi} - \mathbf{MC})$ correspond to a rapid well-behaved decay of any estimator error, with constraints similar to those encountered in pole shifting (Sec. 4.7).

If Σ is multi-output, a dyadic matrix **M** may be specified, while if Σ is unobservable, only the observable states may be estimated (cf. Sec. 3.8).

The CAD facility described in Sec. 3.8 applies directly to these design methods.

(ii) The Reduced-order Estimator (Luenberger Observer)

The estimator equations (cf. Eqs. (3.56), (3.57)) are:

$$\boldsymbol{\eta}(n+1) = \mathbf{L}\boldsymbol{\eta}(n) + (\mathbf{G} - \mathbf{HD})\mathbf{u}(n) + \mathbf{H}\mathbf{y}(n) \tag{4.25}$$

$$\hat{\mathbf{x}}(n) = \mathbf{N}\boldsymbol{\eta}(n) - \mathbf{MD}\mathbf{u}(n) + \mathbf{M}\mathbf{y}(n) \tag{4.26}$$

where **L**, **G**, **H**, **N**, **M** are designed according to the algorithm outlined in Sec. 3.8(ii) (Fig. 3.22). **L** and **M** are selected and improved by trial and error so that Eqs. (4.25), (4.26) are well conditioned and represent good estimator behavior.

As with continuous time systems, the states estimated by these means can readily be used in state feedback control strategies.

EXAMPLE

Consider the system in the example of Sec. 3.8 (Fig. 3.23) buffered with DACs and ADCs (Fig. 4.4), $T = 0.1$.

The state equations are:

$$\dot{\mathbf{x}} = \begin{bmatrix} -0.46 & 1 \\ -210 & 0 \end{bmatrix}\mathbf{x} + \begin{bmatrix} 0 \\ 210 \end{bmatrix}u$$

$$y = [1 \quad 0]\mathbf{x}.$$

Using the program of Fig. 4.5, the discrete system equations are found:

$$\mathbf{x}(n+1) = \begin{bmatrix} 0.1034 & 0.0669 \\ -14.0587 & 0.1342 \end{bmatrix}\mathbf{x}(n) + \begin{bmatrix} 0.8658 \\ 14.4570 \end{bmatrix}u(n)$$

$$y(n) = [1 \quad 0]\mathbf{x}(n).$$

This has eigenvalues:

$$z = 0.1188 \pm j0.9697.$$

An asymptotic state estimator with eigenvalues of $(\boldsymbol{\phi} - \mathbf{MC})$ at $z = 0.3$, 0.3 gives (Eq. 4.23):

$$\mathbf{M} = (-0.3624 \quad -13.6478)^{\mathrm{T}}$$

$$\hat{\mathbf{x}}(n+1) = \begin{bmatrix} 0.4658 & 0.0669 \\ -0.4109 & 0.1342 \end{bmatrix} \hat{\mathbf{x}}(n) + \begin{bmatrix} 0.8658 \\ 14.4570 \end{bmatrix} u(n) + \begin{bmatrix} -0.3624 \\ -13.6478 \end{bmatrix} y(n).$$

To shift the system poles to $z = 0.4 \pm j0.3$, the (not quite correct) assumption is made that $\hat{\mathbf{x}}(n) = \mathbf{x}(n)$, and the method outlined in Sec. 4.7 is used. This results in the state feedback matrix:

$$\mathbf{K} = (-0.7404 \quad 0.0054).$$

The entire control strategy is illustrated in block form in Fig. 4.11(a). The open-loop system response to a unit step function input (Fig. 4.11(b)), contrasts with the closed-loop response (Fig. 4.11(c)).

The concluding comments of Sec. 3.8, which refer to continuous time systems, apply equally to the discrete case. The choices of eigenvalues for the asymptotic estimator ($\boldsymbol{\phi} - \mathbf{MC}$) matrix, or Luenberger $\mathbf{L}$ matrix are constrained by the need to generate estimator algorithms with reasonable parameters, and this is affected by the sampling period T: the longer T, the more difficult the problem. As with the pole shifting problem (Sec. 4.7), the selection of too short a sampling period can result in heavy computer loading and even resolution difficulties. The change in steady-state gain when state feedback is applied is noteworthy.

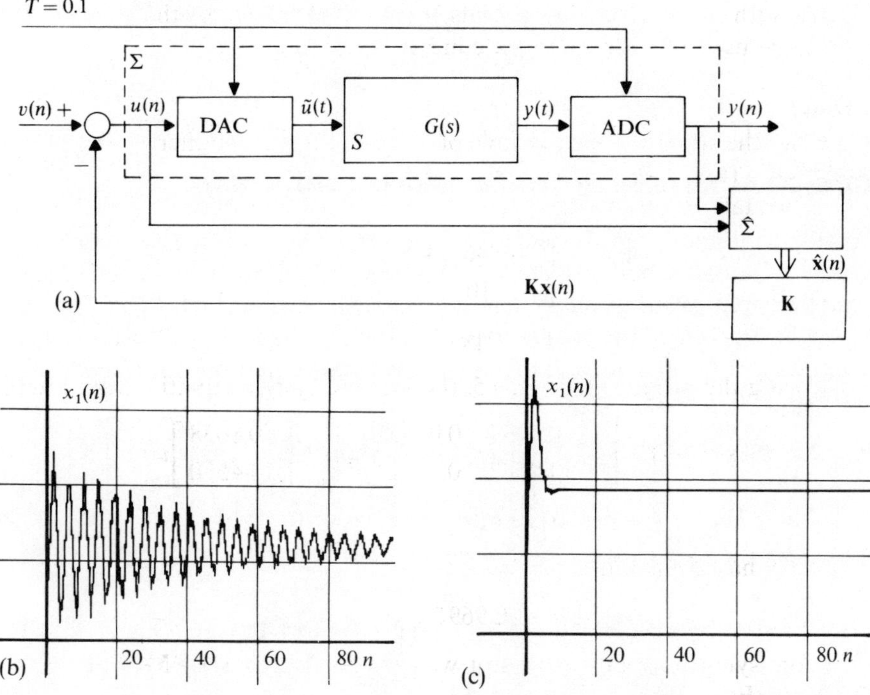

Fig. 4.11 Pole shifting using estimated state feedback (example) (a) Closed-loop system. (b) Open-loop response. (c) Closed-loop response.

4.10 MODELING AND BEHAVIOR OF STOCHASTIC SYSTEMS (REF. 7)

As in the continuous time case (Sec. 3.9) stochastic or noisy linear systems may be modeled by an extension of the state equations, Eqs. (4.2), (4.3):

$$\mathbf{x}(n+1) = \boldsymbol{\phi}(n)\mathbf{x}(n) + \boldsymbol{\psi}(n)\mathbf{u}(n) + \mathbf{w}_d(n) \tag{4.27}$$

$$\mathbf{y}(n) = \mathbf{C}(n)\mathbf{x}(n) + \mathbf{D}(n)\mathbf{u}(n) + \mathbf{v}_d(n). \tag{4.28}$$

$\mathbf{w}_d(n)$ and $\mathbf{v}_d(n)$ are usually vectors of discrete Gaussian white noise with known means and covariances (Sec. 1.22):

$$\mathscr{E}(\mathbf{w}_d(n)) = \mathbf{0}$$
$$\text{cov}(\mathbf{w}_d(n)) = \mathscr{E}(\mathbf{w}_d(n)\mathbf{w}_d^{\mathrm{T}}(n+k)) = \mathbf{Q}_d\delta(k)$$

where $\delta(k)$ is Kronecker delta function (App. C.2).

$$\mathscr{E}(\mathbf{v}_d(n)) = \mathbf{0}$$
$$\text{cov}(\mathbf{v}_d(n)) = \mathscr{E}(\mathbf{v}_d(n)\mathbf{v}_d^{\mathrm{T}}(n+k)) = \mathbf{R}_d\delta(k).$$

$\mathbf{w}_d(n)$ and $\mathbf{v}_d(n)$ represent a noise forcing function vector and a measurement noise vector respectively.

Noisy LTI systems may be modeled by the equations:

$$\mathbf{x}(n+1) = \boldsymbol{\phi}\mathbf{x}(n) + \boldsymbol{\psi}\mathbf{u}(n) + \mathbf{w}_d(n) \tag{4.29}$$

$$\mathbf{y}(n) = \mathbf{C}\mathbf{x}(n) + \mathbf{D}\mathbf{u}(n) + \mathbf{v}_d(n) \tag{4.30}$$

where $\boldsymbol{\phi}$, $\boldsymbol{\psi}$, $\mathbf{C}$, $\mathbf{D}$ are constant matrices.

(i) Systems Affected by Colored Noise (cf. Sec. 3.9(ii))

As with continuous time systems, if a discrete system is affected by nonwhite or colored Gaussian noise, it may be modeled by an augmented set of state equations representing the system illustrated in Fig. 4.12 (cf. Fig. 3.26).

The algebra of the model for an LTI system is identical to that of the continuous time system, and the augmented state equations for the system Σ (cf. Eqs. (3.62), (3.63)) are:

$$\begin{bmatrix} \mathbf{x}(n+1) \\ \mathbf{x}_{\mathrm{w}}(n+1) \\ \mathbf{x}_{\mathrm{v}}(n+1) \end{bmatrix} = \begin{bmatrix} \boldsymbol{\phi} & \mathbf{C}_{\mathrm{w}} & 0 \\ 0 & \boldsymbol{\phi}_{\mathrm{w}} & 0 \\ 0 & 0 & \boldsymbol{\phi}_{\mathrm{v}} \end{bmatrix} \begin{bmatrix} \mathbf{x}(n) \\ \mathbf{x}_{\mathrm{w}}(n) \\ \mathbf{x}_{\mathrm{v}}(n) \end{bmatrix} + \begin{bmatrix} \mathbf{D}_{\mathrm{w}} \\ \boldsymbol{\psi}_{\mathrm{w}} \\ 0 \end{bmatrix} \mathbf{w}(n) + \begin{bmatrix} 0 \\ 0 \\ \boldsymbol{\psi}_{\mathrm{v}} \end{bmatrix} \mathbf{v}(n) + \begin{bmatrix} \boldsymbol{\psi} \\ 0 \\ 0 \end{bmatrix} \mathbf{u}(n) \tag{4.31}$$

$$\mathbf{y}(n) = [\mathbf{C} \quad 0 \quad \mathbf{C}_{\mathrm{v}}] \begin{bmatrix} \mathbf{x}(n) \\ \mathbf{x}_{\mathrm{w}}(n) \\ \mathbf{x}_{\mathrm{v}}(n) \end{bmatrix} + \mathbf{D}\mathbf{u}(n) + \mathbf{D}_{\mathrm{v}}\mathbf{v}(n). \tag{4.32}$$

In these equations, the constant matrices $\boldsymbol{\phi}_{\mathrm{w}}$, $\boldsymbol{\psi}_{\mathrm{w}}$, $\mathbf{C}_{\mathrm{w}}$, $\mathbf{D}_{\mathrm{w}}$, $\boldsymbol{\phi}_{\mathrm{v}}$, $\mathbf{C}_{\mathrm{v}}$, $\mathbf{D}_{\mathrm{v}}$ describe the noise filters Σ_{w}, Σ_{v}, whose forms parallel those of Sec. 3.9(ii).

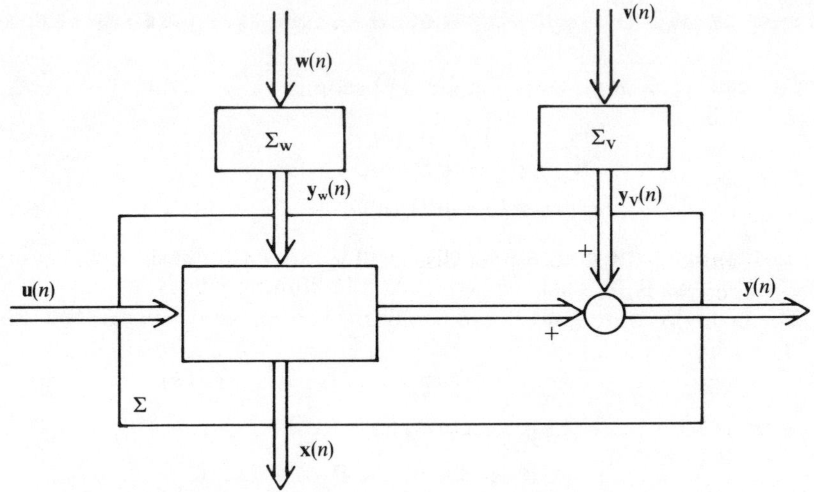

Fig. 4.12 Augmented system modeling colored noise.

Since the noise terms are white, these equations have the same essential form as Eqs. (4.29), (4.30) and this allows the same analysis to be used, with (as in the continuous time case) the difference that the effective white noise forcing function and measurement noise vectors in Eqs. (4.31) and (4.32) are not independent. This feature must of course be remembered where appropriate.

(ii) The Mean and Covariance of the State and Output

Unlike the equivalent continuous time Eqs. (3.58) to (3.61), Eqs. (4.27) to (4.30) are directly meaningful, and the mean and covariance of $\mathbf{x}(n)$, $\mathbf{u}(n) = \mathbf{0}$, behave according to the difference equations (Sec. 1.22):

$$\mathbf{m}_x(n+1) = \boldsymbol{\phi}(n)\mathbf{m}_x(n) \tag{4.33}$$

$$\bar{\mathbf{R}}_x(n+1) = \boldsymbol{\phi}(n)\bar{\mathbf{R}}_x(n)\boldsymbol{\phi}^{\mathrm{T}}(n) + \mathbf{Q}_d \tag{4.34}$$

$$\bar{\mathbf{R}}_x(n, n+k) = \boldsymbol{\phi}(n+k, n)\bar{\mathbf{R}}_x(n) \tag{4.35}$$

using, for clarity, the following symbols, which are slightly fuller than those in Sec. 1.22:

$$\mathbf{m}_x(n) = \mathscr{E}(\mathbf{x}(n))$$

$$\bar{\mathbf{R}}_x(n,\, n+k) = \operatorname{cov}(\mathbf{x}(n)) = \mathscr{E}[(\mathbf{x}(n) - \mathbf{m}_x(n))(\mathbf{x}(n+k) - \mathbf{m}_x(n+k))^{\mathrm{T}}]$$

and $\qquad \bar{\mathbf{R}}_x(n) = \bar{\mathbf{R}}_x(n, n).$

Note that Eq. (4.34) has one term fewer than the 'equivalent' continuous time

Eq. (3.65), a feature which arises from the more complex analysis of the latter case.

For an LTI system, these equations become:

$$\mathbf{m}_x(n+1) = \boldsymbol{\phi}\mathbf{m}_x(n) \tag{4.36}$$

$$\bar{\mathbf{R}}_x(n+1) = \boldsymbol{\phi}\bar{\mathbf{R}}_x(n)\boldsymbol{\phi}^{\mathrm{T}} + \mathbf{Q}_{\mathrm{d}} \tag{4.37}$$

$$\bar{\mathbf{R}}_x(n, n+k) = \boldsymbol{\phi}^k\bar{\mathbf{R}}_x(n). \tag{4.38}$$

Clearly, as in the continuous case, $\mathbf{x}(n)$ is not stationary, though if Σ is stable, $\mathbf{m}_x(n)$ and $\bar{\mathbf{R}}_x(n, n+k)$ approach stable limiting values, $n \to \infty$.

The behavior of the mean and covariance of $\mathbf{y}(n)$, $\mathbf{u}(n) = \mathbf{0}$, can be found from Eq. 4.30.

$$\begin{aligned} \mathbf{m}_y(n) &= \mathbf{C}\mathbf{m}_x(n) \\ \bar{\mathbf{R}}_y(n, n+k) &= \begin{cases} \mathbf{C}\bar{\mathbf{R}}_x(n, n+k)\mathbf{C}^{\mathrm{T}}, & k \neq 0 \\ \mathbf{C}\bar{\mathbf{R}}_x(n, n+k)\mathbf{C}^{\mathrm{T}} + \mathbf{R}_{\mathrm{d}}, & k = 0 \end{cases} \end{aligned} \tag{4.39}$$

where

$$\begin{aligned} \mathbf{m}_y(n) &= \mathscr{E}(\mathbf{y}(n)) \\ \bar{\mathbf{R}}_y(n, n+k) &= \operatorname{cov}(\mathbf{y}(n)) \\ &= \mathscr{E}[\mathbf{y}(n) - \mathbf{m}_y(n))(\mathbf{y}(n+k) - \mathbf{m}_y(n+k))^{\mathrm{T}}]. \end{aligned}$$

(iii) Sampled Data Stochastic Systems

As with deterministic systems (Sec. 4.6), sampled data systems in which continuous time systems are 'viewed' through DACs and ADCs (Fig. 4.4) are of particular interest since on-line computers are now in widespread use in the control of such systems.

The state equations of a continuous time LTI stochastic system are (Eqs. (3.60), (3.61)):

$$\begin{aligned} \dot{\mathbf{x}} &= \mathbf{A}\mathbf{x} + \mathbf{B}\mathbf{u} + \mathbf{w} \\ \mathbf{y} &= \mathbf{C}\mathbf{x} + \mathbf{D}\mathbf{u} + \mathbf{v} \end{aligned}$$

where

$$\begin{aligned} &\mathscr{E}(\mathbf{w}) = \mathbf{0},\ \operatorname{cov}(\mathbf{w}) = \mathbf{Q}.\delta(\tau) \qquad \text{(Secs. (1.21), (3.9))} \\ &\mathscr{E}(\mathbf{v}) = \mathbf{0},\ \operatorname{cov}(\mathbf{v}) = \mathbf{R}.\delta(\tau). \end{aligned}$$

The state equations of the sampled data system derived from this, sampling period T, are (Eqs. (4.29), (4.30)):

$$\begin{aligned} \mathbf{x}(n+1) &= \boldsymbol{\phi}\mathbf{x}(n) + \boldsymbol{\psi}\mathbf{u}(n) + \mathbf{w}_{\mathrm{d}}(n) \\ \mathbf{y}(n) &= \mathbf{C}\mathbf{x}(n) + \mathbf{D}\mathbf{u}(n) + \mathbf{v}_{\mathrm{d}}(n) \end{aligned}$$

where

$$\begin{aligned} &\mathscr{E}(\mathbf{w}_{\mathrm{d}}(n)) = \mathbf{0},\ \operatorname{cov}(\mathbf{w}_{\mathrm{d}}) = \mathbf{Q}_{\mathrm{d}}\delta(k) \\ &\mathscr{E}(\mathbf{v}_{\mathrm{d}}(n)) = \mathbf{0},\ \operatorname{cov}(\mathbf{v}_{\mathrm{d}}) = \mathbf{R}_{\mathrm{d}}\delta(k) \qquad \text{(Sec. 1.22).} \end{aligned}$$

It is not difficult to show that in this case:

$$\left.\begin{aligned}\boldsymbol{\phi} &= e^{\mathbf{A}T}\\ \boldsymbol{\psi} &= \int_0^T e^{\mathbf{A}\lambda}\mathbf{B}\,\mathrm{d}\lambda\\ \mathbf{Q}_\mathrm{d} &= \int_0^T e^{\mathbf{A}\lambda}\mathbf{Q}(e^{\mathbf{A}\lambda})^T\,\mathrm{d}\lambda\\ \mathbf{R}_\mathrm{d} &= \mathbf{R}\end{aligned}\right\} \tag{4.40}$$

In the simpler case where T is small compared with the time constants and cycle periods corresponding to the eigenvalues of **A**, the following approximations (cf. Eqs. (4.17)) may be used:

$$\left.\begin{aligned}\boldsymbol{\phi} &\approx \mathbf{I}+\mathbf{A}T\\ \boldsymbol{\psi} &\approx T\mathbf{B}\\ \mathbf{Q}_\mathrm{d} &\approx \mathbf{Q}T\\ \mathbf{R}_\mathrm{d} &= \mathbf{R}.\end{aligned}\right\} \tag{4.41}$$

4.11 THE KALMAN FILTER (REFS. 2, 6, 7)

A *Kalman filter* is a state estimator designed to estimate the state of a stochastic system using the system input and output as data, as do deterministic system state estimators (Sec. 4.9). Various forms of Kalman filter exist; the basic optimal form applies to linear time variant systems while suboptimal forms apply to LTI systems and various extended forms have been successfully applied to nonlinear systems.

A discrete linear, time variant stochastic system Σ with a Kalman filter $\hat{\Sigma}$

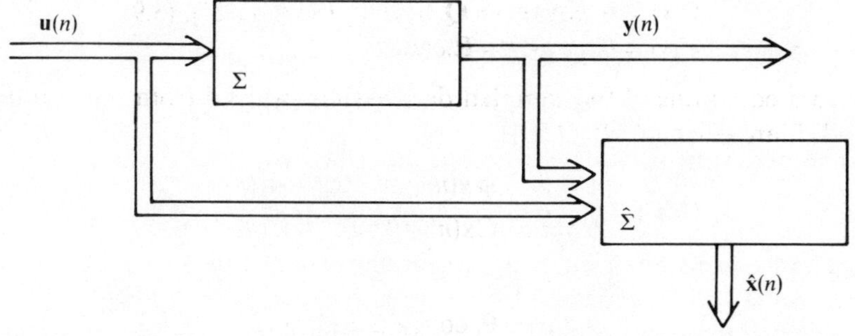

Fig. 4.13 A stochastic system Σ and Kalman filter $\hat{\Sigma}$.

which generates an estimate $\hat{\mathbf{x}}(n)$ of the system state $\mathbf{x}(n)$ is shown in block form in Fig. 4.13.

(i) Basic Kalman Filter for Linear Time-variant Systems (ref. 2)

If the system Σ is observable, all the states may be estimated by the Kalman filter $\hat{\Sigma}$.

The state equations of Σ are (Eqs. (4.27), (4.28)):

$$\mathbf{x}(n+1) = \boldsymbol{\phi}(n)\mathbf{x}(n) + \boldsymbol{\psi}(n)\mathbf{u}(n) + \mathbf{w}_d(n)$$
$$\mathbf{y}(n) = \mathbf{C}(n)\mathbf{x}(n) + \mathbf{D}(n)\mathbf{u}(n) + \mathbf{v}_d(n)$$

where $\mathbf{w}_d(n)$, $\mathbf{v}_d(n)$ represent Gaussian white noise vectors whose properties, in contrast to $\mathbf{x}(n)$, are 'time invariant' or stationary.

$$\mathscr{E}(\mathbf{w}_d(n)) = \mathbf{0}, \qquad \operatorname{cov}(\mathbf{w}_d(n)) = \mathbf{Q}_d\delta(k)$$
$$\mathscr{E}(\mathbf{v}_d(n)) = \mathbf{0}, \qquad \operatorname{cov}(\mathbf{v}_d(n)) = \mathbf{R}_d\delta(k).$$

The Kalman filter algorithm is based on the deterministic system state estimator (Sec. 4.9). In certain cases (ref. 7) the Luenberger observer model (Eqs. (4.25), (4.26)) is used, but more commonly the asymptotic state estimator equation (Eq. (4.23)) forms the basis of the Kalman filter:

$$\hat{\mathbf{x}}(n+1) = \boldsymbol{\phi}(n)\hat{\mathbf{x}}(n) + \mathbf{M}(n)[\mathbf{y}(n) - \mathbf{C}(n)\hat{\mathbf{x}}(n) - \mathbf{D}(n)\mathbf{u}(n)] + \boldsymbol{\psi}(n)\mathbf{u}(n). \tag{4.42}$$

A measure of the estimator 'error' is the *error covariance matrix*, $\operatorname{cov}(\mathbf{x}(n) - \hat{\mathbf{x}}(n))$, $k = 0$ (Eq. (1.76)):

$$\mathbf{G}(n) = \mathscr{E}[(\mathbf{x}(n) - \hat{\mathbf{x}}(n))(\mathbf{x}(n) - \hat{\mathbf{x}}(n))^{\mathrm{T}}]. \tag{4.43}$$

The Kalman filter design is optimal in that it selects the value of $\mathbf{M}(n)$ which minimizes the cost function:

$$\operatorname{tr}\mathbf{G}(n) = \mathscr{E}[(\mathbf{x}_1(n) - \hat{\mathbf{x}}_1(n))^2 + (\mathbf{x}_2(n) - \hat{\mathbf{x}}_2(n))^2 + \cdots + (\mathbf{x}_m(n) - \hat{\mathbf{x}}_m(n))^2].$$

For convenience, a matrix $\mathbf{P}$ is defined:

$$\mathbf{P}(n) = \mathbf{R}_d + \mathbf{C}(n)\mathbf{G}(n)\mathbf{C}^{\mathrm{T}}(n). \tag{4.44}$$

The optimal value of $\mathbf{M}(n)$ is defined by the equation:

$$\mathbf{M}(n) = \boldsymbol{\phi}(n)\mathbf{G}(n)\mathbf{C}^{\mathrm{T}}(n)\mathbf{P}^{-1}(n) \tag{4.45}$$

and the behavior of $\mathbf{G}(n)$ with time is described by:

$$\mathbf{G}(n+1) = [\boldsymbol{\phi}(n) - \mathbf{M}(n)\mathbf{C}(n)]\mathbf{G}(n)\boldsymbol{\phi}^{\mathrm{T}}(n) + \mathbf{Q}_d. \tag{4.46}$$

The basic Kalman filter algorithm consists of Eqs. (4.44), (4.45), (4.42), (4.46), in that order, repeated for $n = 0, 1, 2, \ldots, \infty$.

Some comments on the Kalman filter equations are appropriate at this stage.

1. The Kalman filter algorithm selects dynamically a matrix $\mathbf{M}(n)$ (Eq. (4.42)) which positions the eigenvalues of $(\boldsymbol{\phi} - \mathbf{MC})$ so that the sum of the variances of the estimator errors,
$$\operatorname{tr} \mathbf{G}(n) = \mathscr{E}[(\mathbf{x}(n) - \hat{\mathbf{x}}(n))^{\mathrm{T}}(\mathbf{x}(n) - \hat{\mathbf{x}}(n))],$$
is minimized (cf. Sec. 4.9).
2. Provided Σ is stable, the filter settles down after an initial transient to behavior which is independent of $\mathbf{G}(0)$ and $\hat{\mathbf{x}}(0)$, the elements of which may in principle be set to any values, typically zero. If information is available which may be used to set $\hat{\mathbf{x}}(0)$, $\mathbf{G}(0)$ more accurately, the amplitude and duration of this initial transient can be reduced quite markedly.
3. The filter gain, $\mathbf{M}(n)$ is affected by $\mathbf{Q}_{\mathrm{d}}$ and $\mathbf{R}_{\mathrm{d}}$ which are basically measures of the amplitudes of noise affecting the system Σ. The algorithm fails if $\mathbf{Q}_{\mathrm{d}} = \mathbf{0}$ or $\mathbf{R}_{\mathrm{d}} = \mathbf{0}$, since there is then insufficient data from which to develop an optimal $\mathbf{M}(n)$.
4. $\mathbf{Q}_{\mathrm{d}}$ can, where appropriate, be regarded as a measure of ignorance or uncertainty about the system input, $\mathbf{u}(n)$, rather than as a measure of the noise affecting it. Consequently, $\mathbf{Q}_{\mathrm{d}}$ may be altered as a matter of strategy even during the running of the filter, to improve the filter performance.
5. The gain of the optimal filter, $\mathbf{M}(n)$, develops dynamically as $\boldsymbol{\phi}(n)$, $\mathbf{C}(n)$ change (Eqs. (4.44), (4.45)). In the case of an LTI system, this does not happen after an initial transient; this is considered in comment 6.
6. Where a Kalman filter is used to estimate the states of an LTI system (Eqs. (4.29), (4.30)), the algorithm equations, which are simply a special case of Eqs. (4.42), (4.44), (4.45), (4.46), are:

$$\hat{\mathbf{x}}(n+1) = \boldsymbol{\phi}\hat{\mathbf{x}}(n) + \mathbf{M}(n)[\mathbf{y}(n) - \mathbf{C}\hat{\mathbf{x}}(n) - \mathbf{D}\mathbf{u}(n)] + \boldsymbol{\psi}\mathbf{u}(n) \tag{4.47}$$

$$\mathbf{P}(n) = \mathbf{R}_{\mathrm{d}} + \mathbf{C}\mathbf{G}(n)\mathbf{C}^{\mathrm{T}} \tag{4.48}$$

$$\mathbf{M}(n) = \boldsymbol{\phi}\mathbf{G}(n)\mathbf{C}^{\mathrm{T}}\mathbf{P}^{-1}(n) \tag{4.49}$$

$$\mathbf{G}(n+1) = (\boldsymbol{\phi} - \mathbf{M}(n)\mathbf{C})\mathbf{G}(n)\boldsymbol{\phi}^{\mathrm{T}} + \mathbf{Q}_{\mathrm{d}}. \tag{4.50}$$

It can be seen from Eqs. (4.48), (4.49), (4.50) that $\mathbf{M}(n)$ develops independently of $\hat{\mathbf{x}}(n)$ (Eq. (4.47)). In fact $\mathbf{P}(n)$, $\mathbf{G}(n)$, $\mathbf{M}(n)$ approach limiting values if the LTI system is stable, and there is no reason why these values, $\mathbf{P}^*$, $\mathbf{G}^*$, $\mathbf{M}^*$, cannot be developed off-line. Equation (4.47) can then be used alone, with $\mathbf{M}(n) = \mathbf{M}^*$, as a state estimator, or 'suboptimal' Kalman filter whose performance is identical with the true filter (for an LTI system) except during an initial transient after switch-on. A simple CAD facility can be constructed to find $\mathbf{M}^*$, $\mathbf{G}^*$ (which is a measure of the filter error) under stationary, nontransient, conditions.

CAD Facility

The interactive CAD algorithm in the flow diagram of Fig. 4.14 calculates $\mathbf{M}^*$ and $\mathbf{G}^*$.

The operation of the algorithm is as follows.

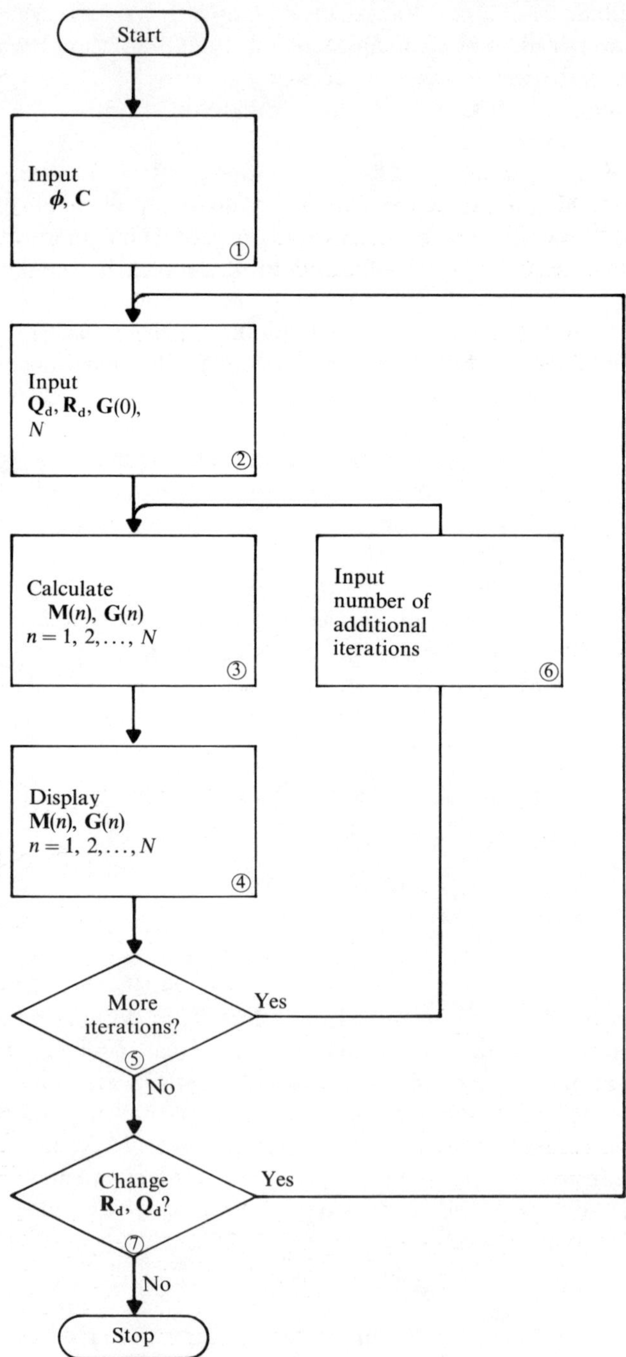

Fig. 4.14 Suboptimal Kalman filter design CAD algorithm.

BLOCKS 1, 2

The system parameters $\boldsymbol{\phi}, \mathbf{C}, \mathbf{Q}_d, \mathbf{R}_d$ and the initial condition $\mathbf{G}(0)$ are input, displayed, and corrected if necessary. A required number of iterations, N, is also input.

BLOCKS 3, 4

The values of $\mathbf{M}(n)$, $\mathbf{G}(n)$ are calculated (Eqs. (4.48), (4.49), (4.50)), $n = 1, 2, \ldots, N$, and displayed. This reveals their evolution from $\mathbf{M}(0)$, $\mathbf{G}(0)$ which, though not perhaps essential, can be of considerable interest.

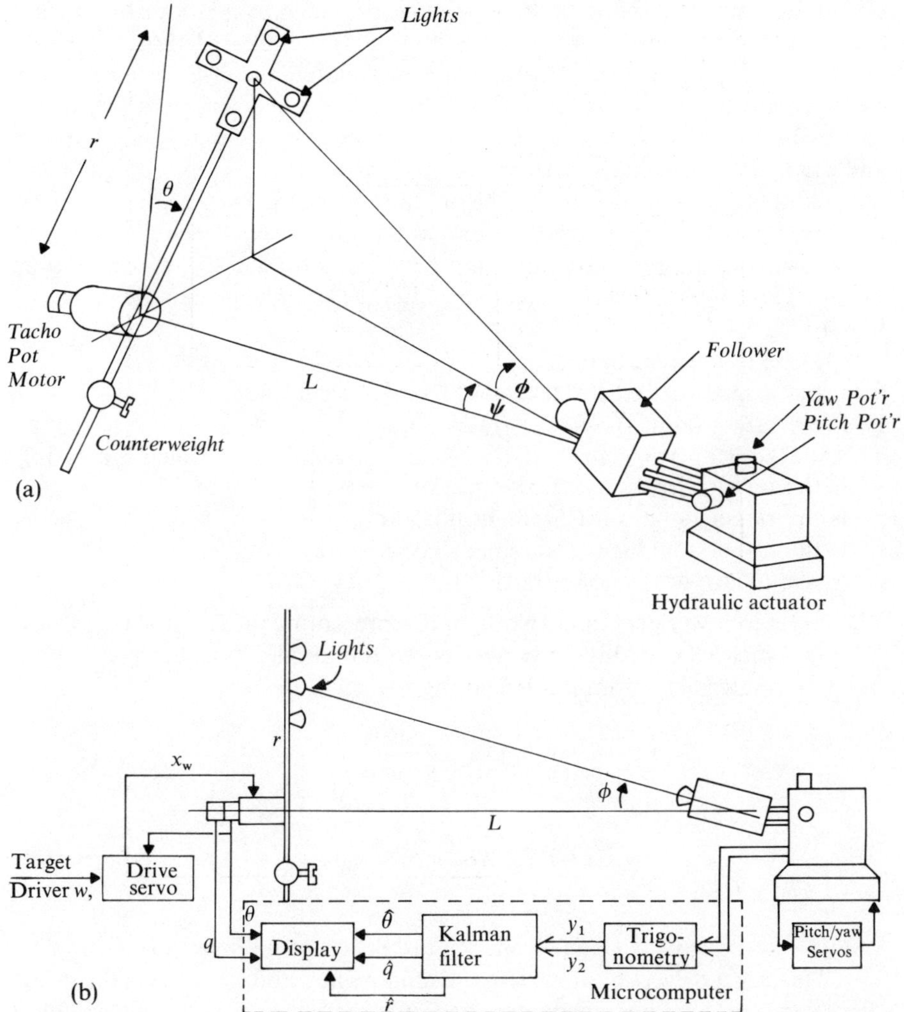

Fig. 4.15 Laboratory target and follower. (a) Target and follower. (b) System schematic.

BLOCKS 5, 6
The operator can demand further iterations, which may be necessary to establish steady-state values $\mathbf{M}(\infty) = \mathbf{M}^*$, $\mathbf{G}(\infty) = \mathbf{G}^*$.

BLOCK 7
The operator can change the parameters $\mathbf{Q}_d$, $\mathbf{R}_d$, $\mathbf{G}(0)$. This helps considerably in 'tuning' the suboptimal filter, i.e. altering its performance by manipulating $\mathbf{Q}_d$.

EXAMPLES
Useful illustration of many of these points is provided by a laboratory equipment comprising a pseudo-randomly (App. B.5) illuminated target array of lights mounted on a vertical rotatable arm situated about 2 meters from a hydraulically actuated target follower. This carries a lens and photocell array and 'locks' on to the 'center of gravity' of the lights via two independent (pitch and yaw) servos, so that the detected position of the target is represented by the outputs of two potentiometers on the follower. The target is actuated either by hand or by a signal generator via an electrically actuated angular position servo. An on-line microcomputer is used to interpret the follower output signals, and to realize various Kalman filters. The arrangement is shown in Fig. 4.15.

In the following analysis,

θ is the target angular displacement from vertical (rad),
q is the target angular velocity (rad s^{-1}),
r is the (constant) target radial displacement from the experiment's axis (m),
x_w is the target disturbance noise (rad s^{-2}),
v_r is the target radial measurement noise (m),
v_θ is the target angular measurement noise (rad),
w is (Gaussian) white noise (rad s^{-2}).

The target follower generates two signals representing pitch φ (rad), and yaw, ψ (rad), which can readily be converted to the 'model' outputs y_1, y_2 by a microcomputer algorithm implementing the relationships:

$$y_1 = \theta + v_\theta = \tan^{-1}\left(\frac{\sin\psi}{\tan\varphi}\right)$$

$$y_2 = r + v_r = L\sqrt{\tan^2\psi + \frac{\tan^2\varphi}{\cos^2\psi}}$$

where L is the axial distance from the target to the follower (m).

The target drive, unknown to the Kalman filter, and applied via the target drive servo, is modeled adequately as Gaussian white noise seen through a first-order filter. The target measurement noise is regarded as Gaussian white noise in each of the two independent axes (App. B5(i), (ii), (iii)), θ, r.

Together with the target equations of motion, these considerations give a system model:

$$\begin{bmatrix} \dot{\theta} \\ \dot{q} \\ \dot{r} \\ \dot{x}_w \end{bmatrix} = \begin{bmatrix} 0 & 1 & 0 & 0 \\ 0 & -0.07 & 0 & 1 \\ 0 & 0 & 0 & 0 \\ 0 & 0 & 0 & -1 \end{bmatrix} \begin{bmatrix} \theta \\ q \\ r \\ x_w \end{bmatrix} + \begin{bmatrix} 0 \\ 0 \\ 0 \\ w \end{bmatrix} \tag{4.51}$$

$$\mathbf{y} = \begin{bmatrix} 1 & 0 & 0 & 0 \\ 0 & 0 & 1 & 0 \end{bmatrix} \begin{bmatrix} \theta \\ q \\ r \\ x_w \end{bmatrix} + \begin{bmatrix} v_\theta \\ v_r \end{bmatrix} \tag{4.52}$$

The covariances of the white noise sources (Sec. 1.21) are:

$$\mathscr{E}(w(t)w(t+\tau)) = 2.5\delta\tau \qquad \text{or} \qquad \mathbf{Q} = \begin{bmatrix} 0 & 0 & 0 & 0 \\ 0 & 0 & 0 & 0 \\ 0 & 0 & 0 & 0 \\ 0 & 0 & 0 & 2.5 \end{bmatrix}$$

$$\mathscr{E}\left[\begin{bmatrix} v_\theta(t) \\ v_r(t) \end{bmatrix} \begin{bmatrix} v_\theta(t+\tau) \\ v_r(t+\tau) \end{bmatrix}^{\mathrm{T}}\right] = \begin{bmatrix} 0.01 & 0 \\ 0 & 0.01 \end{bmatrix} \delta\tau \qquad \text{or} \qquad \mathbf{R} = \begin{bmatrix} 0.01 & 0 \\ 0 & 0.01 \end{bmatrix}$$

$$\mathscr{E}([0 \quad 0 \quad 0 \quad w(t)]^{\mathrm{T}}[v_\theta(t+\tau)v_r(t+\tau)]) = 0\delta\tau.$$

(*a*) *A Suboptimal Kalman Filter*

First, a suboptimal Kalman filter can be designed and tested. For a short sampling period, $T = 0.1$, Eqs. (4.51), (4.52) can be discretized (Eq. (4.17)) to give a model of the system seen through ADCs and DACs (Fig. 4.4). (A longer sampling period could be accommodated using the algorithm of Fig. 4.5).

$$\mathbf{x}(n+1) = \begin{bmatrix} 1 & 0.1 & 0 & 0 \\ 0 & 0.993 & 0 & 0.1 \\ 0 & 0 & 1 & 0 \\ 0 & 0 & 0 & 0.9 \end{bmatrix} \mathbf{x}(n) + \begin{bmatrix} 0 \\ 0 \\ 0 \\ w(n) \end{bmatrix}$$

$$\mathbf{y}(n) = \begin{bmatrix} 1 & 0 & 0 & 0 \\ 0 & 0 & 1 & 0 \end{bmatrix} \mathbf{x}(n) + \begin{bmatrix} v_\theta(n) \\ v_r(n) \end{bmatrix}$$

where $\mathbf{x}(n) = [\theta(n) \quad q(n) \quad r(n) \quad x_w(n)]^{\mathrm{T}}$.

Since these equations are LTI, a suboptimal Kalman filter is appropriate (Comment 6). This will perform identically to a true Kalman filter except during an initial transient after switch-on, and indeed its performance during such a transient will usually be superior to that of the optimal filter (which uses this period to establish an optimal $\mathbf{M}(n)$).

Since the filter is expected to operate without knowledge of the input $w(n)$,

$\mathbf{Q}_d$ is selected to accommodate this ignorance (comment 4).

$$\mathbf{Q}_d = \mathscr{E}(\mathbf{w}(n)\mathbf{w}^T(n)) \approx \mathbf{Q}T \qquad \text{(Eq. 4.41)}.$$

In addition, an element (3,3) is added to accommodate ignorance of r:

Set:

$$\mathbf{Q}_d = \begin{bmatrix} 0 & 0 & 0 & 0 \\ 0 & 0 & 0 & 0 \\ 0 & 0 & 0.25 & 0 \\ 0 & 0 & 0 & 0.25 \end{bmatrix}.$$

The measurement noise, generated by the array of target lights, is represented by:

$$\mathbf{R}_d = \mathbf{R} = \begin{bmatrix} 0.01 & 0 \\ 0 & 0.01 \end{bmatrix}.$$

Substitution of these values in Eqs. (4.48), (4.49), (4.50) allows a gain $\mathbf{M}^*$ to be developed by the algorithm of Fig. 4.14:

$$\mathbf{M}^* = \begin{bmatrix} 0.405 & 0 \\ 0.729 & 0 \\ 0 & 0.776 \\ 0.529 & 0 \end{bmatrix}.$$

This is used in Eq. (4.47) (with $\mathbf{u}(n) = \mathbf{0}$) to realize the suboptimal filter, the performance of which is illustrated in Fig. 4.16 for an input to the target consisting of a constant acceleration of 0.5 rad s^{-2} for 2.5 seconds, followed by a deceleration of equal amplitude for 3.5 seconds.

It can be seen from Fig. 4.16 that in spite of the noise corruption on the measured target position and the absence of any information about target drive, the filter succeeds in establishing quite accurate measurements of target position and (more slowly) velocity. It also estimates the target radial position, r, which is constant, easily and accurately. Different choices of $\mathbf{Q}_d$, of course, result in different filter performances.

(b) An Optimal Filter

Since the optimal filter fully exhibits its properties in estimating the states of a time-varying system (Comment 5), the equipment of Fig. 4.15, which is LTI, must be altered slightly to allow an effective demonstration. This is easily done by removing the target array counterweight so that the target mount now constitutes a motor driven pendulum.

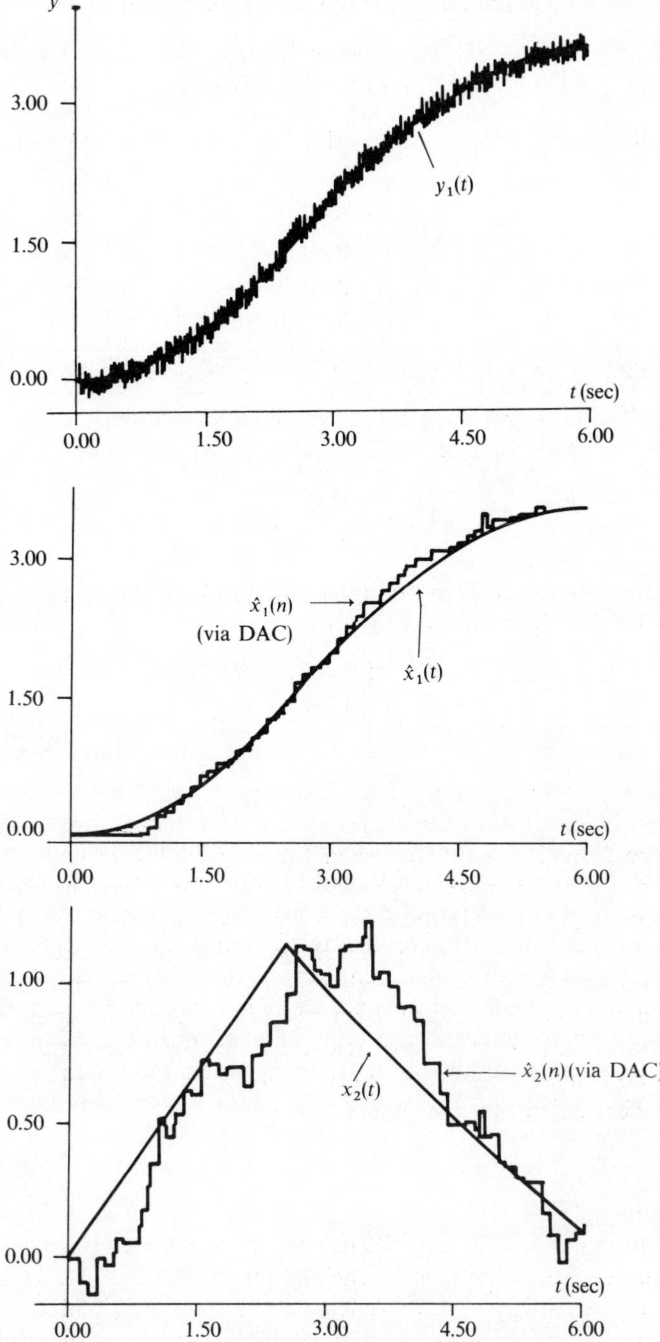

Fig. 4.16 Suboptimal Kalman filter performance (example).

This has model equations:

$$\begin{bmatrix} \dot{\theta} \\ \dot{q} \\ \dot{r} \\ \dot{x}_w \end{bmatrix} = \begin{bmatrix} 0 & 1 & 0 & 0 \\ 0 & -0.07 & -0.5\sin\theta & 1 \\ 0 & 0 & 0 & 0 \\ 0 & 0 & 0 & -1 \end{bmatrix} \begin{bmatrix} \theta \\ q \\ r \\ x_w \end{bmatrix} + \begin{bmatrix} 0 \\ 0 \\ 0 \\ w \end{bmatrix}$$

$$\mathbf{y} = \begin{bmatrix} 1 & 0 & 0 & 0 \\ 0 & 0 & 1 & 0 \end{bmatrix} \begin{bmatrix} \theta \\ q \\ r \\ x_w \end{bmatrix} + \begin{bmatrix} v_\theta \\ v_r \end{bmatrix}.$$

Strictly, this set of equations is nonlinear, but the value of $\sin\theta$ in the $\mathbf{A}$ matrix can be regarded fairly accurately as $\sin y_1$. The resulting system can be treated to all intents and purposes as linear time-variant.

For a short sampling period $T = 0.1$, these equations are discretized to give:

$$\mathbf{x}(n+1) = \begin{bmatrix} 1 & 0.1 & 0 & 0 \\ 0 & 0.993 & -0.05\sin y_1 & 0 \\ 0 & 0 & 1 & 0 \\ 0 & 0 & 0 & 0.9 \end{bmatrix} \mathbf{x}(n) + \begin{bmatrix} 0 \\ 0 \\ 0 \\ w(n) \end{bmatrix}$$

$$\mathbf{y}(n) = \begin{bmatrix} 1 & 0 & 0 & 0 \\ 0 & 0 & 1 & 0 \end{bmatrix} \mathbf{x}(n) + \begin{bmatrix} v_\theta(n) \\ v_r(n) \end{bmatrix}$$

The values of $\mathbf{Q}_d$, $\mathbf{R}_d$ are selected as for the suboptimal filter. Starting with

$$\mathbf{x}(n) = \mathbf{0}, \quad \mathbf{M}(n) = \mathbf{0}, \quad \mathbf{G}(n) = \mathbf{0},$$

and recalculating $\boldsymbol{\phi}(n)$ at every iteration, the Kalman filter consists of Eqs. (4.44), (4.45), (4.42), (4.46) ($\mathbf{u}(n) = \mathbf{0}$, Eq. (4.42)). The performance of this is illustrated in Fig. 4.17 for a target input consisting of a constant torque applied for 2.5 seconds, followed by a reverse torque for 7.5 seconds.

It can be seen from Fig. 4.17 that in spite of the noise corruption on the measured target position, and the absence of information about target drive, the filter establishes reasonable estimates of target position and, more slowly, velocity, following an initial transient lasting some 2 seconds.

Different choices of $\mathbf{Q}_d$ would, of course, result in different filter performances.

(ii) The 'Split' Kalman Filter (refs. 6, 7)

Since most Kalman filters are concerned with estimating the states of sampled continuous time systems, the sampling period T is clearly a factor in determining their performance. If the sampling period is long – perhaps necessarily so for some reason – the estimate $\hat{\mathbf{x}}(n+1)$, which is based on data

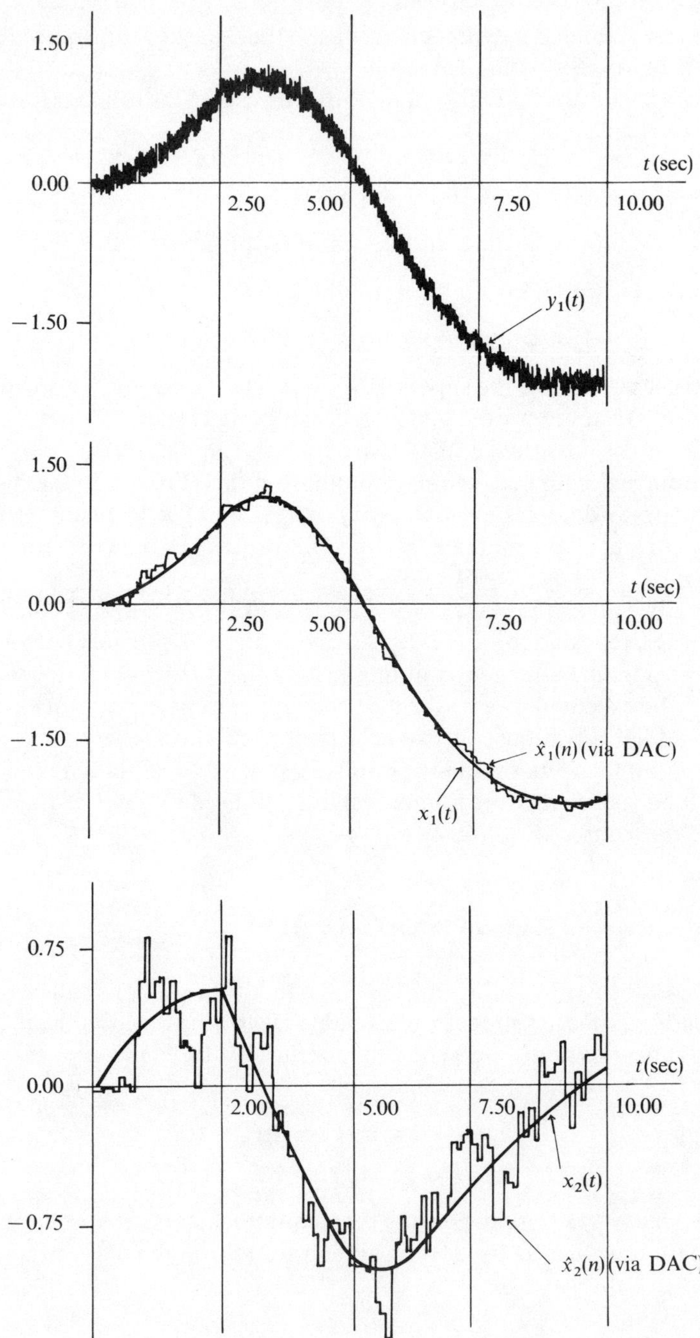

Fig. 4.17 Kalman filter performance (example).

available at time n (Eq. (4.42)) can be quite inaccurate. This situation can be improved by recasting the filter equations so that the most up-to-date data are used to generate the estimated state.

The filter algorithm in this 'split' form consists of the following equations:

$$\hat{\mathbf{x}}^{+}(n) = \hat{\mathbf{x}}^{-}(n) + \mathbf{J}(n)[\mathbf{y}(n) - \mathbf{C}(n)\hat{\mathbf{x}}^{-}(n) - \mathbf{D}(n)\mathbf{u}(n)] \tag{4.53}$$

$$\hat{\mathbf{x}}^{-}(n+1) = \boldsymbol{\phi}(n)\hat{\mathbf{x}}^{+}(n) + \boldsymbol{\psi}(n)\mathbf{u}(n) \tag{4.54}$$

$$\mathbf{J}(n) = \mathbf{G}^{-}(n)\mathbf{C}^{\mathrm{T}}(n)[\mathbf{R}_{\mathrm{d}} + \mathbf{C}(n)\mathbf{G}^{-}(n)\mathbf{C}^{\mathrm{T}}(n)]^{-1} \tag{4.55}$$

$$\mathbf{G}^{+}(n) = [\mathbf{I} - \mathbf{J}(n)\mathbf{C}(n)]\mathbf{G}^{-}(n) \tag{4.56}$$

$$\mathbf{G}^{-}(n+1) = \boldsymbol{\phi}(n)\mathbf{G}^{+}(n)\boldsymbol{\phi}^{\mathrm{T}}(n) + \mathbf{Q}_{\mathrm{d}}. \tag{4.57}$$

In these equations, the superscript^{-} refers to a time just before, while the superscript^{+} refers to a time just after a sampling instant. Thus $\mathbf{G}^{-}(n)$ is the calculated error covariance just before the nth sample.

The current gain matrix $\mathbf{J}(n)$ is found from this (Eq. (4.55)) and used with the most up-to-date system data, $\mathbf{y}(n)$, $\mathbf{u}(n)$ (Eq. 4.53) to generate the best possible current state estimate, $\hat{\mathbf{x}}^{+}(n)$. The process is repeated for the next sample via Eqs. (4.54), (4.56), (4.57).

The filter algorithm thus consists of Eqs. (4.55), (4.53), (4.54), (4.56), (4.57) in that order, repeated for $n = 0, 1, \ldots, \infty$. Equation (4.53) must be executed as quickly as possible after the sampling instant, while the remaining equations need only be executed in time for the following sample. In certain cases, there may be added advantage in two effective calculation intervals: one corresponding to the system sampling period (Eqs. (4.55), (4.56), (4.53)) and another shorter (and therefore more numerous) interval used to 'predict' the variables for the next sample (Eqs. (4.54), (4.57)).

(iii) The Extended Kalman Filter (refs. 6, 7)

In their split form (subsection (ii)) the Kalman filter equations can be extended fairly readily to accommodate reasonably well-behaved nonlinear systems. Such nonlinear systems are generally continuous time and are described by Eqs. (3.1), (3.2) extended to include noise:

$$\dot{\mathbf{x}} = \mathbf{f}(\mathbf{x}, \mathbf{u}, t) + \mathbf{w}_{\mathrm{c}}(t)$$

$$\mathbf{y} = \mathbf{g}(\mathbf{x}, \mathbf{u}, t) + \mathbf{v}_{\mathrm{c}}(t)$$

where $\mathbf{w}_{\mathrm{c}}(t)$, $\mathbf{v}_{\mathrm{c}}(t)$ represent noise. These equations can be discretized to describe the 'equivalent' sampled data system seen through DACs and ADCs (cf. Sec. 4.6):

$$\mathbf{x}(n+1) = \mathbf{x}(n) + \int_{nT}^{(n+1)T} \mathbf{f}(\mathbf{x}, \mathbf{u}, t)\mathrm{d}t + \mathbf{w}(n)$$

$$\mathbf{y}(n) = \mathbf{g}(\mathbf{x}(n), \mathbf{u}(n), nT) + \mathbf{v}(n).$$

A set of Kalman filter equations can be developed from these using the split form of subsection (ii). Usually these equations are considered in two groups, the 'measurement update' and the 'time update' equations, which occur at and between sampling instants respectively.

The measurement update equations are:

$$\hat{\mathbf{x}}^+(n) = \hat{\mathbf{x}}^-(n) + \mathbf{J}(n)[\mathbf{y}(n) - \mathbf{g}(\hat{\mathbf{x}}^-(n), \mathbf{u}(n), nT)] \tag{4.58}$$

$$\mathbf{G}^+(n) = [\mathbf{I} - \mathbf{J}(n)\mathbf{C}(n)]\mathbf{G}^-(n) \tag{4.59}$$

$$\text{where } \mathbf{C}(n) = \left.\frac{\partial \mathbf{g}}{\partial \mathbf{x}}\right|_{\hat{\mathbf{x}}^+(n), \mathbf{u}(n)}.$$

The time update equations are:

$$\hat{\mathbf{x}}^-(n+1) = \hat{\mathbf{x}}^+(n) + \int_{nT}^{(n+1)T} f(\mathbf{x}, \mathbf{u}, t). \tag{4.60}$$

In principle, the integration in Eq. (4.60) can be performed using a Runge–Kutta algorithm (Sec. 1.2) which may require a number of iterations between sampling instants. More usually, if the sampling period, T, is short compared with the system dynamics,

$$\hat{\mathbf{x}}^-(n+1) = \hat{\mathbf{x}}^+(n) + T\mathbf{f}(\hat{\mathbf{x}}^+(n), \mathbf{u}(n), nT). \tag{4.61}$$

In an alternative form, the basic nonlinear state equations can be linearized around the current 'datum' $\mathbf{x}^+(n)$ (ref. 7) and the linear Kalman filter equations, with appropriate augmentation, then apply.

In all these schemes, it is noteworthy that satisfactory operation of the algorithm depends on the filter generating a reasonably good estimate $\mathbf{x}^+(n)$ – which of course in turn depends on satisfactory operation of the algorithm. In practice, if the system nonlinearities are not too severe, the algorithms generally do converge to give useful results.

The remaining equations are:

$$\mathbf{J}(n) = \mathbf{G}^-(n)\mathbf{C}^{\mathrm{T}}(n)[\mathbf{C}(n)\mathbf{G}^-(n)\mathbf{C}^{\mathrm{T}}(n) + \mathbf{R}_{\mathrm{d}}]^{-1} \tag{4.62}$$

$$\mathbf{G}^-(n+1) = \boldsymbol{\phi}(n)\mathbf{G}^+(n)\boldsymbol{\phi}^{\mathrm{T}}(n) + \mathbf{Q}_{\mathrm{d}} \tag{4.63}$$

where

$$\mathscr{E}(\mathbf{w}(n)\mathbf{w}^{\mathrm{T}}(n+k) = \mathbf{Q}_{\mathrm{d}}\delta(k), \mathscr{E}(\mathbf{v}(n)\mathbf{v}^{\mathrm{T}}(n+k)) = \mathbf{R}_{\mathrm{d}}\delta(k), \boldsymbol{\phi}(n) = \mathbf{I} + \left.\frac{\partial \mathbf{f}}{\partial \mathbf{x}}\right|_{\hat{\mathbf{x}}^+(n), u(n)} \cdot \mathrm{T}$$

The filter algorithm thus consists of Eqs. (4.62), (4.58), (4.59), (4.60) (or (4.61)), (4.63) in that order, repeated for $n = 0, 1, 2, \ldots, \infty$. As in the linear case, Eq. (4.58) is evaluated as quickly as possible after the sampling instant, while the remaining equations need only be evaluated in time for the next sampling instant.

The technique is perhaps best explained by example.

EXAMPLE

The behavior of a surface-to-air missile and its target may be modeled in a simplified form in two dimensions as indicated in Fig. 4.18.

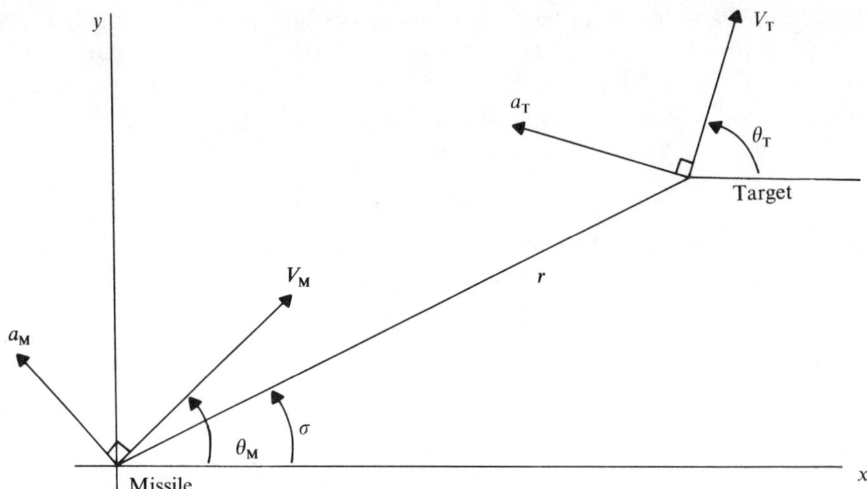

Fig. 4.18 Missile–target system (example).

In Figure 4.18,

a_M is the missile lateral acceleration (m s^{-2}),
a_T is the target lateral acceleration (m s^{-2}),
r is the missile to target distance (m),
θ_M is the missile velocity direction (rad),
θ_T is the target velocity direction (rad),
σ is the 'line of sight' angle (rad),
V_M is the missile forward velocity (m s^{-1}),
V_T is the target forward velocity (m s^{-1}),
T is the sample time (s): $T = 1$ second.

Assuming constant missile and target forward velocities, the following relationships hold:

$$\dot{r} = V_T \cos(\theta_T - \sigma) - V_M \cos(\theta_M - \sigma)$$

$$r\dot{\sigma} = V_T \sin(\theta_T - \sigma) - V_M \sin(\theta_M - \sigma).$$

With the simplifying assumptions that $\theta_M \approx \theta_T \approx \sigma$, $\ddot{r} = 0$, modeling the rate of change of target lateral acceleration as noise, $\dot{a}_T = w_T$, and taking $w_{\dot{\sigma}}$, $w_{\dot{r}}$, w_r, w_T, as measures of ignorance of $\dot{\sigma}$, $\dot{r}$, r, a_T (Comment 4), the following nonlinear state equations may be developed:

$$\dot{\mathbf{x}} = \begin{bmatrix} \dfrac{-2x_1x_2}{x_3} - \dfrac{u}{x_3} + \dfrac{x_4}{x_3} \\ 0 \\ x_2 \\ 0 \end{bmatrix} + \mathbf{w}_c$$

$$y = [1 \quad 0 \quad 0 \quad 0]\mathbf{x} + v_c$$

where

$$u = a_M,$$
$$\mathbf{x} = (\dot{\sigma} \quad \dot{r} \quad r \quad a_T)^T$$

and

$$\mathbf{w}_c = (\omega_{\dot{\sigma}} \quad w_{\dot{r}} \quad w_r \quad w_T)^T.$$

Clearly, for $u = 0$:

$$\phi(n) = \begin{bmatrix} \dfrac{-2T\hat{x}_2^+(n)}{\hat{x}_3^+(n)} + 1 & \dfrac{-2T\hat{x}_1^+(n)}{\hat{x}_3^+(n)} & \dfrac{2T\hat{x}_1^+(n)\hat{x}_2^+(n) - T\hat{x}_4^+(n)}{(\hat{x}_3^+(n))^2} & \dfrac{T}{\hat{x}_3^+(n)} \\ 0 & 1 & 0 & 0 \\ 0 & T & 1 & 0 \\ 0 & 0 & 0 & 1 \end{bmatrix}$$

$$\mathbf{C}(n) = [1 \quad 0 \quad 0 \quad 0]$$

$$\mathbf{Q}_d(n) = \int_0^T e^{\mathbf{A}\lambda}\mathbf{Q}_c(e^{\mathbf{A}\lambda})^T \, d\lambda \qquad \text{(Eq. (4.40))}$$

$$\approx T\mathbf{Q}_c(t)$$

$$\mathbf{Q}_c(t) = \mathscr{E}(\mathbf{w}_c(t)\mathbf{w}_c^T(t)).$$

Set:

$$\mathbf{Q}_d = \begin{bmatrix} \dfrac{360}{r^2} & 0 & 0 & 0 \\ 0 & 20 & 0 & 0 \\ 0 & 0 & 20 & 0 \\ 0 & 0 & 0 & 4800 \end{bmatrix}$$

$$\mathbf{R}_d(n) = \mathbf{R}_c(nT), \qquad \mathbf{R}_c(t) = \mathscr{E}(\mathbf{v}_c(t)\mathbf{v}_c^T(t)).$$

Set:

$$\mathbf{R}_d = \frac{360}{r^2}.$$

Substituting these in Eqs. (4.62) (4.58), (4.59), (4.60), (4.63) and evaluating these in order gives the estimates shown in Fig. 4.19 for a pilot maneuver consisting of a sudden step in lateral target acceleration of $50\,\mathrm{m\,s^{-2}}$.

Comment
It can be seen that even for this very nonlinear system the Kalman filter generates quite accurate estimates of the rate of change of line of sight, $\dot{\sigma}$, and the target lateral acceleration, a_T. The improvements in the estimates at measurement instants are noteworthy.

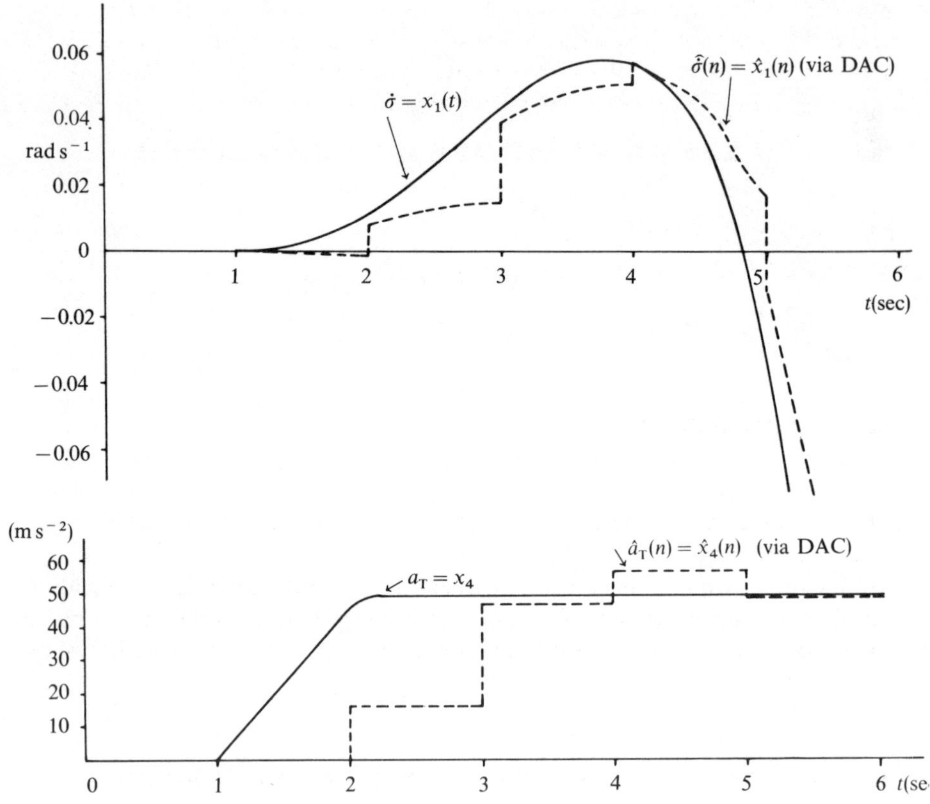

Fig. 4.19 Extended Kalman filter behavior (example). (a) Estimation of $\dot{\sigma}$. (b) Estimation of a_T.

The natural use of such estimates is in a control strategy for the missile which is likely to be an improvement on the proportional navigation technique traditionally used in this context.

REFERENCES

1. Chen, C.T., *Linear System Theory and Design.* Holt, Rinehart and Winston, 1984.
2. Franklin, G.F., and Powell, J.D., *Digital Control of Dynamic Systems.* Addison–Wesley, 1980.
3. Katz, P., *Digital Control using Microprocessors.* Prentice-Hall, 1981.
4. Kuo, B.C., *Digital Control Systems.* Holt Saunders, 1980.
5. Kwakernaak, H., and Sivan, R. *Linear Optimal Control Systems.* Wiley Interscience, 1972.
6. Leondes, C.T., *Theory and Applications of Kalman Filtering.* Agard, 1970 (Agardograph no. 139). *Advances in the Techniques and Technology of the Applications of Nonlinear Filters and Kalman Filters.* Agard, 1982 (Agardograph no. 256).
7. Maybeck, P.S., *Stochastic Models, Estimation and Control*, Vols. I, II, III. Academic Press, 1982.

5 USEFUL COMPUTER TECHNIQUES

5.1 INTRODUCTION

Though usually implemented off-line, the useful techniques of function optimization ('hill-climbing') and system identification hold the promise of significant advances in control engineering (notably adaptive control). A number of basic techniques of the kind are outlined in this chapter. Sections 5.2 to 5.6 deal with function optimization and Secs. 5.7 to 5.10 with system identification.

5.2 GENERAL STATEMENT OF THE OPTIMIZATION PROBLEM

Given an object function $F(\mathbf{x})$ of m independent variables

$$\mathbf{x} = (x_1 \quad x_2 \quad \dots \quad x_m)^T,$$

the problem is to find by a computer search a 'global' extremum $\mathbf{x}^*$ such that:

$$F(\mathbf{x}^*) \leqslant F(\mathbf{x}), \forall \mathbf{x} \qquad \text{for a global minimum, or}$$
$$F(\mathbf{x}^*) \geqslant F(\mathbf{x}), \forall \mathbf{x} \qquad \text{for a global maximum.}$$

If the equality signs in these relationships hold only when $\mathbf{x} = \mathbf{x}^*$, the extremum is 'strong'; otherwise it is 'weak'.

An object function can of course have several 'local' extrema, at which:

$$F(\mathbf{x}^*) \leqslant F(\mathbf{x}), \mathbf{x} \text{ 'near' } \mathbf{x}^* \qquad \text{for a minimum,}$$
$$F(\mathbf{x}^*) \geqslant F(x), \mathbf{x} \text{ 'near' } \mathbf{x}^* \qquad \text{for a maximum.}$$

Developed optimization procedures may incorporate some feature which attempts to ensure that an extremum is global, such as starting a number of searches at various points to check that the same extremum is reached. Such features are mentioned in the following paragraphs.

5.3 THE SIMPLEX METHOD (REFS. 1, 4)

In this method, a regular or 'equilateral' simplex is set up in the m-dimensional space defined by $x_1, x_2, x_3, \ldots, x_m$. In m dimensions this is $(m+1)$ mutually 'equidistant' points. For example, if $m=2$, the simplex is an equilateral triangle ($\bar{A}\bar{B}\bar{C}$, Fig. 5.1); if $m=3$, it is an equilateral tetrahedron, and so on.

For clarity, the procedure is described for $m=2$ and reference made to Fig. 5.1, but the formulas quoted apply to any value of m.

A two-dimensional space is illustrated in Fig. 5.1, with 'contours'

$$F(\mathbf{x}) = F(x_1, x_2) = C.$$

Stage 1

An initial point is selected:

$$\mathbf{x}_0^1 = (x_{01}^1, x_{02}^1, x_{03}^1, \ldots, x_{0m}^1)^{\mathrm{T}}$$

A simplex of edge size s is now set up with vertices at:

$$\mathbf{x}_i^1 = (x_{01}^1 + \delta_1, x_{02}^1 + \delta_1, \ldots, x_{0(i-1)}^1 + \delta_1, x_{0i}^1 + \delta_2, x_{0(i+1)}^1 + \delta_1, \ldots, x_{0m}^1 + \delta_1)^{\mathrm{T}}, \quad i = 1, \ldots, m \tag{5.1}$$

where:

$$\delta_1 = \frac{s[\sqrt{(m+1)} + m - 1]}{m\sqrt{2}} \tag{5.2}$$

$$\delta_2 = \frac{s[\sqrt{(m+1)} - 1]}{m\sqrt{2}}. \tag{5.3}$$

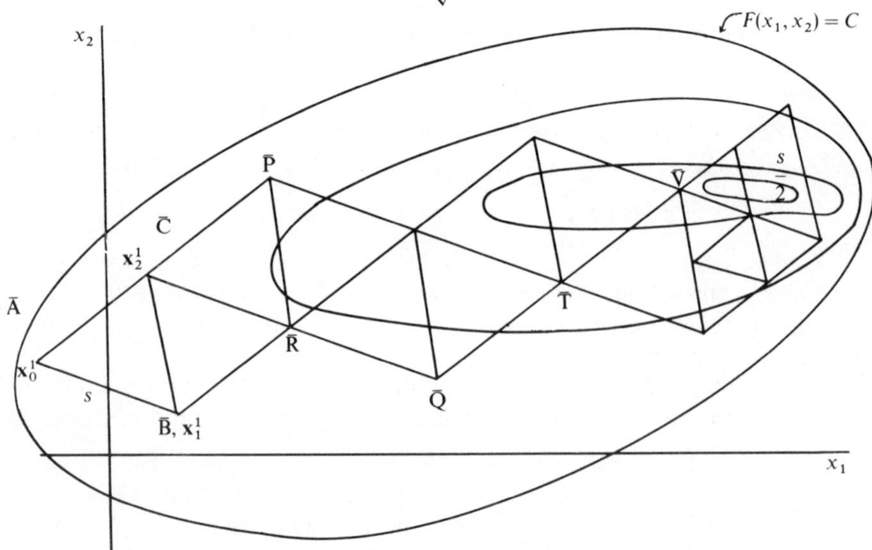

Fig. 5.1 Simplex search ($m = 2$).

Stage 2

$F(\mathbf{x})$ is evaluated at all these vertices and the 'worst' vertex (where $F(\mathbf{x})$ is smallest or largest for maximum or minimum search respectively), say $\mathbf{x}_j$, is found. In Fig. 5.1, $j=0$ for the simplex $\bar{A}\bar{B}\bar{C}$.

$\mathbf{x}_j$ is 'reflected' in the simplex opposite 'side' to give a new simplex ($\bar{B}\bar{R}\bar{C}$) with one vertex:

$$\mathbf{x}_j^2 = \frac{2}{m}\{\mathbf{x}_0^1 + \cdots + \mathbf{x}_{j-1}^1 + \mathbf{x}_{j+1}^1 + \cdots + \mathbf{x}_m^1\} - \mathbf{x}_j^1 . \qquad (5.4)$$

Stage 3

Stage 2 is repeated using the new simplex with appropriately relabeled vertices, and this sequence is repeated until progress comes to a halt.

This occurs when the worst vertex in one simplex is reflected to give the worst vertex in the next simplex (vertices $\bar{P}$, $\bar{Q}$, Fig. 5.1). Reflecting this new worst vertex $\bar{Q}$ would merely regenerate the previous vertex $\bar{P}$.

Stage 4

When this condition is detected, the second worst vertex ($\bar{R}$) is reflected, using the same formula (Eq. (5.4)) to give a new vertex ($\bar{T}$, Fig. 5.1).

The procedure then normally reverts to Stage 3, but where the optimum point is close at hand, the repetition of Stages 3 and 4 results in one vertex ($\bar{V}$) being present in several consecutive simplexes. If any vertex is present in more than M simplices, where

$$M = 1.65\,m + (0.05)m^2, \qquad (5.5)$$

then the search can proceed no further with this simplex size but merely 'rotates' round that vertex.

Stage 5

The simplex side size s is reduced, typically by 50%, and the whole procedure is restarted from Stage 1, using the repeated vertex ($\bar{V}$) as a starting point, $\mathbf{x}_0^{\mathrm{v}}$, say. The other vertices are found from:

$$\mathbf{x}_i = 0.5(\mathbf{x}_0^{\mathrm{v}} + \mathbf{x}_i^{\mathrm{v}}), \quad i = 1, 2, \ldots, m . \qquad (5.6)$$

The search terminates when the side size falls below some prespecified value.

Comment

The simplex method is simple and economical of computing power but it can be slow. Several variations (refs. 1, 4) have been suggested to improve the speed of convergence at the cost of increased complexity.

CAD Facility

A CAD facility is useful for demonstrating the effectiveness (or otherwise) of the simplex routine for a particular function. The function $F(\mathbf{x})$ must be

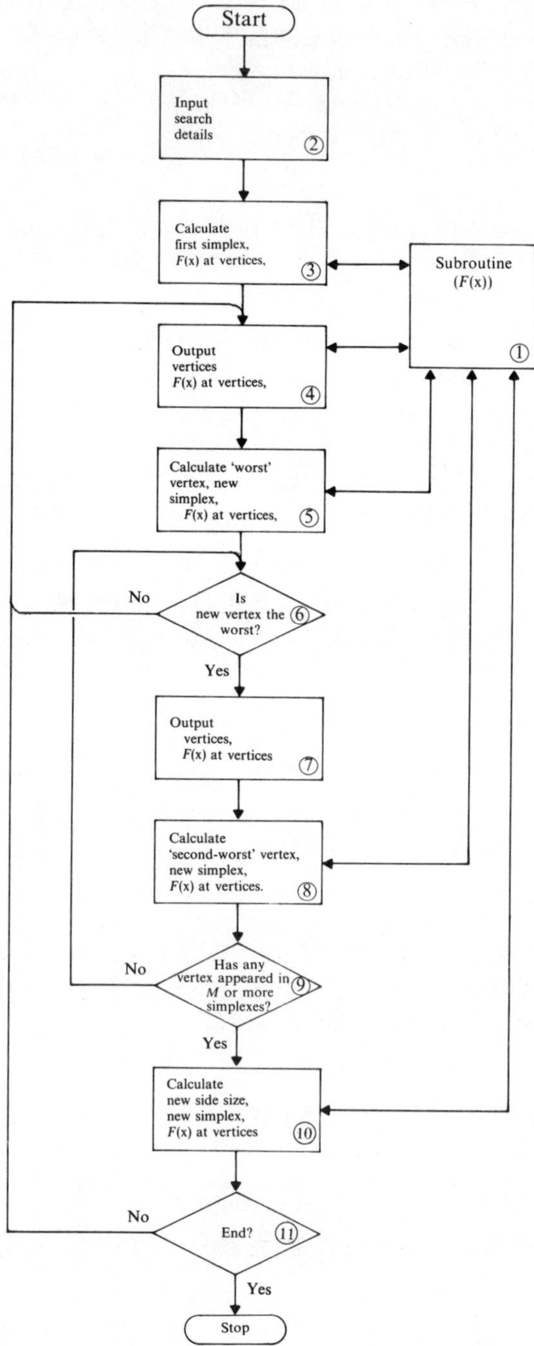

Fig. 5.2 Simple CAD algorithm.

specified by the user, typically in the form of a BASIC or other high-level language subroutine. This is then compiled and linked to the CAD program before the optimization can proceed.

An interactive CAD facility is shown in the flow diagram of Fig. 5.2. The operation of the algorithm is as follows.

BLOCK 1

The object function $F(\mathbf{x})$ is defined in a standardized subroutine of BASIC or other high-level language statements. This is compiled and linked to the main program, which is already in object code.

BLOCK 2

The details of the search are requested and input. These comprise the number of independent variables, m; the starting point $\mathbf{x}_0^1$; the simplex side size s; the resolution required in the search, normally the smallest s which is to be allowed before it is assumed the entremum is reached; whether a maximum or minimum is sought. Opportunity to correct these parameters is provided.

BLOCKS 3, 4

The first simplex vertices are calculated, using Eqs. (5.1), (5.2), (5.3), and $F(\mathbf{x})$ is evaluated at these points, using Block 1. These results are displayed (Block 4).

BLOCKS 5, 6, 7, 8

A sequence of simplexes is generated and $F(\mathbf{x})$ evaluated at their vertices according to the rules outlined above (Stages 2, 3, 4). The vertices and object function values are displayed.

BLOCKS 9, 10, 11

When the search nears an extremum, one vertex appears in more than M simplexes (Eq. (5.5)). In these circumstances a new reduced simplex is calculated (Eq. (5.6)), usually with half the previous side size (Block 10). If the new value s is less than the resolution specified (Block 2), the search is ended. Otherwise the procedure is restarted at Block 4.

EXAMPLES

Some simple examples are examined in Sec. 5.6.

5.4 ALTERNATING VARIABLE METHODS (REFS. 1, 4, 5)

The most obvious method of function optimization is simply to 'climb' in all the independent variables alternately.

A starting point $\mathbf{x}^1 = (x_1 \quad x_2 \quad \cdots \quad x_m)^T$ is selected in the m-dimensional

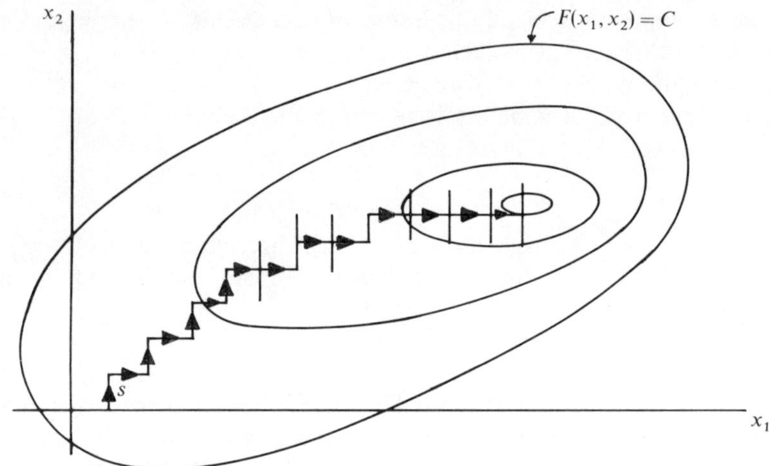

Fig. 5.3 Simple alternating variable search.

space, and a step size s specified. The procedure is illustrated for $m = 2$, in Fig. 5.3.

The independent variables are simply selected in turn and a step of size s, or alternatively no step, is taken in each direction until no further improvement can be achieved with that step length.

In this simple form the method can be exceedingly inefficient, particularly where long narrow 'valleys' or 'ridges' exist, and a number of modifications

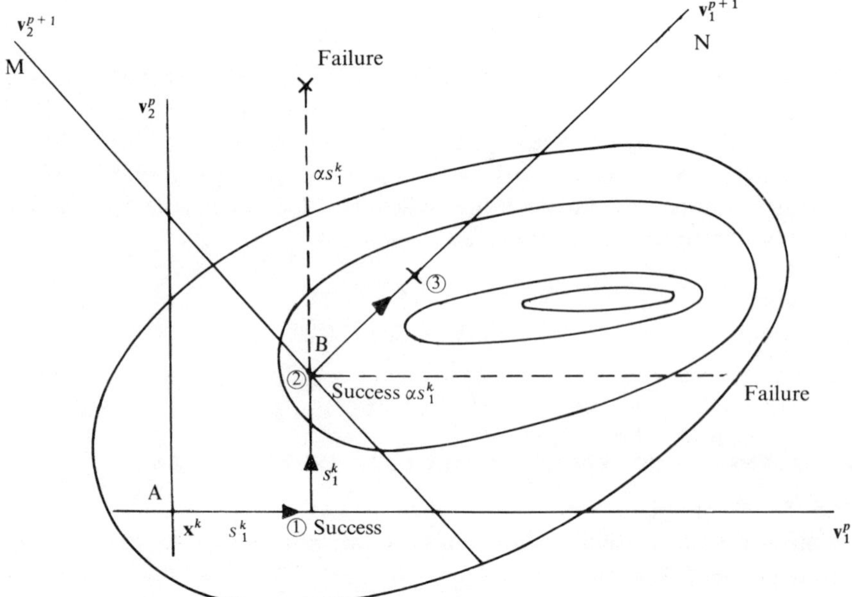

Fig. 5.4 Rosenbrock's alternating variable method ($m = 2$).

have been suggested (refs. 1, 4, 5). One of these, Rosenbrock's method, is described here.

Although the procedure is illustrated for $m = 2$ (Fig. 5.4), the description applies to any number of dimensions.

Stage 1

For the first stage, the directions of search are chosen to be:

$$\mathbf{v}_1^p = (1 \quad 0 \quad \cdots \quad 0)^{\mathrm{T}}$$
$$\mathbf{v}_2^p = (0 \quad 1 \quad \cdots \quad 0)^{\mathrm{T}}$$
$$\mathbf{v}_m^p = (0 \quad 0 \quad \cdots \quad 1)^{\mathrm{T}} \qquad p = 0.$$

An initial point $\mathbf{x}^k$, $k = 0$, typically the origin (A, Fig. 5.4) is selected, $F(\mathbf{x}^k)$ evaluated, and a step size s_i^k chosen for the first attempted step in all directions, $i = 1, 2, \ldots, m$.

Stage 2

A step is taken in the direction $\mathbf{v}_i^p$ to give a new point, $\mathbf{x}^k + s_i^k\mathbf{v}_i^p$.

$F(\mathbf{x}^k + s_i^k\mathbf{v}_i^p)$ is evaluated.

If this shows an improvement on $F(\mathbf{x}^k)$, the attempted step is a success, and:

$$\mathbf{x}^{k+1} = \mathbf{x}^k + s_i^k\mathbf{v}_i^p. \tag{5.7}$$

A new step length, αs_i^k (typically, $\alpha = 3$), is calculated and stored for use as the next step in that direction:

$$s_i^{k+m} = \alpha s_i^k. \tag{5.8}$$

If $F(\mathbf{x}^k + s_i^k\mathbf{v}_i^p)$ shows a deterioration on $F(\mathbf{x}^k)$, the attempted step is a failure, and:

$$\mathbf{x}^{k+1} = \mathbf{x}^k.$$

A new step length, $-\beta s_i^k$ (typically, $\beta = 0.5$), is calculated and stored for use as the next step in that direction:

$$s_i^{k+m} = -\beta s_i^k.$$

Stage 3

Stage 2 is repeated in all directions, $i = 1, 2, \ldots, m$, leaving stored at each stage a new point $\mathbf{x}^{k+1}$ and a new step length s_i^{k+m}, $i = 1, 2, \ldots, m$, for exploration in each direction in the following cycle.

Stage 4

Using the points and step sizes calculated in Stage 3, the sequence Stage 2, Stage 3 is repeated until a success has been followed (not necessarily immediately) by a failure in every direction.

Suppose this occurs when $k = N$.

Stage 5

The search has now proceeded as far as possible with the existing set of axes (B, Fig. 5.4). The axes are now reoriented to align one search direction with the hitherto 'most successful direction of search' (AB, Fig. 5.4):

$$\mathbf{b} = \sum_{k=0}^{N} \sum_{i=1}^{m} s_i^k \mathbf{v}_i^p. \tag{5.9}$$

A set of orthogonal normalized axes based on this are found by the 'Gram–Schmidt' process (ref. 1), as follows:

$$\left.\begin{aligned}
&\mathbf{w}_1 = \mathbf{b}; && \mathbf{v}_1^{p+1} = \frac{\mathbf{w}_1}{|\mathbf{w}_1|}\\
&\mathbf{w}_2 = \mathbf{v}_2^p - [(\mathbf{v}_1^{p+1})^{\mathrm{T}}\mathbf{v}_2^p]\mathbf{v}_1^{p+1}; && \mathbf{v}_2^{p+1} = \frac{\mathbf{w}_2}{|\mathbf{w}_2|}\\
&\mathbf{w}_3 = \mathbf{v}_3^p - [(\mathbf{v}_1^{p+1})^{\mathrm{T}}\mathbf{v}_3^p]\mathbf{v}_1^{p+1} - [(\mathbf{v}_2^{p+1})^{\mathrm{T}}\mathbf{v}_3^p]\mathbf{v}_2^{p+1}; && \mathbf{v}_3^{p+1} = \frac{\mathbf{w}_3}{|\mathbf{w}_3|}\\
&\mathbf{w}_4 = \mathbf{v}_4^p - \cdots - [(\mathbf{v}_3^{p+1})^{\mathrm{T}}\mathbf{v}_4^p]\mathbf{v}_3^{p+1}; && \mathbf{v}_4^{p+1} = \frac{\mathbf{w}_4}{|\mathbf{w}_4|}\\
&\vdots\\
&\mathbf{w}_m = \mathbf{v}_m^p - \cdots - [(\mathbf{v}_{m-1}^{p+1})^{\mathrm{T}}\mathbf{v}_m^p]\mathbf{v}_{m-1}^{p+1}; && \mathbf{v}_m^{p+1} = \frac{\mathbf{w}_m}{|\mathbf{w}_m|}
\end{aligned}\right\}. \tag{5.10}$$

The new axes are illustrated in Fig. 5.4 (BN, BM). After resetting k appropriately, the whole sequence, Stages 2, 3, 4, is repeated until the step sizes fall below some predetermined level, at which point an extremum is assumed to be reached.

CAD Facility

A CAD facility is useful for demonstrating the effectiveness, or otherwise, of this routine for a particular function. The function $F(\mathbf{x})$ must be specified by the user in the form of a BASIC or other high-level language subroutine, which is then compiled and linked to the CAD facility before the optimization procedure is started. An interactive algorithm is shown in Fig. 5.5.

The operation of the algorithm is as follows.

BLOCK 1

The object function $F(\mathbf{x})$ is defined in a standardized subroutine of BASIC or other high-level language statements. This is compiled and linked to the main program which is already in object code.

BLOCK 2

The details of the search are requested and input. These comprise the number of independent variables, m; the starting point $\mathbf{x}^0$; the first set of attempted step sizes s_i^0; the resolution required in the search; whether a maximum or minimum is sought. Opportunity to correct these parameters is provided.

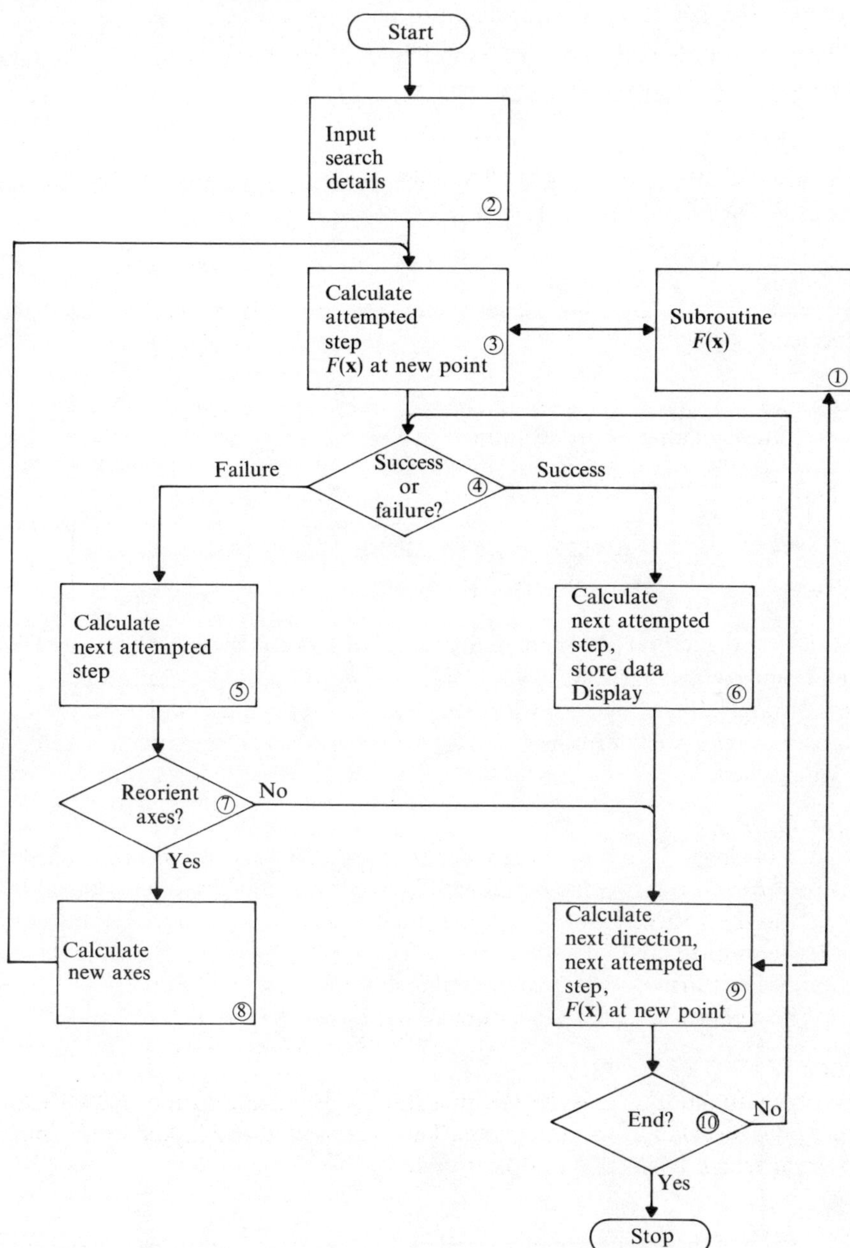

Fig. 5.5 Rosenbrock's alternating variable algorithm.

BLOCKS 3, 4, 5, 6, 9

An attempted step is taken, a new point found, and a new step length calculated according to the criteria outlined in Stage 2 above, and the process is repeated for all axes as described in Stages 2, 3, 4 above. Appropriate data are output or displayed at each stage.

BLOCKS 7, 8

If a success is followed by a failure in all directions, the axes are reoriented using the formulas described in Stage 5 above (Eqs. (5.9), (5.10)).

BLOCK 10

The end of the search is reached when the required resolution has been attained.

EXAMPLES

Some simple examples are examined in Sec. 5.6.

5.5 PATH OF STEEPEST ASCENT OR DESCENT (REFS. 1, 4)

In this type of method (there are many variations) both the object function $F(\mathbf{x})$ and its derivatives w.r.t. $\mathbf{x}$,

$$\bar{\nabla} F(\mathbf{x}) = \left(\frac{\partial F}{\partial x_1} \quad \frac{\partial F}{\partial x_2} \quad \cdots \quad \frac{\partial F}{\partial x_m} \right)^{\mathrm{T}} \tag{5.11}$$

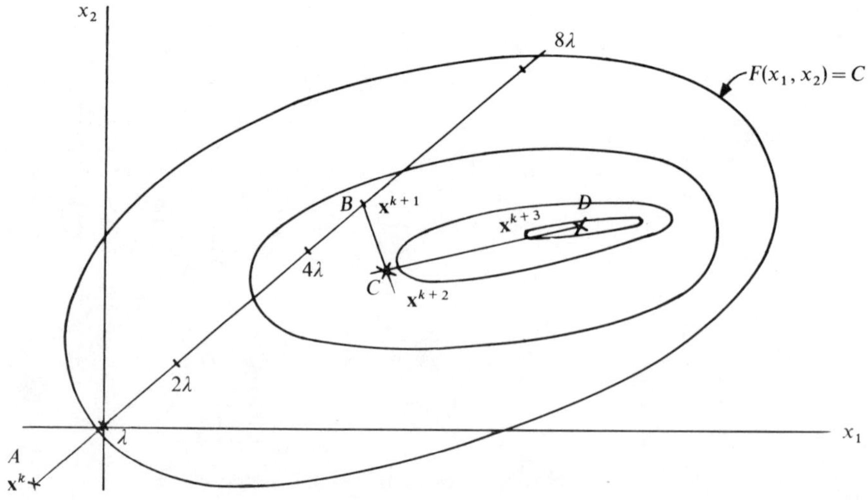

Fig. 5.6 Path of steepest ascent ($m = 2$).

must be available for evaluation. For optimization, steps are taken in the direction $\bar{\nabla}F(\mathbf{x})$. Short steps may be taken in this direction, and $F(\mathbf{x})$, $\bar{\nabla}F(\mathbf{x})$ reevaluated before the next short step is taken, but this leads to a heavy computational load. The method described here involves proceeding along the path of steepest ascent in one direction until the optimum in that direction is reached, then re-evaluating $F(\mathbf{x})$, $\bar{\nabla}F(\mathbf{x})$ and repeating the procedure. The search ends when $\bar{\nabla}F(\mathbf{x}) = \mathbf{0}$, which is the case at saddle points as well as extrema. Consequently the method fails at saddle points which fall across the search path, and several starting points (and search paths) are advised to ensure that a misleading result is avoided.

The procedure is described for $m = 2$ and illustrated in Fig. 5.6, but the formulas quoted apply to any value of m.

In Fig. 5.6 a two-dimensional space is illustrated with contours

$$F(\mathbf{x}) = F(x_1, x_2) = C.$$

Stage 1

An 'initial' small step length, λ, and a starting point (A, Fig. 5.6)

$$\mathbf{x}^k = (x_1^k \quad x_2^k \quad x_3^k \quad \cdots \quad x_m^k)^{\mathrm{T}}, \quad k = 0$$

are selected.

$F(\mathbf{x}^k)$ and $\bar{\nabla}F(\mathbf{x}^k)$ are calculated.

Stage 2

The strategy involves taking a step of length s^k in direction $\bar{\nabla}F(\mathbf{x}^k)$ so that:

$$\mathbf{x}^{k+1} = \mathbf{x}^k \pm s^k \bar{\nabla}F(\mathbf{x}^k) \qquad (5.12)$$

depending on the appropriate direction of search. A single dimensional search is undertaken to find the value of s^k for which $F(\mathbf{x}^{k+1})$ is the best.

This search can be done by a *bracketing procedure* followed by a *Fibonacci search* (refs. 1, 5) as follows:

(a) THE BRACKETING PROCEDURE

Using a small step length λ selected by the operator, $F(\mathbf{x}_i^k)$ is calculated for various step lengths, $(2^i\lambda)$:

$$\mathbf{x}_1^k = \mathbf{x}^k + \bar{\nabla}F(\mathbf{x}^k)2^0\lambda$$

$$\mathbf{x}_2^k = \mathbf{x}^k + \bar{\nabla}F(\mathbf{x}^k)2^1\lambda$$

$$\vdots$$

$$\mathbf{x}_i^k = \mathbf{x}^k + \bar{\nabla}F(\mathbf{x}^k)2^i\lambda$$

$$\vdots$$

This sequence is pursued until an optimum point (in this direction) has been bracketed:

$F(\mathbf{x}_i^k) \lesseqgtr F(\mathbf{x}_{i-1}^k) \gtreqless F(\mathbf{x}_{i-2}^k)$, for a maximum or minimum respectively.

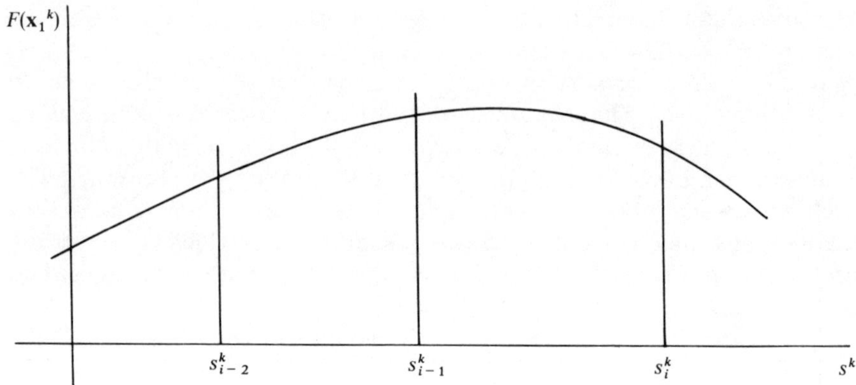

Fig. 5.7 Bracketing in s^k

The bracket containing this optimum is:

$$\{s_{i-2}^k = 2^{i-2}\lambda,\ s_i^k = 2^i\lambda\}$$

This is illustrated in Fig. 5.7.

(b) THE FIBONACCI SEARCH

Having established a bracket $\{s_{i-2}^k, s_i^k\}$ which encloses the optimum point, a more refined search is now undertaken.

For convenience, rename the bracket:

$$s_{i-2}^k = s_1$$
$$s_i^k = s_2, \quad s^k = s.$$

Two new points within the bracket are found:

$$s_3 = (1-\alpha_1)s_1 + \alpha_1 s_2 \tag{5.13}$$

$$s_4 = \alpha_1 s_1 + (1-\alpha_1)s_2 \qquad (\alpha_1 \text{ defined below}). \tag{5.14}$$

This is illustrated in Fig. 5.8.

$F(\mathbf{x}^k + s_3\nabla F(\mathbf{x}^k))$, $F(\mathbf{x}^k + s_4\nabla F(\mathbf{x}^k))$ are evaluated and a new bracket selected from s_1, s_2, s_3, s_4 such that the new bracket contains the optimum point. In the case illustrated,

$$F(\mathbf{x}^k + s_3\nabla F(x^k)) < F(\mathbf{x}^k + s_4\nabla F(\mathbf{x}^k)) > F(\mathbf{x}^k + s_2\nabla F(\mathbf{x}^k)),$$

and the new bracket is $\{s_3, s_2\}$ if a maximum is being sought.

The bracketing procedure is now repeated to generate two new points, s_5, s_6, according to the formulas:

$$s_5 = (1-\alpha_2)s_3 + \alpha_2 s_2 = s_4$$

$$s_6 = \alpha_2 s_3 + (1-\alpha_2)s_2 \qquad (\alpha_2 \text{ defined below}).$$

This procedure is repeated using a sequence of multipliers $\alpha_1, \alpha_2, \cdots, \alpha_j, \cdots$

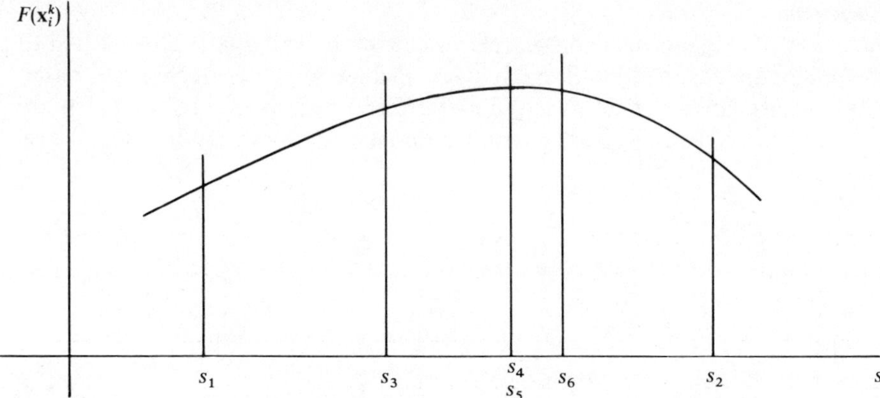

Fig. 5.8 Fibonacci search in s.

selected so that:

$$\alpha_{j+1} = \frac{1 - 2\alpha_j}{1 - \alpha_j}. \tag{5.15}$$

This gives the search the economical property that points in consecutive brackets are shared – in Fig. 5.8, $s_4 = s_5$, etc.

One sequence of multipliers with this property is the Fibonacci series, derived from the sequence of real numbers:

$$R_N = R_{N-1} + R_{N-2}, \quad R_0 = R_1 = 1.$$

(i.e. 1, 1, 2, 3, 5, 8, 13, 21, ...)

$$\alpha_j = \frac{F_{N-j-1}}{F_{N-j+1}} \tag{5.16}$$

(e.g. $\frac{8}{21}, \frac{5}{13}, \frac{3}{8}, \frac{2}{5}, \frac{1}{3}, \frac{1}{2}$).

Alternatively, setting:
$\alpha_j = \alpha_{j+1}$ (Eq. (5.15)), the multipliers have the constant, 'golden section' value:

$$\alpha_1 = \alpha_2 = \cdots = \alpha_j \approx 0.382.$$

The search procedure is repeated until the bracket size falls below some predetermined limit, and the step size s^k has therefore been calculated.

$$\mathbf{x}^{k+1} = \mathbf{x}^k + \mathbf{s}^k \bar{\nabla} F(\mathbf{x}^k)$$

is now evaluated.

Stage 3

Stages 1, 2 are repeated, $k = 0, 1, 2, \ldots$, until no further improvement can be achieved, or until $\bar{\nabla} F(\mathbf{x}^k) = \mathbf{0}$ (steps AB, BC, CD, Fig. 5.6).

Comment

A variety of single-dimension search methods are available as alternatives to the procedure described in Stage 2. These include attractive techniques based on quadratic approximation and undetermined multipliers (ref. 1) which are generally faster, though perhaps more complicated, than that outlined above.

CAD Facility

As with the previous methods, a CAD facility can usefully demonstrate the effectiveness, or otherwise, of this method for a particular object function. The

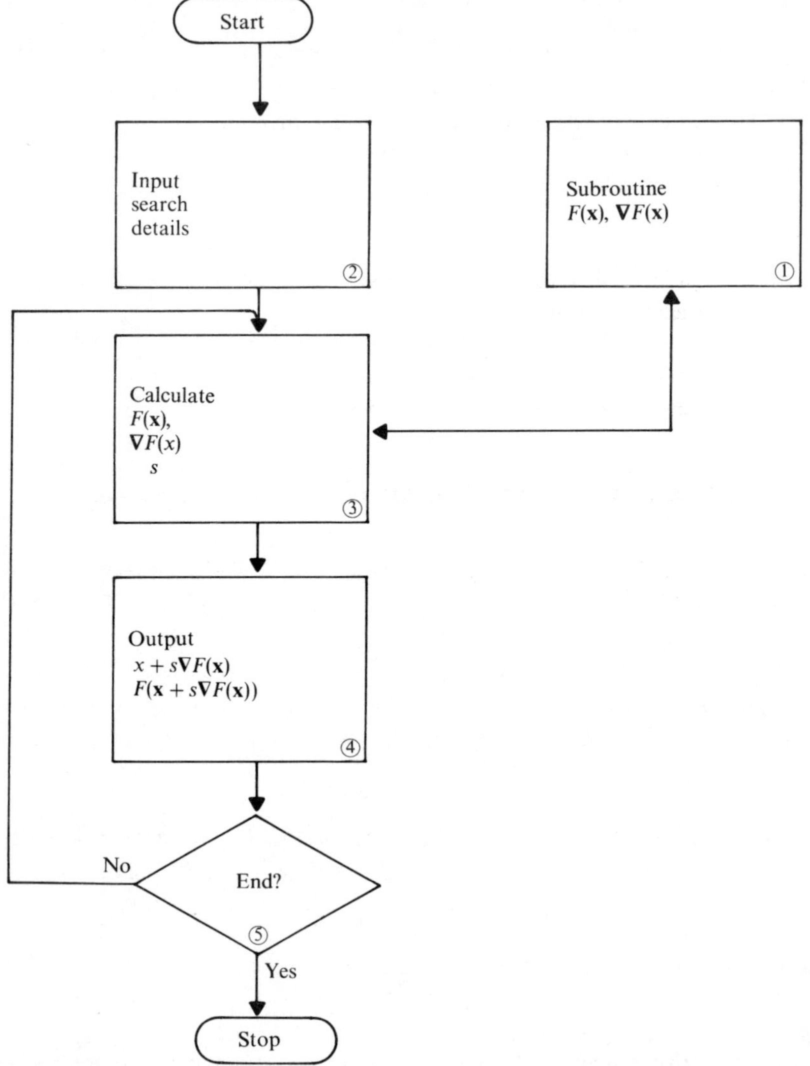

Fig. 5.9 Path of steepest ascent/descent algorithm.

function $F(\mathbf{x})$ and its gradient $\bar{\nabla} F(\mathbf{x})$ must be specified by the user in the form of a subroutine in BASIC or other high-level language, which is compiled and linked to the CAD facility before the optimization is started.

An interactive algorithm is shown in Fig. 5.9.

The operation of the algorithm is as follows.

BLOCK 1
The object function $F(\mathbf{x})$ and its gradient $\bar{\nabla} F(\mathbf{x})$ are defined in a standardized subroutine of BASIC or other high-level language statements. This is compiled and linked to the main program which is already in object code.

BLOCK 2
The details of the search are requested and input with the opportunity to correct the data if necessary. The details comprise the number of independent variables, m; a starting point $\mathbf{x}^0$; a small initial step size λ; the resolution required in the search; whether a maximum or minimum is sought.

BLOCKS 3, 4, 5
The algorithm described above in Stages 1, 2 is effected, the resulting data being output at each stage.

EXAMPLES
Some simple examples are examined in Sec. 5.6.

5.6 SOME USEFUL TWO-DIMENSIONAL OBJECT FUNCTIONS

To assist with the design of CAD packages, three object functions of two independent variables with well-defined global optima are described below.

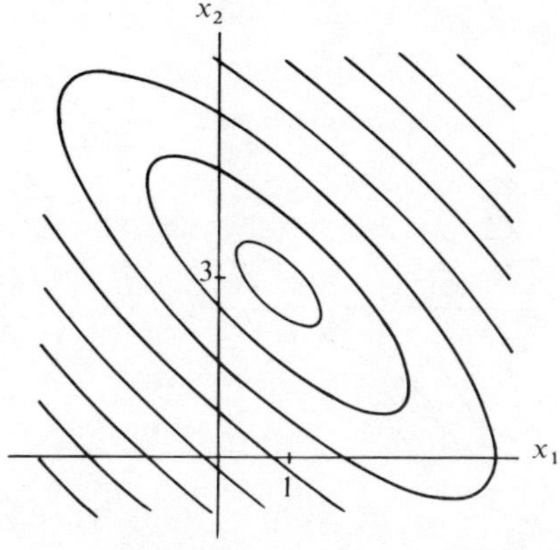

Fig. 5.10 Booth's function (Reproduced with permission from Schwefel, H.P., *Numerical Optimization of Computer Models*. Wiley, 1981).

(i) Booth's Function (ref. 5)

$$F(\mathbf{x}) = (x_1 + 2x_2 - 7)^2 + (2x_1 + x_2 - 5)^2.$$

This function (Fig. 5.10) has a minimum at:

$$\mathbf{x}^* = [1 \quad 3]^{\mathrm{T}}, \quad F(\mathbf{x}^*) = \mathbf{0}.$$

All three methods (Secs. 5.3, 5.4, 5.5) work well with this function, the simplex method and path of steepest ascent being perhaps faster than the alternating variable method.

(ii) Zettl's Function (ref. 5)

$$F(\mathbf{x}) = (x_1^2 + x_2^2 - 2x_1)^2 + 0.25x_1.$$

This function (Fig. 5.11) has a minimum at:

$$\mathbf{x}^* = [-0.02990 \quad 0]^{\mathrm{T}}, \quad F(\mathbf{x}^*) = -0.003791].$$

There is also a local minimum at:

$$\mathbf{x}^* = [1.0630 \quad 0]^{\mathrm{T}}, \quad F(\mathbf{x}^*) = 1.2580$$

and a saddle point at:

$$\mathbf{x} = [1.9670 \quad 0]^{\mathrm{T}}, \quad F(\mathbf{x}) = 0.4962.$$

All three methods work quite well from most starting points for this function, though the path of steepest descent (two paths illustrated) can terminate falsely at the saddle point.

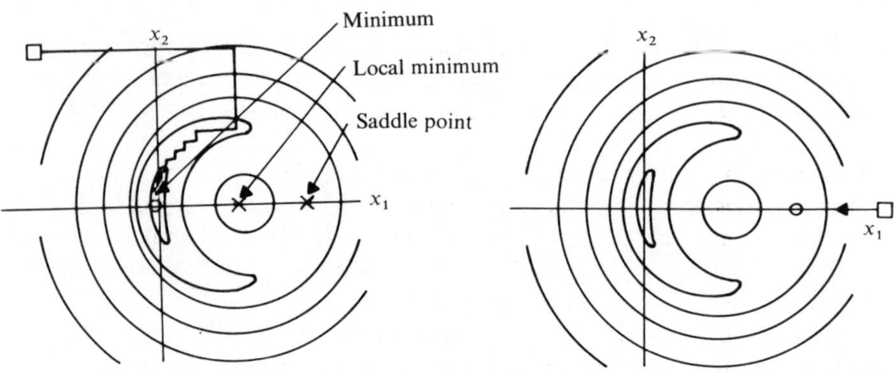

Fig. 5.11 Zettl's function. (Reproduced with permission from Schwefel, H.P., *Numerical Optimization of Computer Models.* Wiley, 1981.)

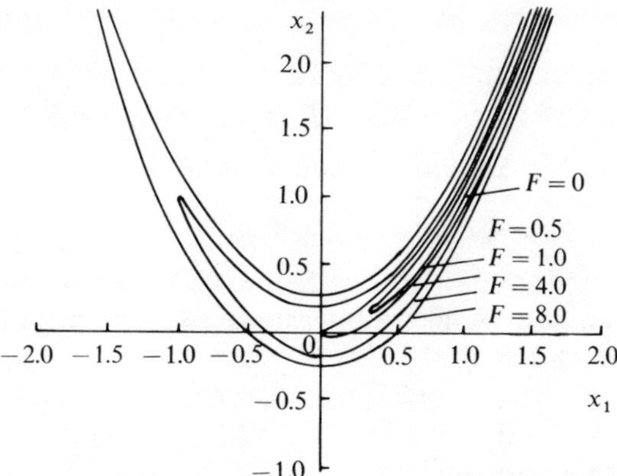

Fig. 5.12 Rosenbrock's function (Reproduced with permission from Schwefel, H.P., *Numerical Optimization of Computer Models,* Wiley, 1981.)

(iii) Rosenbrock's Function (ref. 5)

$$F(\mathbf{x}) = 100(x_2 - x_1^2)^2 + (x_1 - 1)^2 .$$

This function (Fig. 5.12) has a minimum at:

$$\mathbf{x}^* = [1 \quad 1]^T, \quad F(\mathbf{x}^*) = 0.$$

The simplex method has difficulty in negotiating the 'long narrow valley' in this function, the simplex size rapidly becoming very small. The alternating variable method and path of steepest descent are more successful.

5.7 SYSTEM IDENTIFICATION BY COMPUTER PROGRAM

While transfer functions (and so models) for many systems can be developed from experiment-based frequency responses, such methods are in some cases unsuitable (e.g. for ships, submarines, aircraft), and alternatives must be considered.

Clearly, a transfer function (and so a model) for an LTI system can in principle be deduced from time domain input–output data collected from that system (evidently the more data the better), and if the data reflects all the system modes and is practically noise-free, this turns out to be a fairly simple task. Otherwise it can be surprisingly difficult.

The concepts underlying the two possibly best established classes of method of dealing with this problem are outlined in Secs. 5.8, 5.9. The task itself, which is known as *system identification* and which in practice relates to discrete systems, is briefly defined here.

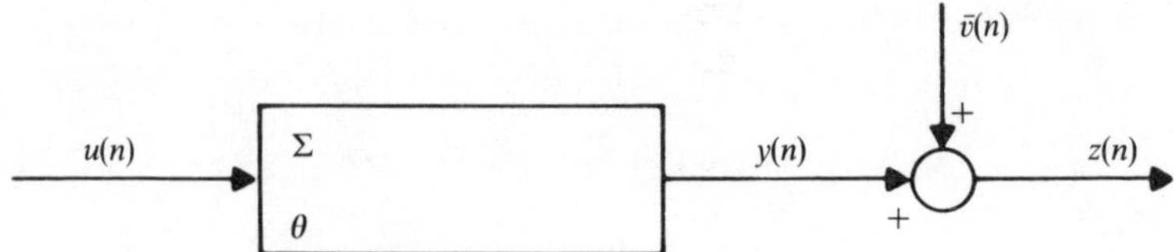

Fig. 5.13 Basic single-input, single-output discrete LTI system.

A basic single-input, single-output discrete LTI system to be identified in this way is shown in Fig. 5.13.

The discrete system Σ (which may consist of a continuous time system with DACs and ADCs (Fig. 4.4)) is subjected to an input sequence $u(n)$, and generates an output sequence $z(n)$:

$$z(n) = y(n) + \bar{v}(n) \tag{5.17}$$

where:

$\bar{v}(n)$ is noise (usually but not necessarily white, Gaussian), and the system behavior is described by an 'auto-regressive-moving-average', or *ARMA*, model, usual in this context:

$$\begin{aligned} y(n) = u(n)\theta_0 + u(n-1)\theta_1 + \cdots + u(n-m)\theta_m \\ + y(n-1)\theta_{m+1} + \cdots + y(n-m)\theta_{2m}. \end{aligned} \tag{5.18}$$

The system 'parameters', $\theta_0, \theta_1, \ldots, \theta_{2m}$ are constants (Secs. 1.9, 1.10).

It is the task of the system identification procedure to estimate θ_0, $\theta_1, \ldots, \theta_{2m}$ given the input and output data sequences $\{u(n)\}$, $\{z(n)\}$, and perhaps some knowledge of the properties of the noise sequence $\{\bar{v}(n)\}$.

5.8 LEAST SQUARES ESTIMATION OF SYSTEM PARAMETERS (REF. 3)

For the sake of neatness, Eqs. (5.17), (5.18) may be written:

$$z(k) = \bar{\mathbf{h}}^{\mathrm{T}}(k)\boldsymbol{\theta} + \bar{v}(k) \tag{5.19}$$

where $\bar{\mathbf{h}}(k) = [u(k) \quad u(k-1) \quad u(k-m) \quad y(k-1) \quad \ldots \quad y(k-m)]^{\mathrm{T}}$.

$\bar{\mathbf{h}}(k)$ includes data which occurred at times $(k-1), (k-2), \ldots, (k-m)$. The independent variable k is used in this argument (rather than n) to help clarify the symbols; $\bar{\mathbf{h}}(k)$ represents data available, but not necessarily occurring, at time k.

$$\boldsymbol{\theta} = [\theta_0 \quad \theta_1 \quad \cdots \quad \theta_m]^{\mathrm{T}},$$

$\bar{v}(k)$ is noise (generally white, Gaussian).

Equation (5.19) may be rewritten:

$$z(k) = \mathbf{h}^{\mathrm{T}}(\mathbf{k})\boldsymbol{\theta} + v(k) \tag{5.20}$$

where $\mathbf{h}(k)$ is used to denote the system input and output information available at time k:

$$\mathbf{h}(k) = [u(k) \quad u(k-1) \quad \cdots \quad u(k-m) \quad z(k-1) \quad \cdots \quad z(k-m)]^{\mathrm{T}} \tag{5.21}$$

$$v(k) = \bar{v}(k) - \bar{v}(k-1)\theta_{m+1} - \bar{v}(k-2)\theta_{m+2} - \cdots - \bar{v}(k-m)\theta_{2m}.$$

Here, $v(k)$ is (almost certainly) nonwhite noise whose properties are partly determined by the system dynamics.

Equations of this kind (Eq. (5.20)) may also be written for the information available at times $(k-1), (k-2), \ldots, (k-L)$:

$$\begin{aligned} z(k-1) &= \mathbf{h}^{\mathrm{T}}(k-1)\boldsymbol{\theta} + v(k-1) \\ z(k-2) &= \mathbf{h}^{\mathrm{T}}(k-2)\boldsymbol{\theta} + v(k-1) \\ &\vdots \\ z(k-L) &= \mathbf{h}^{\mathrm{T}}(k-L)\boldsymbol{\theta} + v(k-L). \end{aligned}$$

These are usually collected or 'concatenated' to give the 'measurement' equation:

$$\mathbf{Z}(k) = \mathbf{H}(k)\boldsymbol{\theta} + \mathbf{V}(k) \tag{5.22}$$

where

$$\mathbf{Z}(k) = [z(k) \quad z(k-1) \quad z(k-2) \quad \ldots \quad z(k-L)]^{\mathrm{T}}$$

$$\mathbf{V}(k) = [v(k) \quad v(k-1) \quad v(k-2) \quad \cdots \quad v(k-L)]^{\mathrm{T}}.$$

$$\mathbf{H}(k) = \begin{bmatrix} \mathbf{h}^{\mathrm{T}}(k) \\ \mathbf{h}^{\mathrm{T}}(k-1) \\ \vdots \\ \mathbf{h}^{\mathrm{T}}(k-L) \end{bmatrix}.$$

The least squares estimation strategy for identifying the 'best' estimate $\hat{\boldsymbol{\theta}}(k)$, which is the 'best' estimate of $\boldsymbol{\theta}$ based on the input–output information available at time k, is shown in diagrammatic form in Fig. 5.14.

The system model, $\hat{\Sigma}$, is fed with the same information, $\mathbf{H}(k)$, as the system Σ, and generates an output $\hat{\mathbf{Z}}(k)$. This is compared with the system output $\mathbf{Z}(k)$ and the difference, $\tilde{\mathbf{Z}}(k)$, calculated:

$$\tilde{\mathbf{Z}}(k) = \mathbf{Z}(k) - \hat{\mathbf{Z}}(k) \tag{5.23}$$

where

$$\hat{\mathbf{Z}}(k) = \mathbf{H}(k)\hat{\boldsymbol{\theta}}(k)\cdot \tag{5.24}$$

$\tilde{\mathbf{Z}}(k)$ is used to drive an identification strategy which selects $\hat{\boldsymbol{\theta}}(k)$ so that a cost function $J(\hat{\boldsymbol{\theta}}(k))$ is minimized:

$$J(\hat{\boldsymbol{\theta}}(k)) = \tilde{\mathbf{Z}}^{\mathrm{T}}(k)\mathbf{W}(k)\tilde{\mathbf{Z}}(k) \tag{5.25}$$

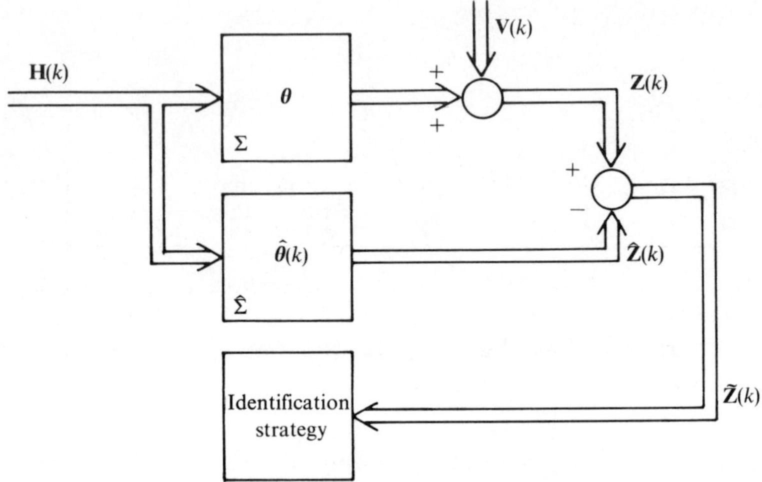

Fig. 5.14 Least squares estimation strategy.

where $\mathbf{W}(k)$ is a positive definite (usually diagonal) weighting matrix:

$$\mathbf{W}(k) = \begin{bmatrix} w_0(k) & 0 & \cdots & 0 \\ 0 & w_1(k) & \cdots & 0 \\ \vdots & & & \\ 0 & 0 & \cdots & w_L(k) \end{bmatrix}.$$

By substituting Eqs. (5.23), (5.24) into (5.25) and finding $\partial J(\hat{\boldsymbol{\theta}}(k))/\partial(\hat{\boldsymbol{\theta}}(k))$, it is not difficult to show that the 'best' (least squares) estimate of $\hat{\boldsymbol{\theta}}(k)$ is:

$$\hat{\boldsymbol{\theta}}(k) = (\mathbf{H}^{\mathrm{T}}(k)\mathbf{W}(k)\mathbf{H}(k))^{-1}\mathbf{H}^{\mathrm{T}}(k)\mathbf{W}(k)\mathbf{Z}(k). \tag{5.26}$$

This is the generalized least squares estimate of $\boldsymbol{\theta}$ based on input–output information available at k, $k-1, \ldots, k-L$.

Usually, the $(L+1) \times (L+1)$ matrix $\mathbf{W}(k)$ is chosen to be the identity matrix, and in this case Eq. (5.26) yields the least squares estimate of $\boldsymbol{\theta}$. However, the elements of $\mathbf{W}(k)$ may be chosen to weight earlier or later measurement vectors $\mathbf{h}(k)$ (ref. 3), and this can be useful in one-line identification where a transfer function may evolve with time. Typically in this case, $w_i(k) = \mu^{k-L+i}$; earlier measurements are weighted if $\mu > 1$, later measurements if $\mu < 1$.

EXAMPLE

The idea is best illustrated by a simple example. Consider the system illustrated in Fig. 5.15 (cf. Fig. 4.4).

The values of R, C are unknown, and an experiment is conducted in which an input series of numbers:

$$u(n) = \{-1, 1, 2, 2, 0, -1, 0, 2, 2, \ldots\}$$

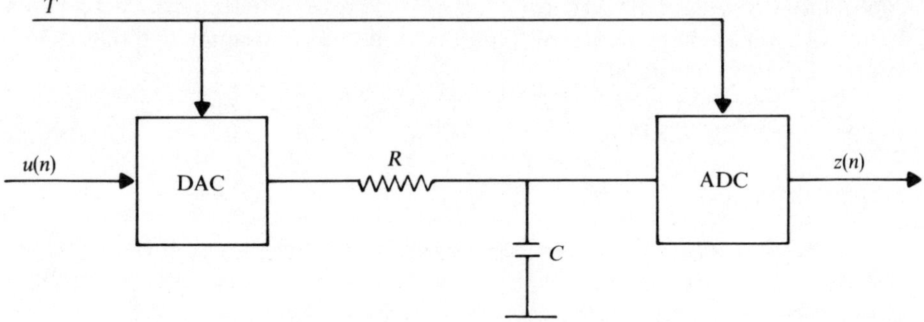

Fig. 5.15 Least squares system identification (example).

gives rise to an output series:

$$z(n) = \{0, -0.63, 0.40, 1.41, 1.79, 0.66, -0.39, -0.14, 1.21, \ldots\}.$$

The transfer function of the continuous time RC network is of the form:

$$G(s) = \frac{a}{1 + sb} \qquad a, b \text{ constants.}$$

Consequently, the z transfer function of the system, Σ, is of the form (Sec. 1.11):

$$\overline{GH}(z) = \frac{\theta_0 + \theta_1 z^{-1}}{1 - \theta_2 z^{-1}} \qquad (\theta_0 = 0).$$

Reverting to k to represent discrete time, this is equivalent to (Eq. (1.32), (1.33)):

$$z(k) = u(k)\theta_0 + u(k-1)\theta_1 + z(k-1)\theta_2. \tag{5.27}$$

To illustrate the method, consider $k = 4$, $L = 2$.

The task is to identify θ_1, θ_2 using the data available at $k = 4$. The input–output data are concatenated (Eq. (5.22) *et seq.*):

$$\mathbf{Z}(4) = [1.79 \quad 1.41 \quad 0.40]^{\mathrm{T}}$$

$$\mathbf{H}(4) = \begin{bmatrix} 0 & 2 & 1.41 \\ 2 & 2 & 0.40 \\ 2 & 1 & -0.63 \end{bmatrix}.$$

Consider the simple case:

$$\mathbf{W}(k) = \begin{bmatrix} 1 & 0 & 0 \\ 0 & 1 & 0 \\ 0 & 0 & 1 \end{bmatrix}.$$

Substituting these in Eq. (5.26):

$$\hat{\boldsymbol{\theta}}(4) = (0 \quad 0.6337 \quad 0.3667)^{\mathrm{T}}.$$

A check on the values of $z(k)$ (Eq. (5.20), $v(k) = 0$) reveals that this gives a good match of estimated and actual output data, and it may be concluded that $\hat{\boldsymbol{\theta}}(4)$ is an accurate estimate of $\boldsymbol{\theta}$.

The larger the quantity of data used in Eq. 5.26 (i.e. the larger L), the more reliable is the estimate $\hat{\boldsymbol{\theta}}(k)$, but it is clearly impractical to deal with matrices in which the number of rows, $(L+1)$, is very large. Consequently, it is common to cast Eq. 5.26 in recursive form:

$$\hat{\boldsymbol{\theta}}(k+1) = \hat{\boldsymbol{\theta}}(k) + \mathbf{P}(k+1)\mathbf{h}(k+1)w(k+1)[z(k+1) - \mathbf{h}^{\mathrm{T}}(k+1)\hat{\boldsymbol{\theta}}(k)] \tag{5.28}$$

where:

$$\mathbf{P}(k+1) = [\mathbf{h}(k+1)\cdot w(k+1)\cdot \mathbf{h}^{\mathrm{T}}(k+1) + \mathbf{P}^{-1}(k)]^{-1}. \tag{5.29}$$

$\mathbf{h}(k+1)$ represents the new information available at $(k+1)$, and $w(k+1)$ is a scalar weighting factor associated with it.

The recursive process is started with $(L+1)$ equations (Eq. (5.26)) and:

$$\mathbf{P}(k) = (\mathbf{H}^{\mathrm{T}}(k)\mathbf{W}(k)\mathbf{H}(k))^{-1}. \tag{5.30}$$

EXAMPLE

Consider the previous example for $(k+1) = 5$.

$$\mathbf{P}(4) = \left[\begin{bmatrix} 0 & 2 & 1.41 \\ 2 & 2 & 0.40 \\ 2 & 1 & -0.63 \end{bmatrix}^{\mathrm{T}} \begin{bmatrix} 1 & 0 & 0 \\ 0 & 1 & 0 \\ 0 & 0 & 1 \end{bmatrix} \begin{bmatrix} 0 & 2 & 1.41 \\ 2 & 2 & 0.40 \\ 2 & 1 & -0.63 \end{bmatrix}\right]^{-1}$$

$$= \begin{bmatrix} 8 & 6 & -0.46 \\ 6 & 9 & 2.99 \\ -0.46 & 2.99 & 2.55 \end{bmatrix}^{-1}$$

$$\hat{\boldsymbol{\theta}}(4) = (0 \quad 0.6337 \quad 0.3667)^{\mathrm{T}}$$

$$\mathbf{h}(5) = (-1 \quad 0 \quad 1.79)^{\mathrm{T}}.$$

Substituting these in Eqs. (5.29), (5.28), with $w(k+1) = 1$:

$$\hat{\boldsymbol{\theta}}(5) = \hat{\boldsymbol{\theta}}(4) + \begin{bmatrix} 0 \\ -0.00045 \\ 0.00135 \end{bmatrix} = \begin{bmatrix} 0 \\ 0.6333 \\ 0.3681 \end{bmatrix}.$$

Clearly, this procedure can be repeated to yield $\hat{\boldsymbol{\theta}}(6)$, $\hat{\boldsymbol{\theta}}(7), \ldots$, until a stable and satisfactory estimate is attained. In fact, in the example above, the input–output data were derived using $\boldsymbol{\theta} = [0 \quad 0.632 \quad 0.368]^{\mathrm{T}}$, and it can be seen that $\hat{\boldsymbol{\theta}}(5)$ represents a slight improvement on an already good $\hat{\boldsymbol{\theta}}(4)$.

CAD Facility

A useful CAD facility depends on the input–output data being presented to it in a workable form – typically on a floppy disk carrying a file of data collected by a compatible computer during experiment on the system concerned. This experiment may involve excitation of the system by a signal such as a random binary signal, or Gaussian white noise, and the collection and recording of this data and the corresponding system responses (ref. 2).

The real-time computer facility required for this operation is connected to the system under test via suitable DACs and ADCs as indicated in Fig. 5.16(a), the necessary experiment specified, and the input–output data collected on disk. An off-line CAD program is then used to process this. A facility of this kind, shown in Fig. 5.16(b), operates as follows.

BLOCK 1

A basic clock period, T, selected to suit the dynamics of the system under test: an excitation sequence type, typically Gaussian or pseudo-random binary white noise (App. B.5) and its variance; and the name of the data file to be generated, are all input, displayed, and corrected if necessary.

BLOCK 2

The excitation sequence and corresponding system output data are collected and filed on disk.

BLOCK 3

The off-line system identification program requests details of the transfer function order m; the number of input–output records to be considered in a preliminary estimate ($L+1$, Eq. (5.26)), any weighting matrix $\mathbf{W}(k)$ required.

BLOCK 4

A preliminary least squares estimate $\hat{\boldsymbol{\theta}}(k)$ is calculated (Eq. (5.26)) and displayed.

BLOCKS 5, 6

The estimate $\hat{\boldsymbol{\theta}}(k)$ may be improved with more data taken in specified quantities from the disk record, using the recursive Eqs. (5.28), (5.29), (5.30). Specification of additional weighting factors may be of interest, though usually $w(k+1)=1$. The updated result is output and the recursive process repeated as required.

The identification process may be repeated using different files of data which may perhaps correspond to different excitation sequences tried in on-line experiments (ref. 2).

Comment

It is easy to show that the method described in this section gives biased results if $v(k)\neq 0$, which is, of course, usually the case. Thus, if the noise level is small, good results are obtained, but if not, the method is unreliable. A surprisingly large amount of data is required to give dependable results, 1000 input–output pairs being by no means excessive.

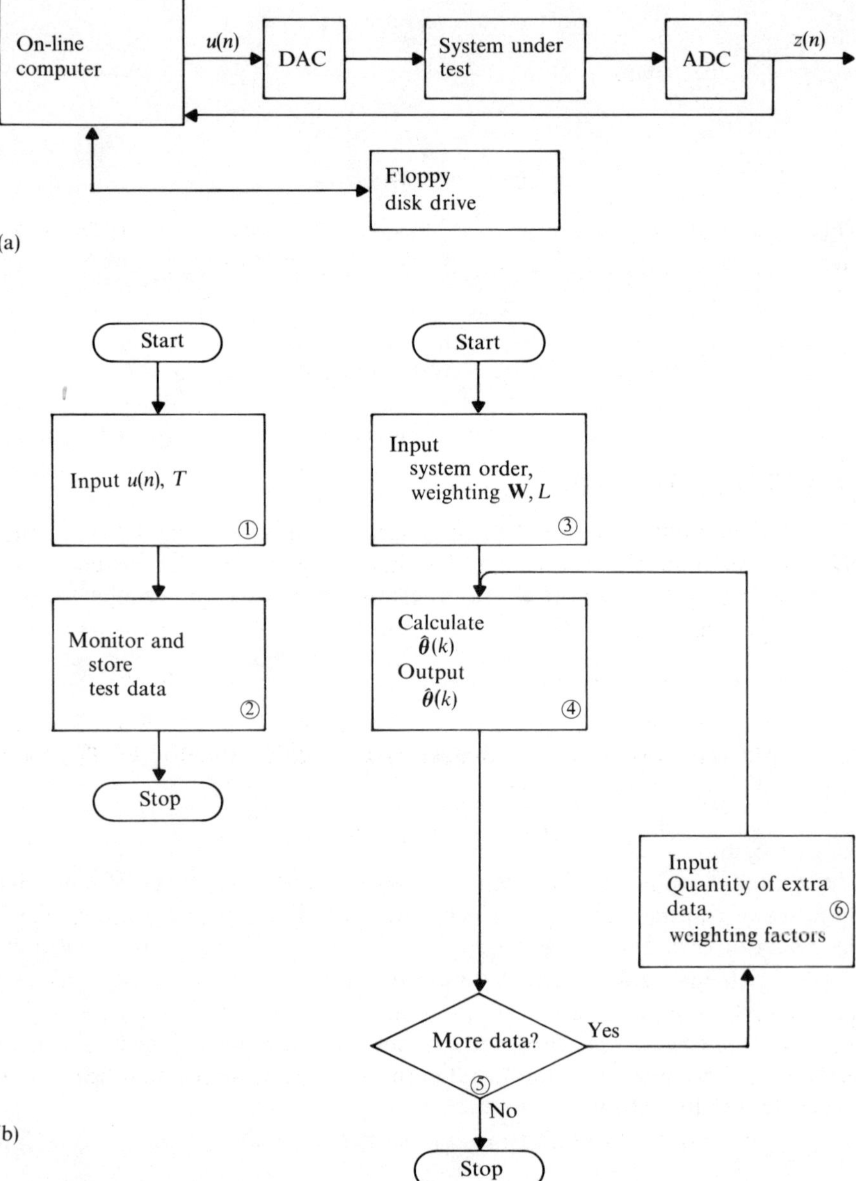

Fig. 5.16 System identification facility. (a) On-line test facilities. (b) Test and data-processing algorithms.

5.9 MAXIMUM LIKELIHOOD ESTIMATION OF SYSTEM PARAMETERS (REF. 2)

This method of parameter estimation uses the information collected in the concatenated measurement equation (Eq. (5.22)) in a way different from the least squares estimation method.

Once more the basic data are expressed in the form:

$$\mathbf{Z}(k) = \mathbf{H}(k)\boldsymbol{\theta} + \mathbf{V}(k).$$

The reasonable assumption is made that $\mathbf{V}(k)$, the noise vector, is Gaussian, and it is not difficult to derive the $(L+1)$-dimensional Gaussian joint probability density function (App. B.3):

$$f(\mathbf{Z}(k)) = \frac{1}{\sqrt{(2\pi)^{L+1}\det \mathbf{R}(k)}} \times \exp\{-\tfrac{1}{2}[\mathbf{Z}(k) - \mathbf{H}(k)\hat{\boldsymbol{\theta}}(k)]^{\mathrm{T}}\mathbf{R}^{-1}(k)[\mathbf{Z}(k) - \mathbf{H}(k)\hat{\boldsymbol{\theta}}(k)]\}$$

where $\mathbf{R}(k) = \mathscr{E}(\mathbf{V}(k)\mathbf{V}^{\mathrm{T}}(k))$

The maximum likelihood method seeks to select a value $\hat{\boldsymbol{\theta}}(k)$ such that $f(\mathbf{Z}(k))$ is maximized, given $\mathbf{Z}(k)$, $\mathbf{H}(k)$. This means that $\hat{\boldsymbol{\theta}}(k)$ is chosen so that the observed system output $\mathbf{Z}(k)$ is as likely as possible, given $\mathbf{H}(k)$.

Maximizing $f(\mathbf{Z}(k))$ w.r.t. $\hat{\boldsymbol{\theta}}(k)$ is equivalent to minimizing:

$$L(\mathbf{Z}(k)) = ((\mathbf{Z}(k) - \mathbf{H}(k)\hat{\boldsymbol{\theta}}(k))^{\mathrm{T}}\mathbf{R}^{-1}(k)(\mathbf{Z}(k) - \mathbf{H}(k)\hat{\boldsymbol{\theta}}(k).$$

The algebra of this is identical to the least squares estimation (LSE) case (Eq. (5.25)) with the weighting matrix $\mathbf{W}(k)$ replaced by $\mathbf{R}^{-1}(k)$. The 'best' estimate $\hat{\boldsymbol{\theta}}(k)$ is (cf. Eq. (5.26)):

$$\hat{\boldsymbol{\theta}}(k) = (\mathbf{H}^{\mathrm{T}}(k)\mathbf{R}^{-1}(k)\mathbf{H}(k))^{-1}\mathbf{H}^{\mathrm{T}}(k)\mathbf{R}^{-1}(k)\mathbf{Z}(k). \tag{5.31}$$

The method in this basic form requires knowledge of $\mathbf{R}(k)$, i.e. knowledge of the noise characteristics, and, in contrast to the LSE case, the estimate $\hat{\boldsymbol{\theta}}(k)$ is free from bias.

Knowledge of $\mathbf{R}(k)$ is of course not often readily available, and one advantage of the maximum likelihood concept is that it can be extended to estimate the parameters of the noise in addition to those of the system. One such estension, which is suited to both off- and on-line identification, is described briefly here.*

Casting Eq. (5.20) in a form more suited to this analysis:

$$z(k) = \sum_{j=0}^{m} a_j u(k-j) + \sum_{j=1}^{m} b_j z(k-j) + \sum_{j=0}^{m} c_j w(k-j)$$

*This method is taken from a useful survey paper: G.N. Saridis, 'Comparison of Six On-line Identification Algorithms', *Automatica,* **10**(1974), 69–79.

where $w(k)$ is white Gaussian noise, $\mathscr{E}(w(k))=0$, variance unknown. By defining a best possible estimate $\hat{z}(k)$ of $z(k)$ based on the evidence $\{u(k-j)\}$, $\{z(k-j)\}$, and estimated parameters $\hat{a}_j$, $\hat{b}_j$, $\hat{c}_j$, an 'error' equation is postulated:

$$\begin{aligned}\tilde{z}(k) &= z(k) - \hat{z}(k)\\ &= z(k) - \sum_{j=0}^{m} \hat{a}_j u(k-j) - \sum_{j=1}^{m} \hat{b}_j z(k-j) - \sum_{j=0}^{m} \hat{c}_j \tilde{z}(k-j).\end{aligned} \tag{5.32}$$

Writing, for convenience,

$$\hat{\boldsymbol{\theta}} = (\hat{a}_0 \quad \hat{a}_1 \quad \cdots \quad \hat{a}_m \quad \hat{b}_1 \quad \cdots \quad \hat{b}_m \quad \hat{c}_0 \quad \cdots \quad \hat{c}_m)^{\mathrm{T}},$$

and assuming $m+M$ sets of data (i.e. $k=m+1,\ldots,m+M$), a probability function is set up and its logarithm taken to give a cost function:

$$L(\hat{\boldsymbol{\theta}}, \sigma^2) = G - \frac{M}{2}\ln(\sigma^2) - \frac{1}{2}\sum_{k=m+1}^{m+M} \frac{1}{\sigma^2}(\tilde{z}(k))^2 \tag{5.33}$$

where $\sigma^2 = \mathscr{E}(\tilde{z}(k))^2$ and G is a constant.
Maximizing $L(\hat{\boldsymbol{\theta}}, \sigma^2)$ w.r.t. $\hat{\boldsymbol{\theta}}$ is equivalent to minimizing:

$$J(\hat{\boldsymbol{\theta}}) = \sum_{k=m+1}^{m+M} (\tilde{z}(k))^2 \cdot \tag{5.34}$$

Moreover, when this minimum is attained:

$$\sigma^2 = \frac{1}{M} J(\hat{\boldsymbol{\theta}}) = \mathscr{E}(w(k))^2.$$

In the algorithm described here, $J(\hat{\boldsymbol{\theta}})$ is minimized by a version of the path of steepest ascent method (Sec. 5.5). The same data may be used repeatedly by the algorithm or, perhaps more interestingly, more blocks of data (each with M input–output pairs) may be fed in at each iteration of the algorithm. In this latter form the algorithm is recommended for on-line identification. For clarity, however, the version described here uses the same data repeatedly.

Suppose that $(m+M)$ input–output data pairs have been collected (typically, $M=3000$). The first m pairs are only used as preliminary data.

Stage 1
Assume initial values for:

(i) $\hat{\boldsymbol{\theta}} = (\hat{a}_0 \quad \hat{a}_1 \quad \cdots \quad \hat{a}_m \quad \hat{b}_1 \quad \cdots \quad \hat{b}_m \quad \hat{c}_0 \quad \cdots \quad \hat{c}_m)^{\mathrm{T}}$

(ii) $\dfrac{\partial \tilde{z}(k-j)}{\partial \hat{\boldsymbol{\theta}}} = \left[\dfrac{\partial \tilde{z}(k-j)}{\partial \hat{a}_0} \quad \dfrac{\partial \tilde{z}(k-j)}{\partial \hat{a}_1} \quad \cdots \quad \dfrac{\partial \tilde{z}(k-j)}{\partial \hat{c}_m}\right]^{\mathrm{T}}, \quad j=0,1,\ldots,m.$

Stage 2
Evaluate $\tilde{z}(k)$, $k=m+1, m+2,\ldots, m+M$ (Eq. (5.32))

Evaluate (see Eq. (5.32)), using the most recent results from Stage 1, or

Stage 4, for $k = m+1, m+2, \ldots, m+M$:

$$\frac{\partial \tilde{z}(k)}{\partial \hat{a}_j} = -u(k-j) - \sum_{j=0}^{m} \hat{c}_j \frac{\partial \tilde{z}(k-j)}{\partial \hat{a}_j}, \quad j = 0, 1, \ldots, m,$$

$$\frac{\partial \tilde{z}(k)}{\partial \hat{b}_j} = -z(k-j) - \sum_{j=0}^{m} \hat{c}_j \frac{\partial \tilde{z}(k-j)}{\partial \hat{b}_j} \quad j = 1\ 2, \ldots, m,$$

$$\frac{\partial \tilde{z}(k)}{\partial \hat{c}_j} = -\tilde{z}(k-j) - \sum_{j=0}^{m} \hat{c}_j \frac{\partial \tilde{z}(k-j)}{\partial \hat{c}_j}, \quad j = 0, 1, \ldots, m\,.$$

Evaluate (see Eq. (5.34)), using these results:

$$\frac{\partial J}{\partial \boldsymbol{\theta}} = -2 \sum_{k=m+1}^{m+M} \tilde{z}(k) \frac{\partial \tilde{z}(k)}{\partial \boldsymbol{\theta}}.$$

Also evaluate, using the same results:

$$S \approx \frac{\partial^2 J}{\partial \boldsymbol{\theta}^2} = \sum_{k=m+1}^{m+M} \left[\frac{\partial \tilde{z}(k)}{\partial \boldsymbol{\theta}} \left(\frac{\partial \tilde{z}(k)}{\partial \boldsymbol{\theta}} \right)^{\mathrm{T}} \right].$$

Stage 3

Evaluate the next estimate of $\hat{\boldsymbol{\theta}}$.

$$\hat{\boldsymbol{\theta}}_{\text{new}} = \hat{\boldsymbol{\theta}} - S^{-1} \frac{\partial J}{\partial \boldsymbol{\theta}}.$$

Stage 4

Repeat from Stage 2 using the latest calculated values of $\hat{\boldsymbol{\theta}}$, $\partial \tilde{z}(k-j)/\partial \boldsymbol{\theta}$, $k = m+M$, $j = 0, 1, \ldots, m$ instead of the Stage 1 estimates.

Continue this process until no appreciable improvement in $\hat{\boldsymbol{\theta}}$ is achieved.

CAD Facility

As with the LSE method, a useful CAD facility depends on the collected information being presented to it in a convenient form, typically as a file of data on a floppy disk. This is prepared in the same way as that described in Sec. 5.8, and indeed the same experimental data can be presented to several system identification algorithms giving confidence (or otherwise) in the results. The interactive, off-line, CAD facility shown in Fig. 5.17 operates as follows.

BLOCK 1

A clock period, T, selected to suit the apparent dynamics of the system under test; an excitation sequence type, which can be Gaussian or pseudo-random binary white noise (App. B.5) and its variance and duration; and the name of the data file to be generated; are all input, displayed, and corrected if necessary.

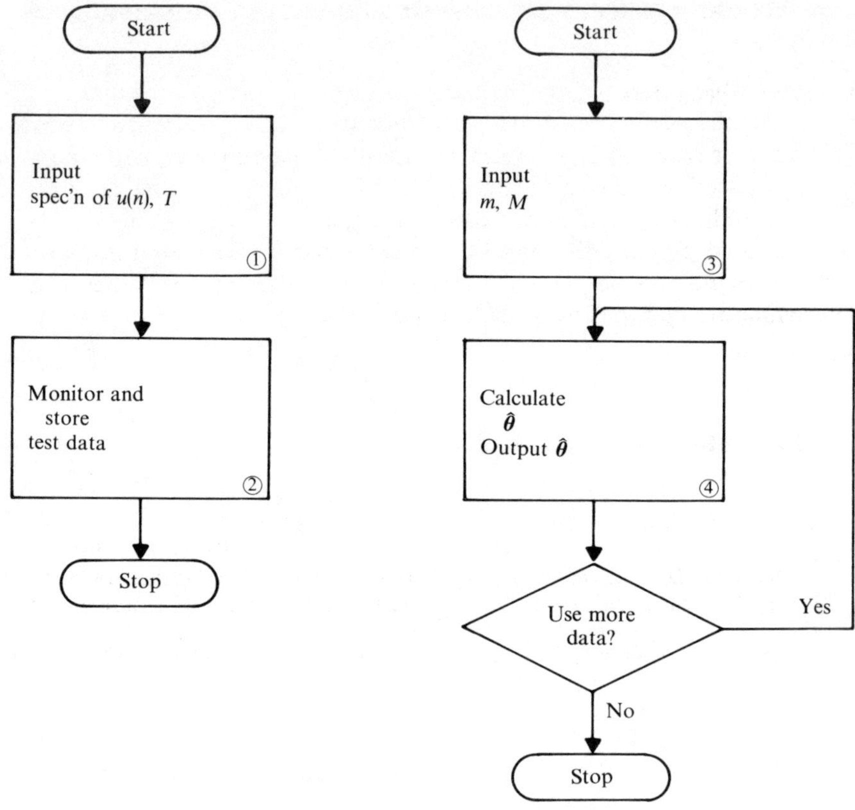

Fig. 5.17 Test and data processing algorithms.

BLOCK 2
The excitation sequence and corresponding system output data are collected and filed on disk.

BLOCK 3
The off-line system identification program requests details of the transfer function order m and the number of input–output records $(m + M)$ on the file.

BLOCK 4
The algorithm proceeds as described in Stages 1–4 above with the latest estimate $\hat{\boldsymbol{\theta}}$ being output at each iteration.

Comment

Maximum likelihood algorithms of this kind generally converge well if the initial estimates, Stage 1, are reasonably good. This may not be easy to achieve, and it might be necessary to use an LSE algorithm, first of all, to obtain initial estimates of sufficient quality.

5.10 OTHER METHODS OF SYSTEM PARAMETER IDENTIFICATION

A considerable variety of developments of the methods outlined in Secs. 5.8, 5.9, and of other techniques, has been proposed (ref. 2). These include use of the Kalman filter algorithm (Sec. 4.11) in a number of extended forms, and also methods based on statistical analyses of input–output data. To achieve reliable results, careful specification of the system excitation signals is recommended (ref. 2); it is unrealistic to expect to identify parameters using excitation signals which merely happen to be available for some other reason. This constraint can, of course, be a major drawback.

REFERENCES

1. Burley, D.M., *Studies in Optimization.* Intertext, 1974.
2. Eykhoff, P., *Trends and Progress in System Identification.* Pergamon, 1981.
3. Mendel, J.M., *Discrete Techniques of Parameter Estimation.* Dekker, 1973.
4. Murray, W., *Numerical Methods for Unconstrained Optimization.* Academic Press, 1972.
5. Schwefel, H.P., *Numerical Optimization of Computer Models.* Wiley, 1977.

APPENDIX A

NOTES ON VECTORS, MATRICES AND DETERMINANTS (REF. 1)

A.1 VECTORS

1. An m-vector is an ordered set of m elements arranged in a column (a *column vector*) or row (a *row vector*):

$$\mathbf{x} = \begin{bmatrix} x_1 \\ x_1 \\ \vdots \\ x_m \end{bmatrix} = [x_1 \quad x_2 \quad \cdots \quad x_m]^{\mathrm{T}}.$$

The suffix 'T' denotes that the row vector is transposed to become a column vector (or vice versa).

2. Like vectors are added or subtracted by adding or subtracting the corresponding elements. A vector is multiplied by a scalar by multiplying all its elements by that scalar. Thus:

$$\mathbf{x} + k\mathbf{y} = \begin{bmatrix} x_1 + ky_1 \\ x_2 + ky_2 \\ \vdots \\ x_m + ky_m \end{bmatrix}.$$

3. An m-vector may be regarded as a point in m-dimensional space. Scalar multiplication results in a new vector with the same 'direction' and different 'length', while the addition (or subtraction) of vectors follows the parallelogram rule.

EXAMPLE
Some points are illustrated, $m = 2$, in Fig. A.1.

$$\mathbf{x} = \begin{bmatrix} 1 \\ 2 \end{bmatrix}; \quad \mathbf{y} = \begin{bmatrix} 3 \\ 1 \end{bmatrix}; \quad k = 2$$

$$\mathbf{z} = \mathbf{x} + k\mathbf{y} = \begin{bmatrix} 7 \\ 4 \end{bmatrix}$$

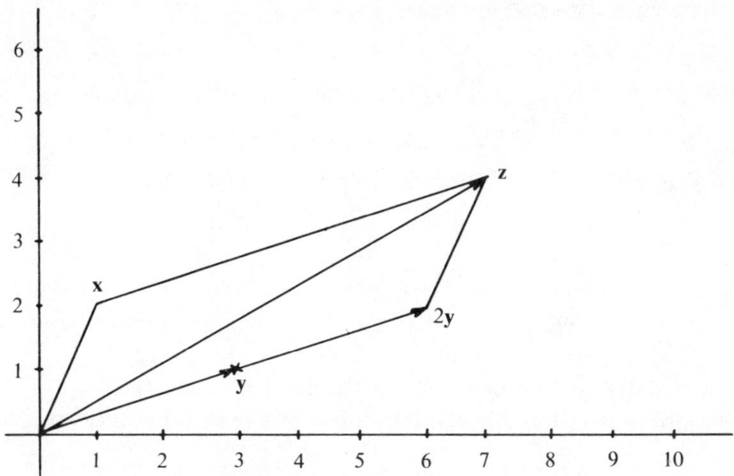

Fig. A.1 Scalar multiplication and addition of vectors (example).

4. A vector may be a function of time (or any other independent variable). In m-dimensional space, $\mathbf{x}(t)$ may be regarded as describing a trajectory as shown in Fig. A.2, $m = 2$.

In the case illustrated in Fig. A.2,

$$\mathbf{x}(0) = \begin{bmatrix} 6 \\ 3.5 \end{bmatrix}, \quad \mathbf{x}(\infty) = \begin{bmatrix} 0 \\ 0 \end{bmatrix}.$$

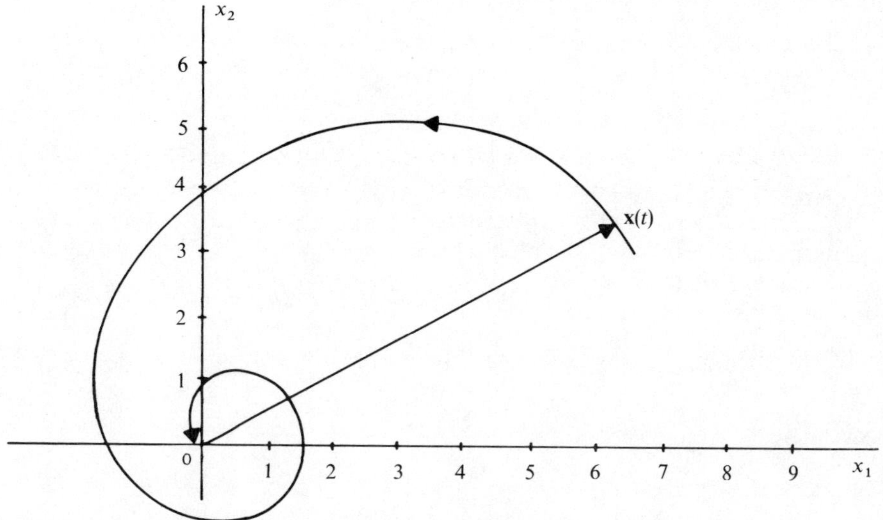

Fig. A.2 Trajectory in two-dimensional space.

A.2 MATRICES – BASIC DEFINITIONS

1. An $m \times n$ *matrix* is an array of m rows and n columns of elements, which is subject to certain rules of operation:

$$\mathbf{A} = \begin{bmatrix} a_{11} & a_{12} & \cdots & a_{1n} \\ a_{21} & a_{22} & \cdots & a_{2n} \\ \vdots & & & \\ a_{m1} & a_{m2} & \cdots & a_{mn} \end{bmatrix}.$$

Two matrices are of *equal order* when they have the same numbers of rows and columns; they are *equal* when all their elements are equal.

A *zero* matrix is one in which all the elements are zero.

A *square* matrix has the same number of rows and columns.

2. Matrices of equal order are added or subtracted by adding or subtracting their corresponding elements.

EXAMPLE

$$\mathbf{A} = \begin{bmatrix} 2 & -3 \\ 4 & 5 \end{bmatrix}, \qquad \mathbf{B} = \begin{bmatrix} 6 & 7 \\ 8 & 9 \end{bmatrix},$$

$$\mathbf{C} = \mathbf{A} + \mathbf{B} = \begin{bmatrix} 8 & 4 \\ 12 & 14 \end{bmatrix}.$$

3. An $m \times n$ matrix may be multiplied by an $n \times r$ matrix, in that order, to give an $m \times r$ matrix according to the following rule:

$$\mathbf{A} = \begin{bmatrix} a_{11} & a_{12} & \cdots & a_{1n} \\ a_{21} & a_{22} & \cdots & a_{2n} \\ \vdots & & & \\ a_{m1} & a_{m2} & \cdots & a_{mn} \end{bmatrix}, \quad \mathbf{B} = \begin{bmatrix} b_{11} & b_{12} & \cdots & b_{1r} \\ b_{21} & b_{22} & \cdots & b_{2r} \\ \vdots & & & \\ b_{n1} & b_{n2} & \cdots & b_{nr} \end{bmatrix}$$

$\mathbf{C} = \mathbf{AB}$

$$= \begin{bmatrix} (a_{11}b_{11} + a_{12}b_{21} + \cdots + a_{1n}b_{n1}) & \cdots & (a_{11}b_{1r} + a_{12}b_{2r} + \cdots + a_{1n}b_{nr}) \\ (a_{21}b_{11} + a_{22}b_{21} + \cdots + a_{2n}b_{n1}) & \cdots & (a_{21}b_{1r} + a_{22}b_{2r} + \cdots + a_{2n}b_{nr}) \\ \vdots & & \\ (a_{m1}b_{11} + a_{m2}b_{21} + \cdots + a_{mn}b_{n1}) & & (a_{m1}b_{1r} + a_{m2}b_{2r} + \cdots + a_{mn}b_{nr}) \end{bmatrix}.$$

EXAMPLE

$$\mathbf{A} = \begin{bmatrix} 1 & -2 & 3 \\ 4 & 5 & -6 \end{bmatrix}, \qquad \mathbf{B} = \begin{bmatrix} 7 & 8 \\ 9 & 10 \\ 11 & 12 \end{bmatrix}$$

$$\mathbf{C} = \mathbf{AB} = \begin{bmatrix} 22 & 24 \\ 7 & 10 \end{bmatrix}.$$

A square matrix may be multiplied by itself.

$$\mathbf{AA} = \mathbf{A}^2$$
$$\mathbf{AAA} = \mathbf{AA}^2 = \mathbf{A}^3.$$

4 The *transpose* of a matrix is found by interchanging the rows and columns.

$$A^{\mathrm{T}} = \begin{bmatrix} a_{11} & a_{21} & \cdots & a_{m1} \\ a_{12} & a_{22} & & a_{m2} \\ \vdots & & & \\ a_{1n} & a_{2n} & & a_{mn} \end{bmatrix}.$$

5. A square matrix $\mathbf{A}$ with elements a_{ij} is *diagonal* if $a_{ij} = 0$, $i \neq j$:

$$\mathbf{A} = \begin{bmatrix} a_{11} & 0 & \cdots & 0 \\ 0 & a_{22} & & \\ \vdots & & & \\ 0 & & & a_{mm} \end{bmatrix}.$$

If $a_{11} = a_{22} = \cdots = a_{mm} = 1$, then $\mathbf{A}$ is an *identity* matrix $\mathbf{I}$:

$$\mathbf{I} = \begin{bmatrix} 1 & 0 & \cdots & 0 \\ 0 & 1 & \cdots & \\ \vdots & & & \\ 0 & & & 1 \end{bmatrix}.$$

6. One useful function of a square matrix $\mathbf{A}$ is:

$$e^{\mathbf{A}} = \mathbf{I} + \mathbf{A} + \frac{1}{2!}\mathbf{A}^2 + \frac{1}{3!}\mathbf{A}^3 + \cdots. \tag{A.1}$$

A.3 THE DETERMINANT OF A SQUARE MATRIX

Only square matrices have determinants.
The determinant of a (1×1) matrix A is $\det(A) = |A| = A$
The determinant of a (2×2) matrix $\mathbf{A}$ is defined as follows:

$$\det(\mathbf{A}) = |\mathbf{A}| = \det \begin{bmatrix} a_{11} & a_{12} \\ a_{21} & a_{22} \end{bmatrix}$$
$$= \begin{vmatrix} a_{11} & a_{12} \\ a_{21} & a_{22} \end{vmatrix}$$
$$= a_{11}a_{22} - a_{21}a_{12}.$$

The determinant of a (3 × 3) matrix is defined as follows:

$$\det(\mathbf{A}) = \begin{vmatrix} a_{11} & a_{12} & a_{13} \\ a_{21} & a_{22} & a_{23} \\ a_{31} & a_{32} & a_{33} \end{vmatrix}$$

$$= a_{11}\begin{vmatrix} a_{22} & a_{23} \\ a_{32} & a_{33} \end{vmatrix} - a_{21}\begin{vmatrix} a_{12} & a_{13} \\ a_{32} & a_{33} \end{vmatrix} + a_{31}\begin{vmatrix} a_{12} & a_{13} \\ a_{22} & a_{23} \end{vmatrix}$$

$$= -a_{21}\begin{vmatrix} a_{12} & a_{13} \\ a_{32} & a_{33} \end{vmatrix} + a_{22}\begin{vmatrix} a_{11} & a_{13} \\ a_{31} & a_{33} \end{vmatrix} - a_{23}\begin{vmatrix} a_{11} & a_{12} \\ a_{31} & a_{32} \end{vmatrix} = \cdots.$$

The determinant of **A** may be found using any row or column in the manner shown, with signs alternating as shown. The terms are found by multiplying each element of a row (or column), with sign positive or negative for odd and even elements respectively, by the determinant of all the elements found by excluding that element's row and column.

The determinant of an $(m \times m)$ matrix is defined according to these rules.

EXAMPLE

$$\det\begin{bmatrix} 1 & -2 & 3 \\ 4 & 5 & -6 \\ -7 & 8 & 9 \end{bmatrix} = 1\begin{vmatrix} 5 & -6 \\ 8 & 9 \end{vmatrix} - (-2)\begin{vmatrix} 4 & -6 \\ -7 & 9 \end{vmatrix} + 3\begin{vmatrix} 4 & 5 \\ -7 & 8 \end{vmatrix}$$

$$= 1(45 + 48) + 2(36 - 42) + 3(32 + 35)$$

$$= 282.$$

A.4 THE MINORS, COFACTORS, ADJOINTS, AND TRACE OF A SQUARE MATRIX

1. The *minors* of an $(m \times m)$ matrix **A** are found by taking the determinants of the $(r \times r)$ square matrices $(r < m)$ formed by removing $(m - r)$ rows and columns from **A**.

EXAMPLE

$$\mathbf{A} = \begin{bmatrix} 1 & -2 & 3 \\ 4 & 5 & -6 \\ -7 & 8 & 9 \end{bmatrix}.$$

The determinants $|1|$, $|8|$, $\begin{vmatrix} 1 & -2 \\ 4 & 5 \end{vmatrix}$, $\begin{vmatrix} 1 & 3 \\ -7 & 9 \end{vmatrix}$

are all examples of minors of **A**.

2. The minor formed by removing the ith row and jth column, with the sign $(-1)^{i+j}$ is a *cofactor* of **A**.

EXAMPLE

$$+\begin{vmatrix} 1 & -2 \\ 4 & 5 \end{vmatrix}, \quad +\begin{vmatrix} 5 & -6 \\ 8 & 9 \end{vmatrix}, \quad -\begin{vmatrix} -2 & 3 \\ 8 & 9 \end{vmatrix}$$

are all cofactors of **A**.

3. The *adjoint* of an $(m \times m)$ square matrix **A** is formed as follows:

$$\operatorname{adj} \mathbf{A} = \begin{bmatrix} \alpha_{11} & \alpha_{21} & \cdots & \alpha_{m1} \\ \alpha_{12} & \alpha_{22} & & \\ \vdots & & & \\ \alpha_{1m} & & & \alpha_{mm} \end{bmatrix}$$

when α_{ij} is the cofactor associated with α_{ij}. The transposition should be noted.

EXAMPLE

The adjoint of **A** (defined above) is:

$$\operatorname{adj} \mathbf{A} = \begin{bmatrix} 93 & +6 & 67 \\ -6 & 30 & -6 \\ -3 & +18 & 13 \end{bmatrix}^{\mathrm{T}} = \begin{bmatrix} 93 & -6 & -3 \\ 6 & 30 & 18 \\ 67 & -6 & 13 \end{bmatrix}.$$

4. The *trace* of a square matrix is the sum of its diagonal elements:

$$\operatorname{tr} \mathbf{A} = a_{11} + a_{22} + \cdots + a_{mm}.$$

A.5 THE RANK OF A MATRIX

From any $(m \times n)$ matrix, several $(r \times r)$ square matrices can be formed by striking out numbers of rows and columns. Determinants of these $(r \times r)$ matrices can be evaluated.

A matrix **A** is said to have rank $\bar{r}$ if there is at least one $(\bar{r} \times \bar{r})$ matrix determinant, derived from **A**, which is not zero, whereas all the determinants of matrices of higher order derived from **A** are zero.

EXAMPLE

$$\mathbf{A} = \begin{bmatrix} 2 & 0 & 2 & 6 \\ 4 & 2 & 0 & -4 \\ -2 & -2 & 2 & 10 \end{bmatrix}.$$

There are four third-order determinants, all of which are zero.
For example

$$\begin{vmatrix} 2 & 0 & 2 \\ 4 & 2 & 0 \\ -2 & -2 & 2 \end{vmatrix} = 0.$$

However, at least one second-order determinant is not zero.
For example,

$$\begin{vmatrix} 2 & 0 \\ 4 & 2 \end{vmatrix} = 4.$$

A therefore has rank 2.

Examining the determinants of all the submatrices of a matrix is not generally a practical method of establishing its rank. Instead, an $(m \times n)$ matrix **A** can be manipulated into an 'echelon' form $(m \times n)$ matrix **B** whose rank is obvious (and equal to that of **A**). This is done by a series of elementary transformations which entail:

(a) interchanging rows;
(b) multiplying a row by a nonzero constant;
(c) adding a multiple of one row to another.

Consider the $(m \times n)$ matrix **A** $(m < n)$:

$$\mathbf{A} = \begin{bmatrix} a_{11} & a_{12} & \cdots & a_{1n} \\ a_{21} & a_{22} & \cdots & a_{2n} \\ \vdots & & & \\ a_{m1} & a_{m2} & \cdots & a_{mn} \end{bmatrix}.$$

If **A** has full rank, m, it can be manipulated into **B**, where:

$$\mathbf{B} = \begin{bmatrix} 1 & b_{12} & b_{13} & \cdots & b_{1n} \\ 0 & 1 & b_{23} & \cdots & b_{2n} \\ 0 & 0 & 1 & \cdots & b_{3n} \\ \vdots & & & & \\ 0 & 0 & 0 & \cdots & b_{mn} \end{bmatrix}.$$

B has echelon form, in which all the elements below the ith element in the ith row are zeros, and the principal diagonal contains 1's.

To achieve this transformation, every row of **A** is divided by its first element and every row, except the first, is then subtracted from the first. This gives a matrix of the form:

$$\bar{\mathbf{A}} = \begin{bmatrix} 1 & \bar{a}_{12} & \bar{a}_{13} & \cdots & \bar{a}_{1n} \\ 0 & \bar{a}_{22} & \bar{a}_{23} & \cdots & \bar{a}_{2n} \\ 0 & \bar{a}_{32} & \bar{a}_{33} & \cdots & \bar{a}_{3n} \\ \vdots & & & & \\ 0 & \bar{a}_{m2} & \bar{a}_{m3} & \cdots & \bar{a}_{mn} \end{bmatrix}.$$

This process is repeated with the appropriate submatrices until the **B** matrix is generated. Naturally, if the first element in a row is zero, the sequence for that row is unnecessary, and if a zero row occurs during the process, it is exchanged with a subsequent non-zero row, if available.

If **A** has full rank, m, a matrix **B** with the form above is in due course generated. If **A** has rank $\bar{r} < m$, **B** has the following echelon form:

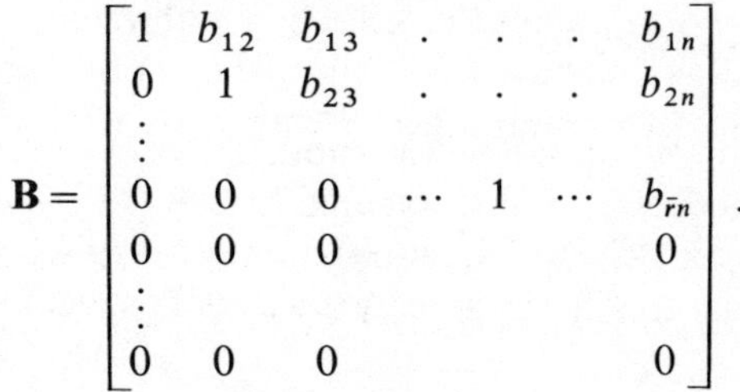

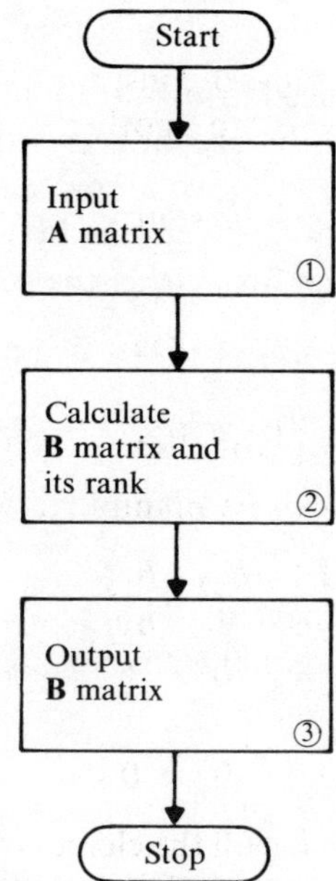

Fig. A.3 Algorithm for matrix rank determination.

The number of nonzero rows, $\bar{r}$, is the rank of **B**, and also, therefore, of **A**.

CAD Facility

The CAD algorithm shown in the flow diagram of Fig. A.3 uses the method described above.

The operation of the algorithm is self-explanatory, with Block 2 performing the calculations outlined above, and Block 3 determining the number of nonzero echelon form rows in **B** (i.e., rank, $\bar{r}$, of **A**). **B** is also displayed.

A.6 THE SOLUTION OF SIMULTANEOUS EQUATIONS (REF. 5)

The Gauss elimination method of solving simultaneous equations, which is easily mechanized as a robust CAD algorithm, is briefly described here.

Consider a set of m simultaneous equations with m unknowns $x_1, x_2, x_3, \ldots, x_m$:

$$\begin{aligned} a_{11}^0 x_1 + a_{12}^0 x_2 + \cdots + a_{1m}^0 x_m &= b_1^0 \\ a_{21}^0 x_1 + a_{22}^0 x_2 + \cdots + a_{2m}^0 x_m &= b_2^0 \\ &\vdots \\ a_{m1}^0 x_1 + a_{m2}^0 x_2 + \cdots + a_{mm}^0 x_m &= b_m^0 \end{aligned}$$

If $a_{11}^0 \neq 0$, this can be reduced to $(m-1)$ equations in $(m-1)$ unknowns. If $a_{11}^0 = 0$, the first equation can be exchanged with another – usually the ith, where $a_{i1}^0 > a_{j1}^0$, $j = 1, 2, \ldots, m$ – to give a new set of equations with $a_{11}^0 \neq 0$.

Multiplying all the equations except the first by a_{11}^0 / a_{i1}^0, $i = 2, 3, 4, \ldots, m$, and subtracting these from the first yields the set:

$$\begin{aligned} a_{11}^0 x_1 + a_{12}^0 x_2 + \cdots + a_{1m}^0 x_m &= b_1^0 \\ a_{22}^1 x_2 + \cdots + a_{2m}^1 x_m &= b_2^1 \\ &\vdots \\ a_{m2}^1 x_2 + \cdots + a_{mm}^1 x_m &= b_m^1 \end{aligned}$$

This process can be repeated, with rows being interchanged where necessary, until a set with diagonal elements all nonzero is developed:

$$\begin{aligned} a_{11}^0 x_1 = a_{12}^0 x_2 + \cdots + a_{1m}^0 x_m &= b_1^0 \\ a_{22}^1 x_2 + \cdots + a_{2m}^1 x_m &= b_2^1 \\ &\vdots \\ a_{mm}^{m-1} x_m &= b_m^{m-1} \end{aligned}$$

Starting at the last equation, the solution is:

$$x_m = \frac{1}{a_{mm}^{m-1}} b_m^{m-1}.$$

Substituting this in the penultimate equation, and repeating the process yields:

$$x_i = \frac{1}{a_{ii}^{i-1}} \left[b_i^{i-1} - \sum_{j=i+1}^{m} a_{ij}^{i-1} x_j \right],$$

$$i = m-1,\ m=2, \ldots, 1.$$

CAD Facility

The CAD algorithm shown in the flow diagram of Fig. A.4 uses the method described above.

The operation of the algorithm is self-explanatory, with Block 2 performing the calculations outlined above.

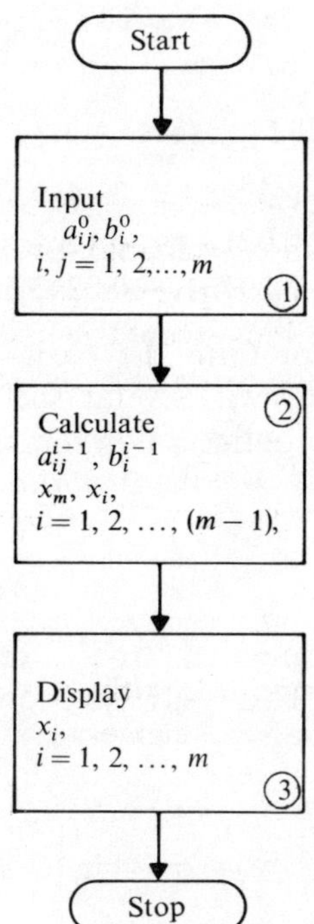

Fig. A.4 Gauss elimination CAD algorithm.

A.7 THE INVERSE OF A SQUARE MATRIX

If **A** and **B** are $(m \times m)$ matrices such that:

$\mathbf{AB} = \mathbf{BA} = \mathbf{I}$, then **B** is the *inverse of* **A** and vice versa.

$$\mathbf{B} = \mathbf{A}^{-1}, \qquad \mathbf{A} = \mathbf{B}^{-1}.$$

It can be shown that:

$$\mathbf{A}^{-1} = \frac{\text{adj } A}{\det A}. \tag{A.2}$$

If $\det \mathbf{A} = 0$, **A** is *singular* and has no inverse.

EXAMPLE
Consider

$$\mathbf{A} = \begin{bmatrix} 1 & -2 & 3 \\ 4 & 5 & -6 \\ -7 & 8 & 9 \end{bmatrix}$$

$$\text{adj } \mathbf{A} = \begin{bmatrix} 93 & 6 & -3 \\ 6 & 30 & 18 \\ 67 & 6 & 13 \end{bmatrix} \qquad \text{(Sec. A.4)}$$

$$\det \mathbf{A} = 282 \qquad \text{(Sec. A.3)}$$

$$\mathbf{A}^{-1} = \begin{bmatrix} 0.3298 & 0.1489 & -0.0106 \\ 0.0213 & 0.1064 & 0.0638 \\ 0.2376 & -0.0213 & 0.0461 \end{bmatrix}.$$

Equation (A.2) does not form the basis of a satisfactory computer algorithm for inverting matrices. Several such algorithms exist and are available in the form of subroutines. For completeness it is appropriate to include a reliable method here, and the *Doolittle algorithm* (ref. 5) is described in the following paragraphs.

Stage 1
The $(m \times m)$ matrix **A** is decomposed into a lower unit triangular matrix and an upper triangular matrix. This is always possible if **A** is nonsingular.

$\mathbf{A} = \mathbf{LU}$, or

$$\begin{bmatrix} a_{11} & a_{12} & \cdots & a_{1m} \\ a_{21} & a_{22} & & a_{2m} \\ \vdots & & & \\ a_{m1} & a_{m2} & & a_{mm} \end{bmatrix} = \begin{bmatrix} 1 & 0 & \cdots & 0 \\ l_{21} & 1 & & 0 \\ \vdots & & & \\ l_{m1} & l_{m2} & & 1 \end{bmatrix} \begin{bmatrix} u_{11} & u_{12} & \cdots & u_{1m} \\ 0 & u_{22} & & u_{2m} \\ \vdots & & & \\ 0 & 0 & & u_{mm} \end{bmatrix}. \tag{A.3}$$

Equating the corresponding first row elements in **A** and **LU**:

$$u_{1j} = a_{1j}, \qquad j = 1, 2, \ldots, m.$$

Equating the first column elements:

$$l_{i1} = \frac{a_{i1}}{u_{11}}, \qquad i = 2, 3, \ldots, m.$$

Equating the remaining second row elements:

$$u_{2j} = a_{2j} - l_{21}u_{1j}, \qquad j = 2, 3, \ldots, m.$$

Equating the remaining second column elements:

$$l_{i2} = \frac{a_{i2} - l_{i1}u_{12}}{u_{22}}, \qquad i = 3, 4, \ldots, m.$$

Proceeding in this manner, the rth row of **U** is computed:

$$u_{rj} = a_{rj} - \sum_{k=1}^{r-1} l_{rk}u_{kj}, \qquad j = r, r+1, \ldots, m. \tag{A.4}$$

Following this, the rth column of **L** is computed:

$$l_{ir} = \frac{a_{ir} - \sum_{k=1}^{r-1} l_{ik}u_{kr}}{u_{rr}}, \qquad i = r, r+1, \ldots, m. \tag{A.5}$$

It may be impossible to complete the decomposition of **A** successfully – for example, if at some stage $u_{rr} = 0$, the subsequent values of l_{ir} become infinite (Eq. A.5). The detailed background to such circumstances is involved (ref. 5), but usually it may be concluded when this happens that **A** is singular and cannot therefore be inverted.

Stage 2

If the factorization of **A** into **LU** is completed successfully, the remaining calculation involves the evaluation of:

$$\mathbf{A}^{-1} = \mathbf{U}^{-1}\mathbf{L}^{-1} \qquad \text{(App. A.8)}.$$

$\mathbf{L}^{-1}$ is a lower triangular matrix:

$$\mathbf{L}^{-1} = \begin{bmatrix} r_{11} & 0 & \cdots & 0 \\ r_{21} & r_{22} & & \\ \vdots & & & \\ r_{m1} & & & r_{mm} \end{bmatrix} \tag{A.6}$$

From Eq. (A.6) and the definition of **L** (Eq. (A.3)),

$$\mathbf{L}_k^{-1}\mathbf{L}_k = 1 = r_{kk}$$

and

$$\mathbf{L}_k^{-1}\mathbf{L}_j = 0 = \sum_{i=j}^{k} r_{ki} l_{ij}, \qquad j = k-1, k-2, \ldots, 1$$

where $\mathbf{L}_i$ is the ith column of $\mathbf{L}$ and $\mathbf{L}_j^{-1}$ is the jth row of $\mathbf{L}^{-1}$.

Using these equations, r_{ki}, $i = k, k-1, \ldots, 1$, $k = 1, \ldots, m$, can be calculated recursively, to give $\mathbf{L}^{-1}$.

$\mathbf{U}^{-1}$ can be calculated using the same formula with rows and columns interchanged:

$$\mathbf{U}^{-1} = \begin{bmatrix} s_{11} & s_{12} & \cdots & s_{1m} \\ 0 & s_{22} & & s_{2m} \\ \vdots & & & \\ 0 & & & s_{mm} \end{bmatrix}. \tag{A.7}$$

From Eq. (A.7) and the definition of $\mathbf{U}$(Eq. (A.3)):

$$\mathbf{U}_k \mathbf{U}_k^{-1} = 1 = s_{kk}$$

and

$$\mathbf{U}_j \mathbf{U}_k^{-1} = 0 = \sum_{j=i}^{k} u_{ji} s_{ik}, \qquad i = k-1, k-2, \ldots, 1,$$

where $\mathbf{U}_j$ is the jth row of $\mathbf{U}$ and $\mathbf{U}_i^{-1}$ is the ith column of $\mathbf{U}^{-1}$.

Using these equations, s_{ik}, $i = k, k-1, \ldots, 1$, $k = 1, 2, \ldots, m$, can be calculated recursively to give $\mathbf{U}^{-1}$.

Stage 3

Having found $\mathbf{U}^{-1}$, $\mathbf{L}^{-1}$, it is a simple matter to calculate:

$$\mathbf{A}^{-1} = \mathbf{U}^{-1}\mathbf{L}^{-1}.$$

CAD Facility

A useful CAD facility can be constructed using this algorithm. This is illustrated in Fig. A.5.

The operation of the algorithm is self-explanatory, with Blocks 2, 3, 4 performing the calculations described in Stages 1, 2, 3 above.

A.8 SOME PROPERTIES OF MATRICES AND DETERMINANTS

The following are true of square matrices:

1. $\det(\mathbf{AB}) = \det \mathbf{A} \det \mathbf{B}$

$$= \det \mathbf{B} \det \mathbf{A}$$

$$= \det(\mathbf{BA}) \tag{A.8}$$

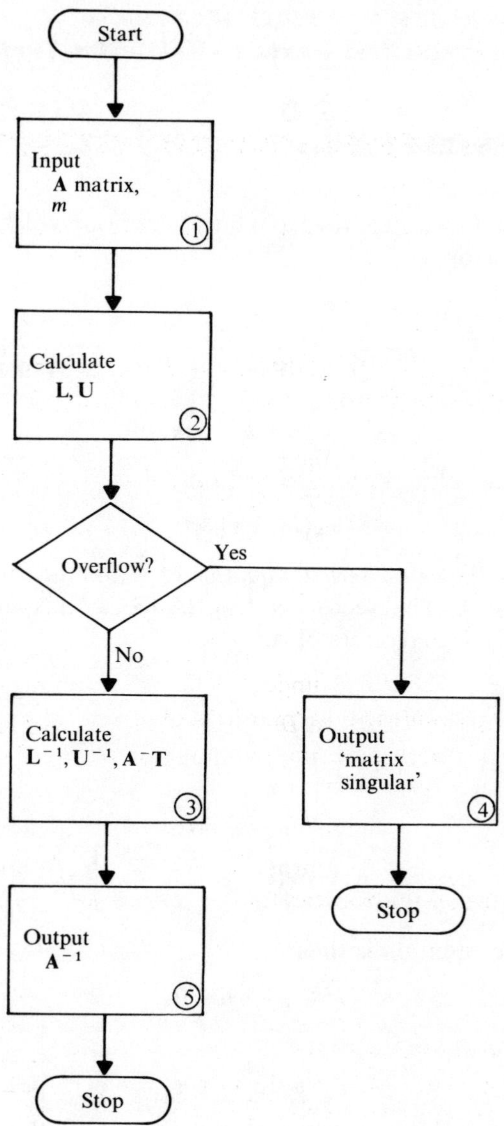

Fig. A.5 Matrix inversion CAD algorithm.

2. $(\mathbf{AB})^{-1} = \mathbf{B}^{-1}\mathbf{A}^{-1}$, if det $\mathbf{A} \neq 0$, det $\mathbf{B} \neq 0$ (A.9)
3. $(\mathbf{AB})^{\mathrm{T}} = \mathbf{B}^{\mathrm{T}}\mathbf{A}^{\mathrm{T}}$ (also true if $\mathbf{A}, \mathbf{B}$ are not square) (A.10)
4. $\mathbf{AB} = \mathbf{0}$ does not imply that $\mathbf{A} = 0$ or $\mathbf{B} = 0$, unless det $\mathbf{A} \neq 0$, det $\mathbf{B} \neq 0$ (A.11)
5. $\mathbf{AB} = \mathbf{AC}$ does not imply that $\mathbf{B} = \mathbf{C}$, unless det $\mathbf{A} \neq 0$. (A.12)

A.9 THE EIGENVALUES AND EIGENVECTORS OF A SQUARE MATRIX; THE CAYLEY–HAMILTON THEOREM

(i) Eigenvalues and Eigenvectors

A square matrix **A** may be regarded as a linear operator which maps a vector, **x**, into another vector, **y**.

$$\text{i.e.} \qquad \mathbf{y} = \mathbf{A}\mathbf{x}.$$

Any vector **x** which is mapped into $\lambda\mathbf{x}$, where λ is a scalar, by **A** is called an *eigenvector* of the matrix **A**.

$$\lambda\mathbf{x} - \mathbf{A}\mathbf{x} = (\lambda\mathbf{I} - \mathbf{A})\mathbf{x} = \mathbf{0} \tag{A.13}$$

This has nontrivial solutions for λ if and only if:

$$\det(\lambda\mathbf{I} - \mathbf{A}) = 0. \tag{A.14}$$

Equation (A.14) is the *characteristic equation* of **A** and the solutions for λ are the *eigenvalues* of **A**. The vectors **x** (Eq. (A.13)) corresponding to those eigenvalues are the eigenvectors of **A**.

EXAMPLE

Consider

$$\mathbf{A} = \begin{bmatrix} 1 & 2 \\ -1 & 4 \end{bmatrix}$$

$$\det(\lambda\mathbf{I} - \mathbf{A}) = \det\begin{bmatrix} \lambda - 1 & -2 \\ 1 & \lambda - 4 \end{bmatrix} = 0.$$

The characteristic equation is thus:

$$\lambda^2 - 5\lambda + 6 = 0.$$

The eigenvalues of **A** are the roots:

$$\lambda = 2, \qquad \lambda = 3.$$

Substitute $\lambda = 2$ in Eq. (A.13) to find an eigenvector:

$$2\begin{bmatrix} x_1 \\ x_2 \end{bmatrix} - \begin{bmatrix} 1 & 2 \\ -1 & 4 \end{bmatrix}\begin{bmatrix} x_1 \\ x_2 \end{bmatrix} = 0.$$

Letting $x_1 = 1$, an eigenvector corresponding to $\lambda = 2$ is $\begin{bmatrix} 1 \\ 0.5 \end{bmatrix}$.

Similarly, an eigenvector corresponding to $\lambda = 3$ is $\begin{bmatrix} 1 \\ 1 \end{bmatrix}$.

Note that when eigenvalues are repeated it may not be possible to find m eigenvectors for an $(m \times m)$ matrix (ref. 1).

(ii) The Cayley – Hamilton Theorem

A square matrix satisfies its own characteristic equation;
i.e. since det $(\lambda\mathbf{I} - \mathbf{A}) = \varphi(\lambda) = 0$ (characteristic equation),

$$\varphi(\mathbf{A}) = 0.$$

EXAMPLE

The characteristic equation of

$$\mathbf{A} = \begin{bmatrix} 1 & 2 \\ -1 & 4 \end{bmatrix} \text{ is } \lambda^2 - 5\lambda + 6 = 0.$$

It is easy to show that:

$$\mathbf{A}^2 - 5\mathbf{A} + 6 = 0.$$

Thoroughly robust modern algorithms for calculating the eigenvalues and eigenvectors of square matrices are beyond the scope of this book (ref. 5). However, a useful CAD facility can be based on *Krylov's method*, which depends on the Cayley–Hamilton theorem.

Consider the characteristic equation of a square $(m \times m)$ matrix:

$$\varphi(\lambda) = \lambda^m + \sum_{i=0}^{m-1} b_i \lambda^i = 0.$$

By the Cayley–Hamilton theorem, for any vector, $\mathbf{y} = [y_1 \quad y_2 \quad \cdots \quad y_m]^T$;

$$\mathbf{A}^m\mathbf{y} + \sum_{i=0}^{m-1} b_i \mathbf{A}^i \mathbf{y} = \mathbf{0}.$$

If the vector $\mathbf{y}$ is first defined, this is a set of m simultaneous equations in m unknowns which may be solved by the algorithm of Fig. A.4 to give b_i, $i = 0, 1, \ldots (m - 1)$.

The roots of the characteristic equation may then be found using the algorithm of Fig. 1.3.

CAD Facility

The CAD algorithm shown in the flow diagram of Fig. A.6 uses this method.

The operation of the algorithm is as follows.

BLOCKS 1, 2

The matrix $\mathbf{A}$ and an arbitrary vector $\mathbf{y}$ are input, displayed, and corrected if necessary.

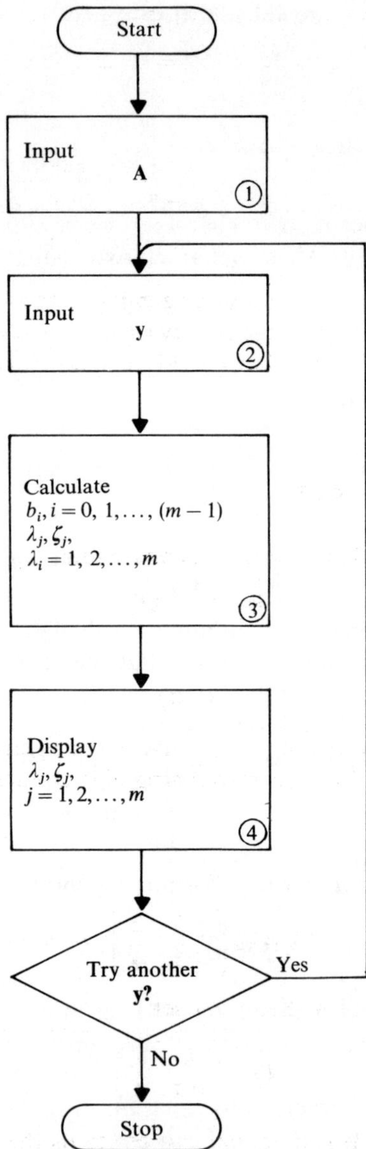

Fig. A.6 Eigenvalue and eigenvector CAD algorithm.

BLOCK 3
The coefficients, b_i, of the characteristic polynomial are calculated using Krylov's method as described above. This involves a subroutine using the algorithm of Fig. A.4.

The eigenvalues, λ_j, are calculated using a subroutine incorporating the algorithm of Fig. 1.3.

The eigenvectors, ξ_j, are calculated using the relationships:

$$\mathbf{A}\xi_j = \lambda\xi_j, \qquad j = 1, 2, \ldots, m$$

where

$$\xi_j = [1 \quad \zeta_{j2} \quad \zeta_{j3} \quad \cdots \quad \zeta_{jm}]^{\mathrm{T}}.$$

This again involves the use of the algorithm of Fig. A.4.

BLOCK 4
The eigenvalues and their corresponding eigenvectors are displayed.

The operator is given the opportunity to select a different arbitrary vector **y** and repeat the calculation. This should, of course, yield the same results.

A.10 SIMILAR MATRICES

Square matrices **A** and **B** are *similar* if a nonsingular matrix **Q** exists such that:

$$\mathbf{B} = \mathbf{Q}^{-1}\mathbf{A}\mathbf{Q}. \tag{A.15}$$

Equivalently:

$$\mathbf{A} = \mathbf{Q}\mathbf{B}\mathbf{Q}^{-1}.$$

If **Q** is a matrix whose columns consist of the eigenvectors of **A**, then **B** is diagonal and consists of the corresponding eigenvalues of **A**.

EXAMPLE
Consider

$$\mathbf{A} = \begin{bmatrix} 1 & 2 \\ -1 & 4 \end{bmatrix}.$$

From the eigenvectors of **A** (Sec. A.9), set:

$$\mathbf{Q} = \begin{bmatrix} 1 & 1 \\ 0.5 & 1 \end{bmatrix}.$$

It is easy to show that **B** is diagonal and contains the eigenvalues of **A**:

$$\mathbf{B} = \mathbf{Q}^{-1}\mathbf{A}\mathbf{Q} = \begin{bmatrix} 2 & 0 \\ 0 & 3 \end{bmatrix}.$$

When eigenvectors (and usually eigenvalues) are repeated, a matrix **Q** comprising these eigenvectors would be singular and Eq. (A.15) would fail ($\mathbf{Q}^{-1}$ would not exist). A *Jordan canonical form* matrix **Q** can be formed, however, in which off-diagonal elements correspond to repeated eigenvalues. The details of this are best pursued in the appropriate literature (ref. 1).

A.11 DEFINITE AND SEMIDEFINITE QUADRATIC FORMS

1. Any $(m \times m)$ symmetric matrix **A** is of *quadratic form* since the homogeneous quadratic polynomial p in the variables $x_1, x_2, \ldots, x_m$ can be written:

$$p = \sum_{i=1}^{m} \sum_{j=1}^{m} a_{ij} x_i x_j$$
$$= \mathbf{x}^T \mathbf{A} \mathbf{x}$$

where $\mathbf{x} = [x_1 \quad x_2 \quad \cdots \quad x_m]^T$

EXAMPLE

Consider the homogeneous polynomial:

$$p = x_1^2 + 3x_2^2 - 5x_2^3 - 6x_1x_2 + 10x_1x_3 - 4x_2x_3$$
$$= [x_1 \quad x_2 \quad x_3] \begin{bmatrix} 1 & -3 & 5 \\ -3 & 3 & -2 \\ 5 & -2 & -5 \end{bmatrix} \begin{bmatrix} x_1 \\ x_2 \\ x_3 \end{bmatrix}.$$

The symmetric metrix is in quadratic form.

2. A symmetric matrix **A** is *positive definite* if for any nontrivial choice of a vector **x** (i.e. $\mathbf{x} \neq \mathbf{0}$):

$$p = \mathbf{x}^T \mathbf{A} \mathbf{x} > 0.$$

This implies that there exists a linear transformation $\mathbf{y} = \mathbf{T}\mathbf{x}$ such that:

$$p = y_1^2 + y_2^2 + \cdots + y_m^2.$$

EXAMPLE

The matrix

$$\mathbf{A} = \begin{bmatrix} 1 & 0 & 0 \\ 0 & 3 & 0 \\ 0 & 0 & 5 \end{bmatrix}$$

is clearly positive definite, since

$$p = \mathbf{x}^T \mathbf{A} \mathbf{x} = x_1^2 + 3x_2^2 + 5x_3^2 > 0 \qquad \forall \mathbf{x} \ (\mathbf{x} \neq \mathbf{0}).$$

3. A symmetric matrix **A** is *positive semidefinite* if for any nontrivial choice of the vector **x** ($\mathbf{x} \neq \mathbf{0}$):

$$p = \mathbf{x}^T \mathbf{A} \mathbf{x} \geqslant 0$$

This implies that there exists a linear transformation $\mathbf{y} = \mathbf{T}\mathbf{x}$ such that:

$$p = y_1^2 + y_2^2 + \cdots + y_r^2, \qquad r < m.$$

EXAMPLE

The matrix

$$\mathbf{A} = \begin{bmatrix} 1 & 0 & 0 \\ 0 & 3 & 0 \\ 0 & 0 & 0 \end{bmatrix}$$

is positive semidefinite, since

$$p = \mathbf{x}^{\mathrm{T}}\mathbf{A}\mathbf{x} = x_1^2 + 3x_2^2 \geqslant 0 \qquad \forall \mathbf{x}\ (\mathbf{x} \neq \mathbf{0}).$$

4. The definitions of *negative* definite and semidefinite quadratic forms follow the obvious pattern with $p < 0$ and $p \leqslant 0$ $\forall \mathbf{x}$ respectively.

5. The terms positive and negative definite and semidefinite generally apply only to symmetric matrices.

APPENDIX B

NOTES ON PROBABILITY FUNCTIONS (REFS. 2, 3)

B.1 BASIC DEFINITIONS

Consider an event with an outcome of value X which is dependent on chance. Then the *probability distribution function* $P(x)$ is defined:

$$\text{prob}(X \leqslant x) = P(x), \quad -\infty \leqslant x \leqslant +\infty.$$

Some properties of the function $P(x)$ are:

1. $0 \leqslant P(x) \leqslant 1$
2. $P(-\infty) = 0$, $P(+\infty) = 1$
3. $\dfrac{dP(x)}{dx} \geqslant 0, \quad \forall x$
4. $\text{prob}\,(x_1 < X \leqslant x_2) = P(x_2) - P(x_1)$.

EXAMPLES

(a) If X is the result of tossing a die, $P(x)$ is as shown in Fig. B.1(a).

(b) If X is a number equally likely to lie anywhere between $+1$ and $+3$, $P(x)$ is as shown in Fig. B.1(b).

In both cases the properties of $P(x)$ listed above can be seen to hold.

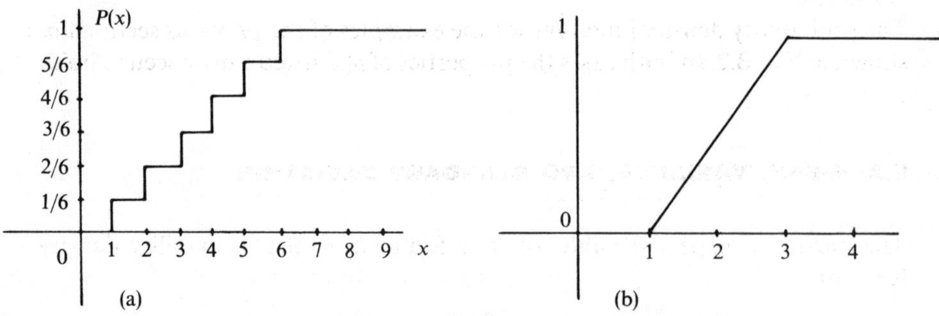

Fig. B.1 Probability distributions (examples).

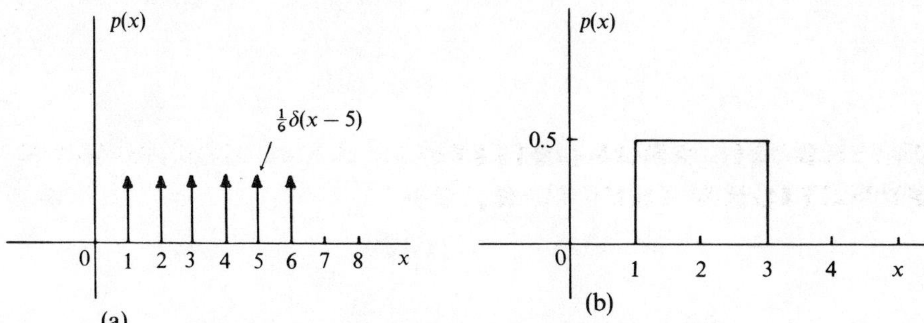

Fig. B.2 Probability density functions (examples).

More usefully, a probability can be described by a *probability density function*:

$$p(x) = \frac{dP(x)}{dx},$$

$$= \lim_{\delta x \to 0} \left[\frac{\text{prob}\,(x < X \leqslant x + \delta x)}{\delta x} \right], \qquad -\infty \leqslant x \leqslant +\infty.$$

Some properties of $p(x)$ are:

1. $p(x) \geqslant 0, \quad \forall x$
2. $p(-\infty) = p(+\infty) = 0$
3. $\text{prob}\,(X \leqslant x_1) = \int_{-\infty}^{x_1} p(x)\,dx$
4. $\text{prob}\,(x_1 < X \leqslant x_2) = \int_{x_1}^{x_2} p(x)\,dx$
5. $\int_{-\infty}^{+\infty} p(x)\,dx = 1.$

EXAMPLES

The probability density functions for the examples of the previous section are shown in Fig. B.2. In both cases the properties of $p(x)$ listed can be seen to hold.

B.2 MEAN, VARIANCE, AND STANDARD DEVIATION

The mean, or 'expected value' of X is found from the probability density function:

$$\mathscr{E}(X) = \int_{-\infty}^{+\infty} x p(x)\,dx.$$

The mean of any function of X, $f(X)$ is:

$$\mathscr{E}(f(X)) = \int_{-\infty}^{+\infty} f(x)p(x)\,dx.$$

Commonly (strictly incorrectly) the event value itself is represented by x, giving the formulas:

$$\bar{x} = \mathscr{E}(x) = \int_{-\infty}^{+\infty} xp(x)\,dx$$

$$\overline{f(x)} = \mathscr{E}(f(x)) = \int_{-\infty}^{+\infty} f(x)p(x)\,dx.$$

Also useful is the *variance*:

$$\sigma^2 = \mathscr{E}(x-\bar{x})^2 = \int_{-\infty}^{+\infty} (x-\bar{x})^2 p(x)\,dx\,.$$

The *standard deviation*, a measure of the 'spread' of a distribution is:

$$\sigma = +\sqrt{\sigma^2}.$$

It can easily be shown that:

$$\sigma^2 = \mathscr{E}(x^2) - (\mathscr{E}(x))^2.$$

EXAMPLES

(a) If x (strictly X) is the result of tossing a die (Fig. B.2(a)):

$$\bar{x} = \mathscr{E}(x) = \int_{-\infty}^{+\infty} \frac{x}{6} \sum_{n=1}^{6} \delta(x-n)\,dx$$

$$= \tfrac{1}{6} + \tfrac{2}{6} + \cdots + \tfrac{6}{6} = 3.5$$

$$\mathscr{E}(x^2) = \int_{-\infty}^{+\infty} \frac{x^2}{6} \sum_{n=1}^{6} \delta(x-n)\,dx = 15.166$$

$$\sigma^2 = \mathscr{E}(x^2) - (\mathscr{E}(x))^2 = 2.916$$

$$\sigma = 1.708.$$

(b) If x (strictly X) is equally likely to take any value between $+1$ and $+3$ (Fig. B.2.(b)):

$$\bar{x} = \mathscr{E}(x) = \int_{1}^{3} x0.5\,dx = 2$$

$$\mathscr{E}(x^2) = \int_{1}^{3} x^2 0.5\,dx = 4.333$$

$$\sigma^2 = \mathscr{E}(x^2) - (\mathscr{E}(x))^2 = 0.333$$

$$\sigma = 0.577.$$

B.3 JOINT AND CONDITIONAL PROBABILITY DENSITY FUNCTIONS

Consider two events X, Y. If Y depends on X, *conditional probability* distributions and density functions are defined:

$$\text{prob}(Y \leqslant y, X = x) = P(y|x), \quad -\infty < y < +\infty$$

$$p(y|x) = \frac{\partial P(y|x)}{\partial y}.$$

Joint probability distribution and density functions are defined whether or not Y depends on X (or vice versa):

$$\text{prob}(X \leqslant x, Y \leqslant y) = P(x, y)$$

$$p(x, y) = \frac{\partial^2 P(x, y)}{\partial x \partial y}.$$

Some properties of these functions are:

1. $p(x, y) \geqslant 0, \qquad \forall x, y.$
2. $\text{prob}(x_1 < X \leqslant x_2, y_1 < Y \leqslant y_2) = \int_{x_1}^{x_2} \int_{y_1}^{y_2} p(x, y)\, dx\, dy$
3. $p(y|x)p(x) = p(y); \qquad p(x|y)p(y) = p(x)$
4. $p(y) = \int_{-\infty}^{+\infty} p(x, y)\, dx; \qquad p(x) = \int_{-\infty}^{+\infty} p(x, y)\, dy.$

The joint probability density function is easily defined for any number of dimensions:

$$\text{prob}(X_1 \leqslant x_1, X_2 \leqslant x_2, \ldots, X_m \leqslant x_m) = P(x_1, x_2, \ldots, x_m)$$

$$p(x_1, x_2, \ldots, x_m) = \frac{\partial^m P}{\partial x_1 \partial x_2 \ldots \partial x_m}.$$

EXAMPLE

The *Gaussian*, or *normal* probability density function in m dimensions may be written:

$$p(\mathbf{x}) = p(x_1, x_2, \ldots, x_m)$$

$$= \frac{1}{\sqrt{(2\pi)^m \det \mathbf{R}}} \exp\left\{\frac{-1}{2}(\mathbf{x} - \bar{\mathbf{x}})^{\mathrm{T}} \mathbf{R}^{-1} (\mathbf{x} - \bar{\mathbf{x}})\right\}$$

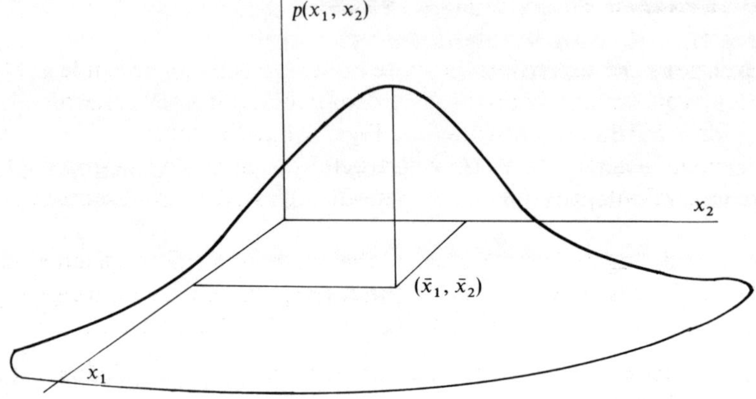

Fig. B.3 Gaussian distribution in two dimensions.

where

$$\mathbf{R} = \mathscr{E}[(\mathbf{x} - \bar{\mathbf{x}})(\mathbf{x} - \bar{\mathbf{x}})^{\mathrm{T}}]$$

$$= \mathscr{E}\begin{bmatrix} (x_1 - \bar{x}_1)^2 & (x_1 - \bar{x}_1)(x_2 - \bar{x}_2) & \cdots & (x_1 - \bar{x}_1)(x_m - \bar{x}_m) \\ (x_2 - \bar{x}_2)(x_1 - \bar{x}_1) & (x_2 - \bar{x}_2)^2 & & \\ \vdots & & & \\ (x_m - \bar{x}_m)(x_1 - \bar{x}_1) & & & (x_m - \bar{x}_m)^2 \end{bmatrix}.$$

For $m = 2$, this is shown in Fig. B.3.

B.4 THE CENTRAL LIMIT THEOREM

If a random process consists of the sum of a large number of independent processes, the distribution of the overall process is Gaussian, or normal.

This is a loose statement of the *Central Limit Theorem*, and since many noise processes fall into this category, it is reasonable to model them with Gaussian distributions.

B.5 PSEUDO-RANDOM SIGNAL GENERATION

Pseudo-random signals of various kinds can easily be generated. Some simple methods are described here.

(i) Pseudo-random Binary Signals (PRBS)

A signal of this kind approximates white noise. It has two amplitude levels. A sample function, amplitude probability density function, and autocorrelation function for a PRBS are illustrated in Figs. B.4(a), (b), (c).

A simple generator can be constructed using a shift register, or its software equivalent, and one or more modulo 2 adders as illustrated in Fig. B.4(d).

If the shift register has N stages, the maximum length of unrepeated sequence which can be generated (without the pattern disappearing) is:

$$L = 2^N - 1.$$

It can be shown that the sequence is of this length if K (the feedback stage) is chosen so that the equation

$$\lambda^N + \lambda^{N-K} + 1 = 0$$

has no real roots.

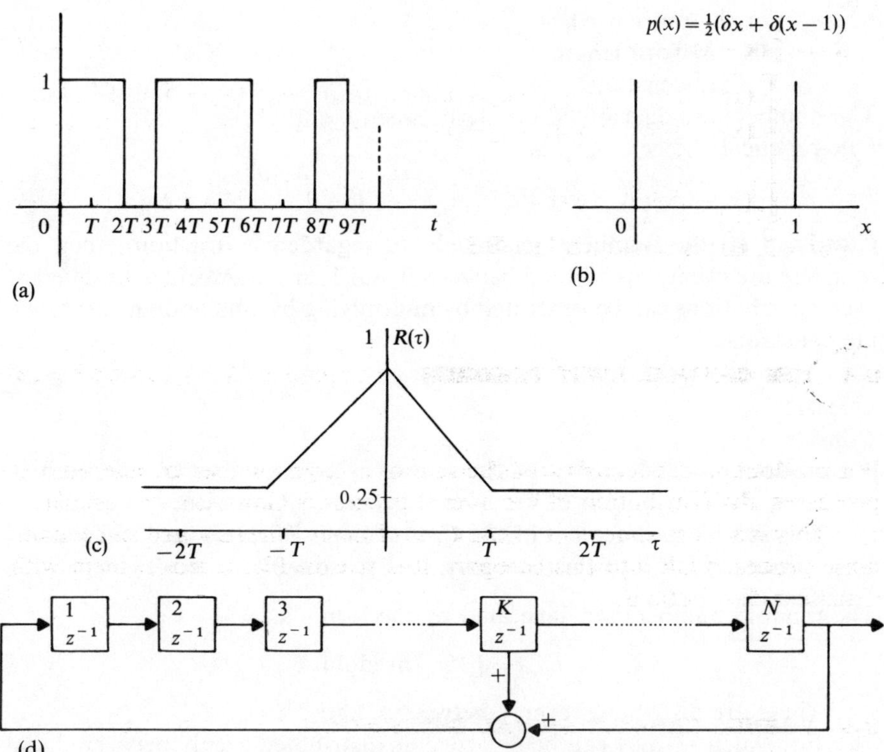

Fig. B.4 PRBS and a schematic generator. (a) PRBS sample function. (b) Probability density function. (c) Autocorrelation function. (d) PRBS generator.

EXAMPLE

For a shift register with $N = 15$, $K = 13$, and clock period T, a pseudo-random binary signal with amplitudes 0, 1 is generated:

$$R(\tau) = \begin{cases} 1, & \tau = 0 \\ 0.25, & |\tau| > T. \end{cases}$$

The sequence is repeated every $(2^{15} - 1)$ clock periods.

(ii) Evenly Distributed Random Numbers (ref. 6)

Evenly distributed random numbers can easily be generated by the *multiplicative congruential* formula:

$$r_{n+1} = [[\alpha r_n + \beta]] \text{ modulo } \gamma^p$$

where γ is the radix used,
p is the word length,
α, β are constants.

The numbers are distributed evenly between 0 and γ^p.

One practical choice of α, β is:

$$\alpha = 2^{p/2+1} + 3, \qquad \beta = 0.$$

Clearly, if all the numbers generated are regarded as fractions, then the numbers are evenly distributed between 0 and 1, and numbers with different even distributions can be generated by multiplying by, and adding, appropriate constants.

EXAMPLE

Consider

$$\gamma = 2$$
$$p = 8.$$
$$\text{Select } \alpha = 2^5 + 3 = 35, \quad \beta = 0.$$

The random numbers are generated by the formula:

$$r_{n+1} = [35 r_n] \text{modulo}_{2^8}.$$

These are distributed evenly between 0 and 2^8.

Equivalently, they can be regarded as distributed evenly between 0 and $(1 - 2^{-8})$. In this case, if, for example, each number is multiplied by 4 and 1 is subtracted (using double length word arithmetic), a new set of numbers evenly distributed between -1 and $+3$ is generated.

(iii) Random Numbers with Gaussian Distribution (ref. 4)

A sequence of uniformly distributed numbers $r(n)$ can conveniently be converted to a set of Gaussian distributed random numbers by a formula which depends on the central limit theorem:

$$y(n) = \frac{1}{N} \sum_{i=0}^{N-1} r(nN - i).$$

N is theoretically large, but $N \geqslant 10$, typically, gives good results. In this case $\mathscr{E} y(n) = \mathscr{E} r(n)$,

$$\operatorname{var}(y(n)) = \frac{1}{N} \operatorname{var}(r(n)).$$

APPENDIX C

NOTES ON DELTA FUNCTIONS (REF. 2)

C.1 THE DIRAC DELTA FUNCTION

Consider the function illustrated in Fig. C.1.

Define:

$$\delta_a(t) = \begin{cases} \dfrac{1}{a}, & 0 \leqslant t \leqslant a \\ 0, & 0 > t,\ t > a\,. \end{cases}$$

The 'area' under $\delta_a(t)$ is always 1, regardless of a.

Then a *Dirac δ function* (or 'impulse function') is:

$$\lim_{a \to 0} (\delta_a(t)) = \delta(t)\,.$$

Some properties of this function are:

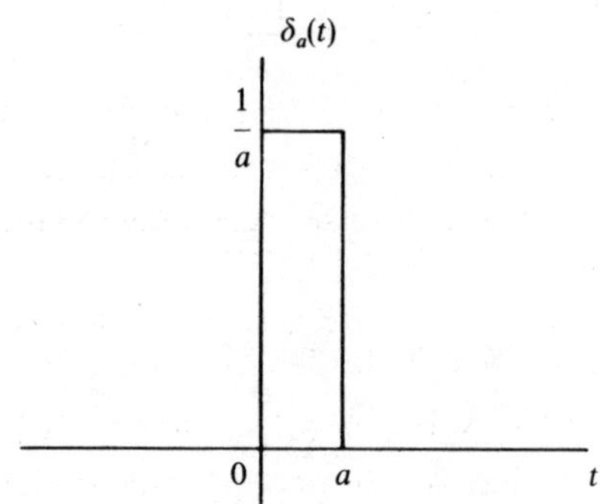

Fig. C.1 Dirac δ-function.

1. Laplace transform:

$$\mathscr{L}\{\delta(t)\} = 1.$$

2. Integration:
   If $f(t)$ is continuous at $t = 0$,

$$\int_{-\infty}^{+\infty} \delta(t) f(t)\,\mathrm{d}t = f(0).$$

3. Strength:
   $\delta(t)$ has a 'strength' (or area) or 1.
   $k\delta(t)$ has a strength (or area) of k.

4. Delay:
   An impulse of strength k at time τ is described by the Dirac δ function,

$$k\delta(t-\tau).$$

C.2 THE KRONECKER DELTA FUNCTION

A *Kronecker δ function* of a discrete independent variable (typically time, n) is defined:

$$\delta(n) = \begin{cases} 1, & n = 0 \\ 0, & n \neq 0. \end{cases}$$

Sometimes it is convenient to define a Kronecker δ-function of two independent variables:

$$\delta(l, k) = \begin{cases} 1, & l = k \\ 0 & l \neq k. \end{cases}$$

REFERENCES

1. Ayres, F., *Matrices*, Schaum Outline Series, 1962.
2. Chirlian, P.M., *Signals, Systems and the Computer*. Intertext, 1973.
3. Lipschutz, S., *Probability*. Schaum Outline Series, 1974.
4. Rabiner, L.R., and Gold, B., *Theory and Application of Digital Signal Processing*. Prentice-Hall, 1975.
5. Ralston, A., and Rabinowitz, P., *A First Course in Numerical Analysis*. McGraw-Hill, 1978.
6. Ralston, A., and Wilf, H.S., *Mathematical Methods for Digital Computers*. Wiley, 1966.

INDEX

A

Adams–Moulton, 5
Alternate variables, 261
Analog to digital converter, 35
Argand diagram, 16
ARMA model, 274
Asymptote (root locus), 79
Asymptotic stability, 158, 222
Asymptotic state estimator, 194, 235
Autocorrelation, 59, 66
Autopilot, 178, 185

B

Bairstow–Hitchcock, 11
Bias (identification), 279
Bilinear transformation, 110
Bode plot, 71, 92, 109, 126
Bounded-input, bounded-output stability, 159, 222
Bracketing, 267
Bromwich–Wagner integral, 9
Bryson and Ho, 232

C

Calculus of variations, 181
Canonical form state equations, 142
Cayley–Hamilton theorem, 301
Central limit theorem, 310
Characteristic
  equation, 152, 221, 300
  polynomial, 152, 221
Closed-loop poles, 79, 115
Companion form, 161, 223
Compensation
  algorithm, 123
  network, 91
Continuous cycling method, 137
Controllability, 155
Controlled variable, 134
Correlation, 59, 66
Cost function, 181
Costate variable, 189
Covariance, 59, 66, 239
Cross-correlation, 59, 66
Crossover frequency, 92, 126

D

Damping ratio, 74, 116
Decibels, 71
Decimation in time algorithm, 53
Delta function, 314, 315
Determinant, 289
Diagonal dominance, 205
Digital to analog converter, 35
Dirac delta function, 314
Discrete
  Fourier transform, 52
  state equation, 215
  system, 215
Discretization, 224
Doolittle algorithm, 296
Dynamic programming, 232

E

Echelon form, 292
Eigenvalue, 300

Eigenvector, 300
Ensemble, 58
Ergodic process, 59
Error covariance, 243
Estimator, 194, 195, 235, 236
Expected value, 307
Extremum, 257

F
Factorization of polynomials, 11
Faddeev algorithm, 154
Fast Fourier transform, 53
Fibonacci search, 267
Final value theorem, 8, 24
Fourier
  spectra, 42, 48
  transform, 42
  transform table, 45

G
Gain margin, 77, 111
Gauss elimination, 294
Gershgorin
  band, 206
  circle, 205
Golden section, 269
Gram–Schmidt, 264

H
Hamiltonian, 189
Hill climbing, 257
Ho and Kalman algorithm, 163

I
Identification, 273
Initial
  conditions, 1
  value theorem, 8, 24
Integral control, 135
Inverse
  Fourier transform, 42
  Laplace transform, 9
  Nyquist diagram, 89, 98
  z transform, 24

J
Jordan form, 303

K
Kalman filter, 241
Kronecker delta function, 315
Krylov's method, 301

L
Laplace
  transform, 7
  transform table, 26
Lead
  algorithm, 128
  network, 92
Least squares estimate, 274
Linear
  quadratic optimal control 181
  time-invariant system, 143, 216
  time-variant system, 143, 216
Linearization, 144
Luenberger observer, 195, 236

M
M contour, 84, 87, 91, 120, 122
Matrix, 288
  cofactor, 290
  covariance, 201, 238
  inverse, 296
  negative def./indef., 305
  positive def./indef., 304
  rank, 291
  singular, 296
Maximum
  likelihood, 281
  overshoot, 81
Mean, 239, 307
Measurement noise, 202, 238
Modified
  z transform, 40
  z transform table, 26
Multivariable
  control, 204
  system, 163

N
N contour, 84, 87, 91, 120, 122
Nichols chart, 86, 98, 122, 132
Noise, 201, 238, 243, 311
Non-minimum phase, 126
Normal distribution, 309

Numerical integration, 3
Nyquist
  diagram, 83, 96, 119, 131
  stability criterion, 84

O

Observability, 157
Observer, 194, 195, 235, 236
Open-loop system, 180
Optimal control, 180, 189, 232
Optimization, 257
Ostrowski circles, 207
Overshoot, 81

P

P matrix, 156
Path of steepest ascent, 266
Phase margin 77, 111
PID controller, 134
Polynomial roots, 11
Pontryagin's minimum principle, 189
Positive
  definite, 304
  semi definite, 304
Potter's solution, 182
Power spectral density, 63
Predictor corrector, 5
Probability
  density function, 307
  function, 306
Process control, 134
Proportional
  band, 135
  control, 135
Pseudo-random signals, 310

Q

Q matrix, 158
Quadratic form, 304

R

Random
  number, 311
  variable, 311
Rank of matrix, 291
Rate time, 136
Reachability, 222
Reaction curve, 136
Realization of state equations, 160, 223
Repeats per unit time, 136
Reset rate, 136
Riccati equation, 182
Root locus, 77, 94, 115, 130
Rosenbrock, 263, 273
Runge–Kutta, 3

S

s plane, 16
Sample
  data systems, 35
  function, 59
  period, 35
  theorem, 50
Separation property, 198
Servomechanism design, 91, 123
Setpoint, 135
Shannon's theorem, 50
Shaping
  algorithm, 123
  network, 91
Simplex, 258
Stability, 77, 84, 86, 91, 113, 120, 122, 158, 222
State
  equation, 142, 215
  feedback, 172, 176, 228
  variable, 142, 215
Stationary process, 60
Stochastic
  process, 58
  system, 201, 238
Switching curve, 189
System
  equations, 142, 215
  identification, 273

T

Taylor series, 3
Transfer function
  Fourier, 46
  Laplace, 14
  z, 31
Transport lag, 141

U

Unbiased estimate, 281

Undamped natural frequency, 74, 116
Unit circle, 34

V
Variance, 308
Vector, 286

W
w domain, 110
w' domain, 113
Weighting matrix, 276
Wiener Kinchine theorem, 63
White noise, 64

Z
z plane, 31
  transform, 22
  transform table, 26